Lecture Notes of the Institute
for Computer Sciences, Social-Informatics
and Telecommunications Engineering   67

Fernando Antonio Basile Colugnati
Lia Carrari Rodrigues Lopes
Saulo Faria Almeida Barretto (Eds.)

Digital Ecosystems

Third International Conference, OPAALS 2010
Aracuju, Sergipe, Brazil, March 22-23, 2010
Revised Selected Papers

 Springer

Volume Editors

Fernando Antonio Basile Colugnati
Instituto de Pesquisas em Tecnologia e Inovação - IPTI
Av. São Luis, 86, cj 192-19° andar
Cep 01046 000 São Paulo - SP, Brazil
fernando@ipti.org.br

Lia Carrari Rodrigues Lopes
Instituto de Pesquisas em Tecnologia e Inovação - IPTI
Av. São Luis, 86, cj 192-19° andar
Cep 01046 000 São Paulo - SP, Brazil
E-mail: lia.carrari@gmail.com

Saulo Faria Almeida Barretto
Instituto de Pesquisas em Tecnologia e Inovação - IPTI
Av. Beira Mar, sem n°, Povoado do Crasto
Cep 49230-000, Santa Luzia do Itanhy, Sergipe, Brazil
E-mail: saulo@ipti.org.br

Library of Congress Control Number: 2010931633

CR Subject Classification (1998): H.4, I.2.6, H.3, H.5, I.2.4

ISSN   1867-8211
ISBN-10   3-642-14858-1 Springer Berlin Heidelberg New York
ISBN-13   978-3-642-14858-3 Springer Berlin Heidelberg New York

springer.com

© ICST Institute for Computer Science, Social Informatics and Telecommunications Engineering 2010
Printed in Germany

Typesetting: Camera-ready by author, data conversion by Scientific Publishing Services, Chennai, India
Printed on acid-free paper   SPIN: 219/3180

Preface

The Third International OPAALS[1] Conference was an opportunity to explore and discuss digital ecosystem research issues as well as emerging and future trends in the field. The conference was organized by IPTI – Instituto de Pesquisas em Tecnologia e Inovação (www.ipti.org.br). IPTI is a member of the OPAALS Framework Programme 7 Network of Excellence, which is led by the London School of Economics and Political Science. OPAALS is a multi-disciplinary research network of excellence for developing the science and technology behind digital ecosystems. The conference was held within the scope of a broader EU–Brazil bilateral workshop hosted by IPTI in cooperation with the Brazilian government and the European Commission and designed to foster EU support of information and communications technologies (ICT) enablement and socio-economic development in Brazil.

The event was held in the city of Aracajú, Sergipe, in the northeast of Brazil, during March 22–23, 2010. Aracajú is the capital of the state of Sergipe and is located on the coast, a tropical region with lush vegetation, rivers and mangroves and an economic landscape dominated by fisheries, tourism and the challenges associated with fostering local economic development in the presence of low ICT penetration.

Digital ecosystems (DEs) in some ways represent the next generation of ICT and Internet usage. Applicable to many contexts, they will perhaps have the greatest effect in enabling small and medium-sized enterprises (SMEs) to compete on the global stage. Based on open, peer-to-peer architectures and open source software (OSS), digital ecosystems integrate emerging trends in the technical, social, and economic aspects of the connected world.

DEs, whose distributed and evolutionary architecture is inspired by biological metaphors, require a process of 'cultivation' in order to be instantiated in a particular social, cultural and economic context. This implies the need for their appropriation by local stakeholders, which comes with the responsibility for their technical maintenance and economic sustainability – as in all OSS projects. Because digital ecosystems are designed to support ICT adoption, collaborative practices between SMEs, social and knowledge networks, and socio-economic development, they naturally extend the familiar governance requirements of OSS projects to an integrated, interdisciplinary, and multi-stakeholder process that strengthens the democratic institutions in any given context.

A research question that has arisen from our work over the past 7 years is whether we can claim that such strengthened democratic institutions provide a structure out of which economic growth and innovation can develop naturally and spontaneously, as 'by-products' of social innovation processes. More work will be required to explore this question, but in the meantime research in Des as represented by the articles in this volume and related publications is tantalisingly beginning to suggest that some such rationalization might be justified.

---

[1] Open Philosophies for Associative Autopoietic Digital Ecosystems, http://www.opaals.eu

Social Science

DEs, therefore, are based on the integration of *bottom–up processes* with *structural principles* of socio-economic action. This can be considered a new 'synthesis' of what traditionally have been labelled as rightist/agent-based/individualist and leftist/ structuralist/collectivist perspec-tives, respectively, on the explanation of socio-economic action, and in this sense this research is very innovative. This new integration has been made possible by adopting a social constructivist perspective in our research, with a strong reflexive component, and has been significantly influenced by Giddens's Structu-ration (Giddens, 1984).

The concept of social construction is very useful to make tangible objects that, in spite of their sometimes abstract character, make up our everyday life, such as money, citizenship, and newspapers: "Money, citizenship and newspapers are transparent social constructions because they obviously could not have existed without societies" (Boghossian, 2001). We generally speak of social constructions as resulting from social processes mediated by language. Thus, social constructions acquire *meaning* through a consensual social process. The fact that such processes create meaning apparently out of nothing and give us a way to talk concretely about concepts that are otherwise difficult to define affords to social constructivism the status of an epistemology.

Thus, the various flavors of social constructivism fall broadly in the middle of a spectrum whose extremes are identified with the radical subjectivism of individual phenomenology at one end and the radical objectivism of physics at the other. Be-cause of their intermediate position between subjectivism and objectivism, social constructivist processes are sometimes called 'inter-subjective' (e.g., Popper, 2002).

Figure 1 can help position these concepts within a broader context of social theory. The figure, inspired by Hollis (1994), summarizes the main analytical traditions in social science over the past few centuries in addressing questions of socio-economic structure and human action. A few indicative and by no means exhaustive names are added to make the table easier to interpret. The left-hand column is generally associ-ated with the rationalistic, deterministic tradition, it is the older of the two, and grew out of naturalistic philosophy. The right-hand column is more recent, it reflects a greater emphasis on the social world for defining our reality (ontology) and the con-struction of knowledge (epistemology). Although interpreting the two columns in terms of an objective–subjective dichotomy can only be a gross oversimplification, the thinkers in the left-hand column could be loosely grouped as sharing a belief in some form of 'objective' reality, whereas a more 'subjective' perspective permeates the ideas found in the right column. The different widths of the columns are meant to reflect the much greater constituency (and funding), within social science, that a criti-cal tradition inspired by naturalistic philosophy still commands.

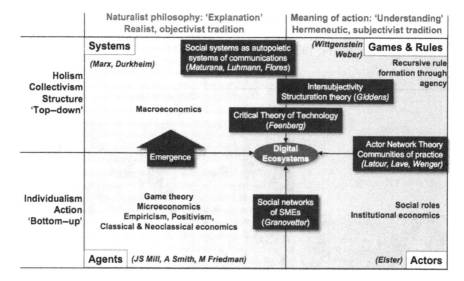

Fig. 1. Map of social science (adapted from Hollis, 1994)

The table can also be understood in terms of different accounts of social systems and therefore human action. The top row favors a view of society and the economy that biases the importance of structures and systems over individuals, whereas the bottom row represents the opposite view. This distinction correlates also with methodology, in the sense that theories in the top row tend to be deductive in deriving behavior from general principles, whereas the bottom row is best associated with the longstanding and currently overriding tradition of empiricism and positivism, where general principles are derived from experience through an inductive process.

The relatively new field of complexity science is proposing new words for describing processes and phenomena that have long been studied in the social sciences, such as 'emergence' to describe the not-so-well-understood relationship between local interactions and global behavior. Part of the excitement felt by practitioners in this new field derives from the development of new conceptual, mathematical, and computational tools for modelling processes that had until recently been considered too difficult for the reductionist scientific approach – and had therefore been mainly studied in the social sciences. DE research is opportunistic with regard to complexity science. Some insights are useful and illuminating, but a vigilant eye needs to be kept on the assumed ontological and epistemological basis when engaging in this particular type of interdisciplinary discussion in order to avoid falling back into the trap of 'monorail' rationalistic thinking.

With this figure in mind, therefore, we can see that both the bottom–up agent-based and the top–down structuralist approaches mentioned above belong to the naturalistic tradition. DE research is capable of reconciling them by bringing a more intersubjective perspective from the right-hand side column, in particular Giddens's Structuration (Giddens, 1984).

As shown in Figure 2, DE computer science research is similarly able to reconcile another important dichotomy, between the view of software as medium of communications (Winograd and Flores, 1987) and software as machine.

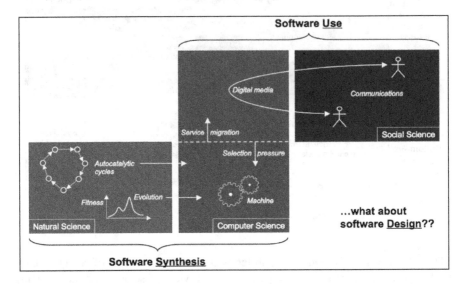

Fig. 2. Interdisciplinary structure of OPAALS

Natural Science

As stated in the paper "A Research Framework for Interaction Computing," in these proceedings,

> [Natural science DE] research is motivated by the fundamental question whether a biological ecosystem, or a subset thereof, could be used as a *model* [instead of merely as a metaphor] from which to derive self-organising, self-healing, and self-protection properties of software. This research question is premised on the assumption that such biological properties can increase the effectiveness of information and communication technologies (ICTs) in various application domains, from ubiquitous computing, to autonomic communications, to socio-economic processes aimed at regional development, simply on the basis of their greater and spontaneous adaptability to user needs.

Relative to the social science perspective, therefore, natural science DE research is functionalist, as succinctly summarized in Figure 2. Its aim is to improve the performance and computational properties of the software as a machine.

Computer Science

Last but not least, DE research is concerned with the realization of distributed architectures for all aspects of the software system. This has been a huge challenge for us, due to the need to reconcile opposing requirements:

1. On the one hand, the (normative) requirement to avoid single points of control and the (engineering) requirement to avoid single points of failure called for a 'pure' peer-to-peer architecture, which is very inefficient for data and message propagation.
2. On the other hand, the (functionalist) requirement for efficient data and message propagation called for a hierarchical network, based on so-called super-peers, which, however, violate the first requirement.

The solution, dubbed "Flypeer", was the concept of *dynamic, virtual* super-peers: 'virtual' means that each super-peer is actually a collection of machines, whereas 'dynamic' refers to the fact that membership in the cluster of virtual super-peer machines is entirely dependent on the resource availability of the participating peers.

Computer science DE research addresses many facets of distributed software systems and of formal/business modelling languages, only some of which are represented in this collection of papers. At a more theoretical level, the normative requirement mentioned above can be reconciled with the social science perspective of the research with the help of Fig. 3, which answers the question "What about software design?" shown in Fig. 2. In particular, the realization that technology embodies our cultural values (Feenberg, 1991, 2002) calls for the assumption of a greater level of responsibility on the part of the technologists. It is with this perspective in mind that the computer scientists of the DE community have embraced the normative requirement of distributing every aspect of the architecture.

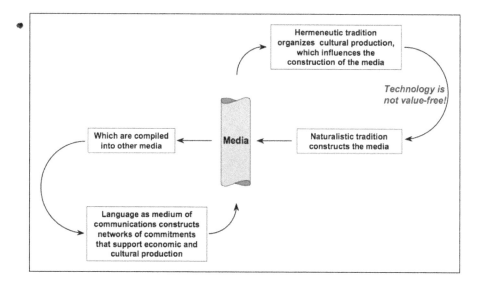

Fig. 3. Autopoiesis of media

Interdisciplinary Paper Map

During the two days of the OPPALS 2010 Conference, 21 multidisciplinary papers were presented by authors from diverse backgrounds on multiple areas and topics related to DE adoption, technologies, theoretical research and models, among others. Figure 4 shows how the papers of these proceedings are approximately distributed across the three disciplinary domains of the project.

It should be understood that the three colors provide only an approximate categorization, since every paper is in fact interdisciplinary to different degrees. As this is the final conference of the project, we can take the overall interdisciplinary mixing of the papers presented at this conference as an encouraging measure of our progress toward the building of an interdisciplinary science of DEs.

Paolo Dini
Anne English

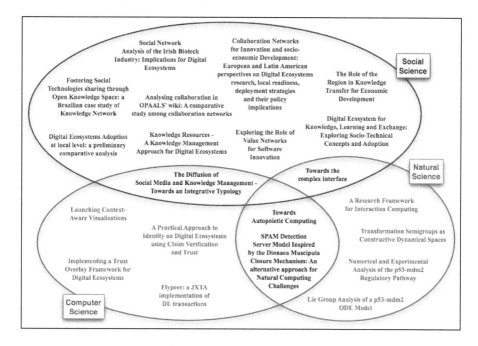

Fig. 4. Interdisciplinary map of the papers presented at OPAALS10

References

Boghossian, P A (2001). "What is Social Construction?", *Times Literary Supplement*.
     Available at: http://philosophy.fas.nyu.edu/docs/IO/1153/socialconstruction.pdf
Feenberg, A (1991). *Critical Theory of Technology*, Oxford.
Feenberg, A (2002). *Transforming Technology: A Critical Theory Revisited*, Oxford.
Giddens, A (1984). *The constitution of society: Outline of the theory of structuration*, Cam-
     bridge: Polity.
Hollis, M (1994). *The philosophy of social science: an introduction*, Cambridge.
Popper, K (2002). *The Logic of Scientific Discovery* (reprinted translation of 1935 original
     Logik der Forschung), London: Routledge.
Winograd, T and Flores, F (1987). *Understanding Computers and Cognition*, Addison-Wesley.

Organization

Chair

Saulo F. A. Barretto   Instituto de Pesquisas em Tecnologia e Inovação (IPTI), Brazil

Organizing Committee

Fernando A.B. Colugnati   Instituto de Pesquisas em Tecnologia e Inovação (IPTI), Brazil

Renata Piazzalunga   Instituto de Pesquisas em Tecnologia e Inovação (IPTI), Brazil

Ossi Nykänen   Technical University of Tampere (TUT), Finland

Paolo Dini   London School of Economics and Political Science (LSE), UK

Frauke Zeller   University of Kassel, Germany

Thomas J. Heistracher   Salzburg University of Arts and Science (SUAS), Austria

Jayanta Chatterjee   Indian Institute of Technology Kanpur (IITK), India

Paul Krause   University of Surrey, UK

Scientific Committee

Neil Rathbone   London School of Economics and Political Science (LSE), UK

Lia Carrari Lopes   Instituto de Pesquisas em Tecnologia e Inovação (IPTI), Brazil

Chris Van Egeraat   NUI Maynooth, Ireland

Declan Curran   NUI Maynooth, Ireland

Thomas Kurz   Salzburg University of Arts and Science (SUAS), Austria

Raimund Eder   Salzburg University of Arts and Science (SUAS), Austria

Jaakko Salonen   Technical University of Tampere (TUT), Finland

Jukka Huhtamäki   Technical University of Tampere (TUT), Finland

Ingeborg van Leeuwen   University of Dundee, UK

Paolo Dini   London School of Economics and Political Science (LSE), UK

Daniel Schreckling   University of Passau, Germany

Francesco Botto   Create-Net, Italy

Marco Bräuer   University of Kassel, Germany

Ingmar Steinicke	University of Kassel, Germany
Oxana Lapteva	University of Kassel, Germany
Amritesh Chatterjee	Indian Institute of Technology Kanpur (IITK), India
Jayanta Chatterjee	Indian Institute of Technology Kanpur (IITK), India
Gabor Horvath	London School of Economics and Political Science (LSE), UK
Attila Egri	University of Hertfordshire (UH)
Paolo Dini	London School of Economics and Political Science (LSE), UK
Chrystopher Nehaniv	University of Hertfordshire (UH), UK
Maria Schilstra	University of Hertfordshire (UH), UK
Lorraine Morgan	University of Limerick (UL), Ireland
Paul Malone	Waterford Institute of Technology, Ireland
Jimmy McGibney	Waterford Institute of Technology, Ireland
Mark McLaughlin	Waterford Institute of Technology, Ireland
Gerard Briscoe	London School of Economics and Political Science (LSE), UK
Paolo Dini	London School of Economics and Political Science (LSE), UK
Lorena Rivera Leon	London School of Economics and Political Science (LSE), UK
Mark McLaughlin	Waterford Institute of Technology, Ireland
Amir reza Razavi	University of Surrey, UK
Paulo Siqueira	Instituto de Pesquisas em Tecnologia e Inovação (IPTI), Brazil
Fabio K Serra	Instituto de Pesquisas em Tecnologia e Inovação (IPTI), Brazil
Anne English	Intel Ireland

Table of Contents

Collaboration Networks for Innovation and Socio-economic
Development: European and Latin American Perspectives on Digital
Ecosystems Research, Local Readiness, Deployment Strategies and
Their Policy Implications 1
    Lorena Rivera León and Rodrigo Kataishi

Exploring the Role of Value Networks for Software Innovation 20
    Lorraine Morgan and Kieran Conboy

Social Network Analysis of the Irish Biotech Industry: Implications for
Digital Ecosystems 31
    Chris van Egeraat and Declan Curran

Digital Ecosystem for Knowledge, Learning and Exchange: Exploring
Socio-technical Concepts and Adoption 44
    Amritesh and Jayanta Chatterjee

The Diffusion of Social Media and Knowledge Management – Towards
an Integrative Typology 62
    Frauke Zeller, Jayanta Chatterjee, Marco Bräuer,
    Ingmar Steinicke, and Oxana Lapteva

Digital Ecosystems Adoption at Local Level: A Preliminary
Comparative Analysis 76
    Francesco Botto, Antonella Passani, and Yedugundla Venkata Kiran

The Role of the Region in Knowledge Transfer for Economic
Development 92
    Neil Rathbone

Fostering Social Technologies Sharing through Open Knowledge Space:
A Brazilian Case Study of Knowledge Network 100
    Lia Carrari R. Lopes, Saulo Barretto, Paulo Siqueira,
    Larissa Barros, Michelle Lopes, and Isabel Miranda

Analysing Collaboration in OPAALS' Wiki: A Comparative Study
among Collaboration Networks 109
    Fernando A. Basile Colugnati and Lia Carrari R. Lopes

Towards the Complex Interface 118
    Renata Piazzalunga and Fernando A. Basile Colugnati

Knowledge Resources - A Knowledge Management Approach for Digital
Ecosystems 131
    Thomas Kurz, Raimund Eder, and Thomas Heistracher

Launching Context-Aware Visualisations   146
  Jaakko Salonen and Jukka Huhtamäki

A Practical Approach to Identity on Digital Ecosystems Using Claim
Verification and Trust   161
  Mark McLaughlin and Paul Malone

Implementing a Trust Overlay Framework for Digital Ecosystems   178
  Paul Malone, Jimmy McGibney, Dmitri Botvich, and
  Mark McLaughlin

SPAM Detection Server Model Inspired by the *Dionaea Muscipula*
Closure Mechanism: An Alternative Approach for Natural Computing
Challenges   192
  Rodrigo Arthur de Souza Pereira Lopes, Lia Carrari R. Lopes, and
  Pollyana Notargiacomo Mustaro

Towards Autopoietic Computing   199
  Gerard Briscoe and Paolo Dini

Flypeer: A JXTA Implementation of DE Transactions   213
  Amir Reza Razavi, Paulo R.C. Siqueira, Fabio K. Serra, and
  Paul J. Krause

A Research Framework for Interaction Computing   224
  Paolo Dini and Daniel Schreckling

Transformation Semigroups as Constructive Dynamical Spaces   245
  Attila Egri-Nagy, Paolo Dini, Chrystopher L. Nehaniv, and
  Maria J. Schilstra

Numerical and Experimental Analysis of the p53-mdm2 Regulatory
Pathway   266
  Ingeborg M.M. van Leeuwen, Ian Sanders, Oliver Staples,
  Sonia Lain, and Alastair J. Munro

Lie Group Analysis of a p53-mdm2 ODE Model   285
  Gábor Horváth and Paolo Dini

Author Index   305

Collaboration Networks for Innovation and Socio-economic Development: European and Latin American Perspectives on Digital Ecosystems Research, Local Readiness, Deployment Strategies and Their Policy Implications

Lorena Rivera León[1] and Rodrigo Kataishi[2]

[1] Department of Media and Communications
The London School of Economics and Political Science
Houghton Street WC2A 2AE. London, UK
l.rivera-leon@lse.ac.uk
[2] Instituto de Industria, Universidad Nacional de General Sarmiento.
J. M. Gutierrez 1150. Buenos Aires, Argentina
rkataish@ungs.edu.ar

Abstract. International cooperation and knowledge transfer among countries has become increasingly important in the last decades, giving opportunity to a set of multiple interaction programs particularly amongst developed and developing regions. This paper discusses the feasibility of the adoption of Digital Ecosystems (DEs) in the Latin American context, based on the experience of deployment of DEs in the European Union. Different deployment experiences in the European context revealed the need of a methodology for planning and implementing DEs that resulted in a set of tools for measuring the maturity grade of localities related to the deployment of DEs and the need of an impact index for understanding its long-term implications of the dynamics of their implementation. This paper proposes a new methodological framework that integrates concepts related to ICT adoption, connectivity and absorption capacities and recognises the strong influence of social capital over these. The paper concludes with the description of a methodological tool oriented towards the mapping, evaluation and modification of scenarios related to ICT adoption process among multiple agents.

Keywords: Collaboration networks; Digital Ecosystems; Latin America; quantitative- qualitative methodological framework; ICT adoption; absorption capabilities; connectivity; social capital; sustainable socio-economic development; policymaking.

1 Introduction: Collaboration Networks and Digital Ecosystems

Digital Ecosystems (DEs) are context-specific socio-technical systems enabling sustainable socio-economic development at the local/regional scale or within a particular industrial sector, driven by networks of social actors and economic agents, and

F.A. Basile Colugnati et al. (Eds.): OPAALS 2010, LNICST 67, pp. 1–19, 2010.
© Institute for Computer Sciences, Social Informatics and Telecommunications Engineering 2010

entirely reliant on distributed architectures for their technical components. DEs can be seen as socio-technical processes that offer ultimately affordable and trustworthy cooperative solutions through investment and engagement by the local stakeholders. DEs are composed by virtual communities that are connected by an open source and low-cost peer-to-peer infrastructure that minimises transaction costs by allowing more efficient participation of stakeholders in the socio-economic system through the integration and sharing of knowledge in a given territory or sector. As a result, DEs maximise the participation of small enterprises with each other and thus allowing and reinforcing their participation in global value networks. This facilitates their access to regional and global markets through the principle of more participation and collaboration for better competition.

Economists commonly describe Digital Ecosystems as an enabler (tool) of development [1] [2]. From an economics empirical perspective, these collaboration networks are:

- A socio-technical system and process
- A link between the 'micro-economy' and the 'macro-economy'

The latter definition implies that the networks minimise transaction costs within clusters at the regional level through knowledge integration and sharing within the region, and thus through more dynamic Regional Innovation Systems. They maximise the benefits to enterprises in participating to Global Value Chains because, when referring to SMEs and distributed markets, evidence shows that more collaboration could lead to better competition and performance.

The findings and results on deployment plans at the European level confirmed that there are some differences in regional needs, requirements and opportunities for DEs. Typically, the regional variations reflect the differences in innovation capabilities, in enterprises' ICT capabilities, and in the characteristics of the social capital of the region. Nevertheless, regions interested in the deployment of DEs are typically characterised by their commitment to regional development and by their support to regional innovative capabilities.

More precisely, regions interested in the implementation and deployment of Digital Ecosystems are characterised by:

- An interest in mechanisms for sharing and for open diffusion of knowledge within local clusters, supported by the interaction and Europe-wide/ international co-operation between regional/local networks;
- A need for easy-to-use services with high user value;
- A shared interest and support for distributed infrastructures and Open Source; and
- An interest for the promotion of the knowledge "embedded" within local territories, and the recognition of the importance of knowledge sharing and best practices through regional innovation programs and plans.

Based on the practical experiences of deployment strategies and plans studied in several EU funded research projects (i.e. DBE project, the DBE Lazio Project and the OPAALS project), an identification of different uses and applications of Digital

Ecosystems emerged for the regional level of intervention. Four different typologies have been identified within a Regional Digital Ecosystem (RE): Digital Business Ecosystems (DBEs), Digital Ecosystem for Public Administration (DE-PA), Digital Ecosystems for Researchers (or DEs of Knowledge – DEK) and Digital Ecosystems for the Labour Market (or DEs of Work - DEW).

Figure 1 is self-explanatory for understanding the possible co-existence of different ecosystems (with different objectives and purposes) within a RDE. The way a RDE would evolve and be composed in terms of the interaction of different 'ecosystems' is context- specific.

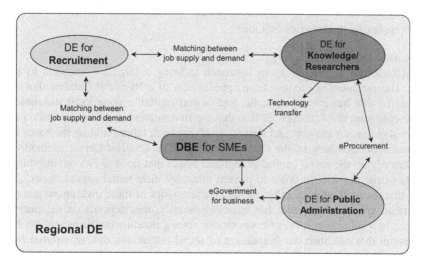

Source: Passani, Rivera León, 2009

Fig. 1. A Regional Digital Ecosystem

DEs are adaptable to different regional applications and needs, and they are thus not exclusive to the business sector. Each region has the opportunity to shape DEs to fit regional and local priorities best.

In the context of their deployment and evaluation, the first concern is to identify potential users and their needs for planning the deployment of these collaboration networks and thus for evaluating and assessing their impact.

2 Planning the Deployment of Digital Ecosystems and Its Evaluation: Past and Present

One of the core objectives of DEs research is to provide regional stakeholders with new approaches and strategies for fostering sustainable regional development.

The following sub-sections discuss different strategies for the introduction of DEs at the regional level, with a focus on practical issues, difficulties and key success factors encountered by regions and key-players that have deployed or are planning the deployment of Digital Ecosystems. The subsection ends with the introduction of the Latin American perspectives on these deployment strategies and the evaluation of the deployment of DEs.

2.1 Existing Tools, Methods and Methodologies for Assessing and Deploying Digital Ecosystems

Feasibility studies are appointed as an essential starting point when planning the deployment of Digital Ecosystems and collaboration networks, in order to relate local needs, plans and strategies to the different technology necessities and requirements.

More complex methodologies include:

The Catalyst-Driver methodology
The methodology is offered as an approach to bring a Digital Ecosystem to local SMEs. The methodology requires the appointment of a Regional Catalyst that is an organisation that has certain strengths and 'social capital' among local stakeholders; and the engagement of 'Driver' SMEs, that are first movers in the construction of the network dynamics. Catalysts and Drivers reinforce each other building the bases for a community of members of the local ecosystem. The Catalyst-Driver methodology relies heavily on the social capital of regional actors that need to be convinced of the benefits of the network in order to protect mutually their social capital 'asset'. This means that the methodology relies on the characteristics of these institutions and their capacity to grow and succeed, but most importantly, this depends on the previous existence of a minimum level of connectivity among institutions, firms and the local government that establish the foundation of social interaction that is required by an efficient collaboration network.

Balanced Scorecard self-assessment system
The Balanced scorecard is a strategic management technique that seeks to make objective and quantitative measures using four headings or 'perspectives' in order to provide a comprehensive and balanced view of an organisation that is able to usefully inform management. In the case of Digital Ecosystems, the four perspectives used are: Financial, User, Business Process, and Development. The aim of the Scorecard is to investigate how the different characteristics of each regional catalyst influence the role and the success of achieving the network objectives. It serves as a complementary tool of readiness and effectiveness of the proposed Regional Catalyst.

Regional Maturity Grade
The Regional Maturity Grade (RMG) serves as a tool for regional analysis. It is a theoretical framework used for interpreting innovative processes at the regional level. It is formed of different techniques of analyses, qualitative and quantitative, that give a complex description of reality, thus becoming a useful instrument for programming and policy interventions.

The RMG function is conformed of three key elements: Social Capital (SC), Innovative Capacity (IC), and the relation between SMEs and ICTs (ICT). The variables analysed under each of the key elements are:

- Social Capital: improvement through networking. The territorial SC has five dimensions and several variables. The first dimension, leadership towards innovation, analyses the entities recognised as innovators in the public and private sectors as a point of reference for obtaining information. The second dimension, Relational Typologies, looks at face-to-face contacts, participation in associations, information sharing, resources sharing, and participation through common plans. Density is the third dimension, analysing the overlaps between various relational typologies, and the rate between potential and real links. The fourth dimension is called Differences, introducing in the network the various typologies of actors (by number and preponderance). Finally, Trust is the fifth dimension, looking at the level of trust per each typology of actors.
- Innovative capacity: change for better competing. Three dimensions are analysed in the IC element: characteristics of the human resources; ability to generate knowledge; and transmission and diffusion of knowledge and access to the market. The first dimension looks at variables such as percentage of the local population with a higher education degree, technical and scientific education, employees in the services sector (high-tech) and employees in the manufacturer sector (medium and high-tech). The second dimension analyses variables like public and private expenditures in R&D; local participation in European projects; international scientific publications; etc. Finally, the variables analysed in the third dimension include percentage of innovative enterprises in the manufacturing and service sectors; total expenditures in innovation in the manufacturing and service sectors; introduction of new products in the market; and the amount and number of investments from early stage venture capitalists.
- Relations between SMEs and ICTs. Two dimensions are analysed. The first one, analyses broadly the characteristics of the business sector. The second one, ICT use by SMEs, analyses variables such as access to networks, and SMEs activities regarding e-business.

The RMG describes the territory from a socio-economic point of view and measures the dynamics and modifications resulting from a development intervention. It is useful for understanding the regional background and it is usually undertaken within feasibility studies to the deployment of the collaboration network.

Social Network Analysis
Social Network Analysis (SNA) is generally used as a preliminary phase when planning the deployment of DEs at the regional level. It helps in characterising and measuring the socio-economic dynamics of a network at the local level in order to evidence relations and the potentials to collaboration, the structure of leadership, the flows of communication and transmission of knowledge, the nature and intensity of the inner links to the territory, and the social capital. The visualisation of the network allows an initial understanding of the processes of collaboration and the temporal changes on them. SNA serves as a tool for understanding collaboration capabilities on an ecosystem-like environment.

2.2 The Construction of an Impact Index for Digital Ecosystems

As the research became more complex there was an increasing interest from policy-makers and researchers on understanding the socio-economic impact of Digital Eco-systems. An impact index of DEs deployments was then defined [4] as an aggregated composite indicator, formed of four evaluation accounts: financial, user, economic development and social. It is an open and scalable tool for assessing the socio-economic impact of DEs at local/regional level; sharing the principles of three different methodologies: impact assessment studies through the 'before-after' approach of project impact assessment (1), methods of valuation of tangible and intangible goods from Value Network Analysis (2), and Multiple Account Cost- Benefit Analysis (3).

In the framework of the construction of the impact index, a multiple-account CBA global methodological approach was used. Multiple-Account CBA evaluation frameworks capture all of the factors considered in a socioeconomic CBA, but present the results in several distinct evaluation accounts. The use of different evaluation accounts allows having a clear description on what the consequences and trade-offs, for instance of the deployment of DEs are.

Four evaluation accounts were designed in order to provide an overall assessment related DEs deployment:

- (Public/Private) Financial Account: net return or cost to 'investors'.
- User/Consumer Account: net benefit to users as 'consumers' of what the collaboration network provides.
- Economic Development Account: micro-economic effects and macro-economic net benefits related to productivity, growth and employment.
- Social Account: Community and social impacts, mainly net benefits on social capital.

Although its complexity and its completeness, the impact index revealed to be ahead as a theoretical framework if compared to real cases of deployment. For instance, the framework did not consider at that point any progress evaluation or ex-ante evaluations or impact assessments. This revealed to be a limitation when trying to implement it.

2.3 The Latin American Context and the Adaptation of the Analytical Framework

The research above was further extended to the Latin American context. The activities undertaken were mainly oriented towards partnership building, knowledge transfer and sharing of experiences in the area of DEs research for development. The central focus of the activities is the extension of existing European networks to Latin American and Caribbean (LAC) research communities. The aim is to stimulate the collaboration between communities and networks engaged in research on multi-disciplinary approaches concerned with the role ICTs and new media have on the evolution and dynamics of the Information and Knowledge Society in the European Union and LAC countries.

One pilot case study was planned in Argentina, province of Buenos Aires, and more specifically in the city of Morón; with applicability to the metal-mechanical industry. When meeting policymakers and researchers of the region, and presenting the 'European' methods for planning the deployment of DEs and assessing/evaluating their implementation, it was revealed that these methods were rather ahead on applicability as no real use cases were already taking place, and thus no real benchmark case (or zero case) existed.

Based on concepts more suitable to international cooperation projects common to th e Latin American context, a new conceptual framework was introduced, taking into account the bases of methodologies for the implementation of projects for development such as the *'Marco Lógico"* approach [5]. This framework also situated all the previous methods and tools developed by previous projects. Figure 2 presents the 'adapted' conceptual framework.

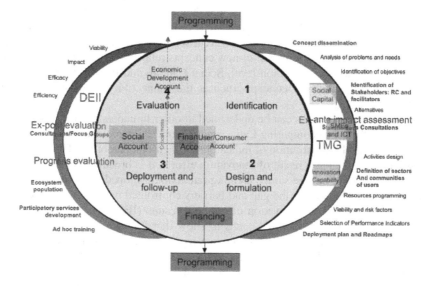

Source: Rivera León, 2009

Fig. 2. Four phases for planning the deployment of Digital Ecosystems

Figure 2 shows four phases related to the deployment of Digital Ecosystems: Identification (1); Design and formulation (2); deployment and follow-up (3); and Evaluation (4). Several deployment steps and requirements complement each phase. The right side of the figure is more related to ex-ante actions previous to the actual deployment of the network. This right side, joined by the red semi-circle, is the 'area' of application of the previously developed Regional/Territorial Maturity Grade methodology. The left side of the figure, joined by the green semi-circle corresponds to the phases of applicability of the DEs Impact Index. Also to note, is that this methodological framework introduces ex-ante impact assessments (i.e. through Stakeholders consultations); progress evaluations; and ex- post evaluations (i.e. through consultations and focus groups). The four accounts of the Impact Index introduced before are situated in the 4 main phases of the deployment process. Their location is related to the moment where data for its completion can be obtained. For instance, data for the

User/Consumer Account and Financial Account can be obtained from the beginning of the process, but only completed until Phase four is finished; the Social Account between phases 3 and 4; whereas the Economic Development Account requires the completion of the deployment process. The arrow pointing upwards in the Economic Development account refers to the time spam between post evaluation of the impact assessment of economic variables (that usually requires a long-period of time to be concretised).

This conceptual framework then raised the question on how could a Latin American region be situated in this framework and theoretically and empirically compared ex-ante to any European deployment case in terms of readiness and possibility of successful deployment, given the considerable structural differences in both regions. This stimulated a debate on the role of collaboration among regional socio-economic stakeholders in Latin America, linkages, and their impact on innovation activities, as well as the role of intermediate institutions in speeding up innovation processes.

The mentioned problems regarding different scales, absence or inadequacy of certain problems enforced the creation of new concepts related to collaboration networks not only in the triad of innovation level, Social Capital, and ICT-SMEs relationship, but also in introducing new concepts such as the connectivity level and absorptive capacities [6] [7] [8].

Two representative cases can be underlined in order to understand the heterogeneity and the difficulty in comparing Latin America and EU deployment cases, particularly considering the constitution of indicators for their comparability. The first one is the use of patent data as a result of innovation efforts. It is broadly known that developing countries have not the same patenting rate than developed countries, especially if one takes in consideration the gap and barriers confronted by SMEs. As a result of using patents to measure innovation efforts and results (IER) in Latin America, an analysis on innovation capabilities tends to show that the patenting level is near to zero in most cases, which means that there is very low level of IER. However, several studies [9] underlined that in spite of the comparative lack of IER in LA, it is possible to measure multiple gradients of efforts and results that not necessarily points to obtaining a new a patent, but impact into the innovation level of the firm (specially in incremental innovations at firm and local market level). The same analysis can be applied to R&D investments: while the raw indicator is near to zero in the private sector in Latin America, it is known that informal R&D teams are more commonly found in developing countries and that their results generally are oriented to incremental solutions to "every day" problems.

Former discussions focused the problem of inter-regional measurement in efforts to develop a set of indicators that could be both broadly comparable but also sensitive enough to collect differences between regions at a qualitative level.

3   New Approach on Measuring Collaboration Networks through Digital Ecosystems

The pre-feasibility analysis of a Digital Ecosystems implementation in a region relies on the idea that the technology involved in a network is not a mechanism that *generates* innovation and connectivity in firms by itself. The main vision, instead, is that

ICTs are a booster of pre-existent relationships and innovative behaviours, allowing the users of these technologies to enrich their productive activities.

This is very important when evaluating the role of ICTs in the implementation of a collaboration network. How does a policy maker know if a region is prepared to transform technology into a boosting mechanism of a local innovation system? What are the priorities for each region to enforce this process?

The methodological framework developed is based on four dimensions that summarise what is considered as the main aspects of ICTs as a networking and innovation booster in a specific context: the absorption capacities level, the connectivity, ICT adoption and the social capital.

3.1 Absorption Capacities

Absorption capabilities are usually defined [6] as the ability of given actors to recognise external information, assimilate it, and apply it within the system these actors are part of. These capacities are not just related to the access to knowledge and information, but also on the ability of identifying useful knowledge and generate new knowledge. Absorption capabilities are not developed automatically. They require time and the development of previous competencies, and are consequently influenced by the framework conditions of the region.

The absorption capacity dimension concentrates indicators of those aspects of the local environment that allow agents to understand and modify their context, such as innovation capabilities and results, human resources qualifications and characteristics, quality management on firms and organizations, among other.

The absorption capacities are key for the firms and organisations in order to understand their environment. This understanding makes them able to achieve their objectives relatively more efficiently. This set of indicators could point out "innovation based agents", able to develop learning process under dynamic and adapting strategies, by analysing their context. On the other hand, very low absorption capabilities will show firms with lack of strategy, with no interest on innovation activities and without qualifications.

3.2 Connectivity

This axe is related to the potential of the relevant stakeholders in the region to establish relationships and linkages with other relevant stakeholders (internal or external to the region). As was the case for the absorption capacities axe, 'connectivity' is not just the simple interaction with other agents, but rather selected exchanges and the capacity to prioritise relationships according to the 'use' that these could give to the (network of) stakeholders.

This axe takes into account not only firm-to-firm linkages, but also institutional and organisational relationships. Also, it is interesting to underline that linkages could be analysed beyond the dual indicator based on the existence of the relationship, for example considering the quality or the impact of that interaction.

Recent literature [10] points that there is a strong relationship between absorption capabilities and connectivity, since the good development of the first dimension in many cases could enrich and empower the frequency and quality on the second one.

3.3 ICT Adoption

ICT adoption takes the form of a requisite for establishing collaboration networks among enterprises through Digital Ecosystems. Although it is a pre-requisite, this approach to development is not a technocentric approach, since the success relies on the capabilities of the region of reference and its actors/agents. Technology thus, through Digital Ecosystems, plays a 'boosting' role of dynamic cooperative processes highly influenced by the interaction of absorption capacities, connectivity and social capital.

This set of indicators is oriented to measure the existence and uses of different technologies related to information and communication, such as basic infrastructures as broadband diffusion and its characteristics, and more complex ones, such as the properties of ICT use in production processes among enterprises.

It is interesting to note that these three dimensions could be analysed separately, but because of the dynamics of their interaction, they should be analysed jointly. This is particularly important when evaluating policies related to the adoption of new technologies.

For instance, first approaches on supporting ICT use in developing countries only considered the ICT adoption dimension, leaving behind both connectivity and absorption capacities. The result is an increase of the *stock* of available technology, but without a business and networking enforcement, because of the lack of previous absorptive capacities (i.e. firms that do not understand the benefits of ICT use) and connectivity (very weak relationships and links between firms and between firms and institutions).

3.4 Social Capital

In the world of economics [7] social factors have been increasingly recognised as central to the competitive challenges of the knowledge-based economy. The density and structure of social networks is crucial for acquiring democratic organisation in a society, and thus also in business interactions. The OECD [8] defines social capital as *"networks together with shared norms, values and understanding that facilitate cooperation within or among groups"*. Following Steinmueller lines, 'social capital' considers formal organisations in the society and social networks, including 'communities of practice'. The concept of 'social capital' is thus central for a complete analysis of collaboration networks for innovation.

Under this conceptual framework, a region (political geographical concept) is understood as a localised system of interaction in which networks of stakeholders play a central role. This recognition supports the analysis of the role of social capital and knowledge exchange as relevant elements for local innovation. Generation and accumulation of social capital is mostly based on trust and made possible as actors share norms, values and understanding [9].

The effect of stronger social capital tends to generate inertia towards the convergence of a successful implementation of ICTs in boosting regional development. A weak social capital creates obstacles in achieving this objective.

The above discussion can be graphically represented as shown in Figure 3. Each of the axes, or determinants, described above corresponds to one of the spheres in the figure. The arrows try to express a quantitative maximisation exercise, as it will be discussed in the next section. The graph shows that in the maximisation of the three axes (i.e.. where the three spheres overlap) there is a 'successful' area related to the deployment of DEs for innovation in a given region. As it will be discussed below, this framework allows the graphical identification of studied stakeholders (i.e. enterprises) in the region. The equilibrium, or 'success area', would be when most of the enterprises in the region are located in the intersection of the 3 axes.

Figure 3 would be better represented in a tri-dimensional way (this is why social capital is expressed geometrically as a cylinder). For this effect, social capital would act as an 'elevator' to the maximisation process of the three axes. For example, a strong social capital in the region would 'push up' the three axes, and thus accelerate reaching maximum values in each axe. The contrary would happen with a low level or weak social capital.

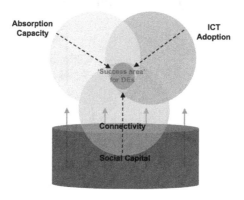

Source: Rivera León, Kataishi, 2009

Fig. 3. Determinants of the 'successful' deployment of collaboration networks through Digital Ecosystems

4 How to Map These Complex Socio-economic Interactions?

As it was mentioned in the previous section, the presented conceptual framework is able to map local readiness and the structure of socio-economic interactions in a given region towards the deployment of DEs. This is possible thanks to a (mathematic) maximisation process through the use of several indicators serving as proxy variables for the definition of the concepts descripted above.

Several secondary and primary sources exist in the EU to approximate quantitatively the axes of the conceptual framework. An exhaustive list of variables, indicators and data sources are presented in the Appendix A of this paper.

In Latin America, however, the statistical systems are generally less developed, generating comparability problems not only between Latin American countries and the EU, but also amongst many LA countries. These problems are particularly important when looking at S&T indicators, use and diffusion of ICTs and the performance of institutions.

Regarding this lack of systemic and periodic official information in many Latin American countries, particularly in a number of aspects that concerns the sets of indicators that were proposed before, the main methodological challenges are: to generate primary data from specially-build case studies and surveys; and to study the availability of secondary sources including those available from other research projects.

Three main surveys were developed in the scope of the Latin American case study undertaken in the city of Morón, province of Buenos Aires, Argentina. One of them was oriented to understanding ICT use in the manufacturing industry in Morón. The second focused on understanding the general characteristics of the metal mechanical industry; and the third one was based on an in-depth interviews methodology in order to map the behavior, the relationships and the influence over other manufacturing

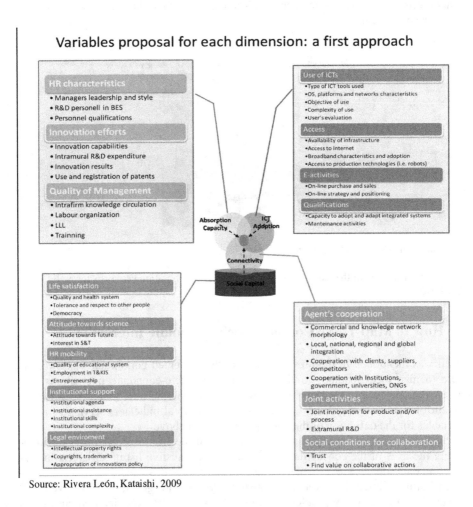

Source: Rivera León, Kataishi, 2009

Fig. 4. A proposal of the set of indicators and variables used to map complex socio-economic interactions through Digital Ecosystems

industries of organisations and institutions (NGOs, chambers of commerce, public bodies). Additionally, these interviews also captured the interests and business strategies of enterprises in the industry (i.e. through interviews with the CEOs of the main companies). Regarding secondary data for understanding the structure of the region, the study focused on analysing the results of 15 in-depth interviews of existent research projects, mainly focused on NGOs.

Figure 4 presents the complete list of indicators used to map each dimension of the study. This is a schematic view focused on showing how this methodology works in practice. The set of indicators can be amended depending on the case of study and data availability.

There is also a sort of hierarchy among indicators. Primary indicators are represented in the figure with bold names and are set as the 'head' of sub-sets of indicators. Those sub-sets of variables are usually ones that cannot be replaced because of their importance to the dimension considered.

As a result of the normalisation and addition of the different sub-sets of indicators, subtotals for each sub-dimension can be built. This process allows to make a normalized 0-1 value for each dimension considered, that consequently allows the representation of these several aspects in a three axes graph (figure 5).

Mapping quantitatively the axes above is very much related to mathematical integration. For example, taking an integral can also be described as a process of dimensional reduction, i.e. going from many variables to few variables, and thus losing some information of the studied phenomenon. This is particularly common in all statistical exercises and in economic models trying to model reality.

The analytical framework is a very expressive qualitative tool (the 3 domains/axes: ICT adoption, connectivity, absorption capacity, influenced by social capital); and quantitative (the triangular coordinates of the maximum of the same three axes) representation of data that characterises well the properties and characteristics of the business collaborative networks in the studied region.

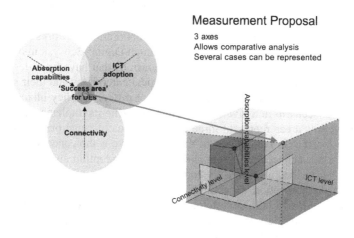

Source: Rivera León, Kataishi,2009

Fig. 5. Mapping the characteristics of enterprises in a given region

The benefits are on the relatively simple way to achieve a graphical quick view of the relative positioning of each region studied. Also, this scheme allows researchers and/or policymakers to see where the strengths and weaknesses of the region are. Figure 5 presents an example of the mapping exercise and characterisation of the enterprises in a studied region.

In this figure, the green point located at the maximum values of the three axes is also found in the 3D cube on the right of the picture, explaining the tri-dimensional aspects of the methodological framework. This green spot is in fact representing 'one given' enterprise. Examples of other enterprises are the red spots on the cube. While one has similar values in terms of the indicators in the axes of ICT adoption and connectivity, while having lower levels of absorption capabilities; the other has high levels of connectivity and absorption capabilities while having lower ICT adoption.

The average of all the dots, which can be taken exactly or which can be easily estimated by the human eye aided by the three circles, gives a single number and graphical representation that not only sums up the contributions of all the enterprises in a clear way, but, by so doing, gives a very significant global indicator for the region that is comparable no matter the differences of the framework conditions (i.e. between the developed and the developing world) as the framework in itself represents them internally already. In other words, it presents macroscopic information from the aggregation of microscopic information.

5 Policy Mixes for Enterprises and Regions

In the previous sections an approach and measurement proposal for following an ICT implementation process through DEs was introduced. This however should be complemented with a set of policy actions related to the identified stages on the deployment process. The objective of this section is to describe the linkages between the presented tools and specific policy actions that are able to modify certain scenarios.

Being able to map enterprises in the way described above also has strong implications in terms of policymaking and the implementation of different policy mixes in the studied regions towards the goal of building collaboration networks for innovation and sustainable socio-economic development through Digital Ecosystems.

Figure 6 represents graphically the three dimensions that have been described before together with the representation of hypothetical enterprises and their situation in the framework. This is shown together with sets of different policy mixes applicable to each possible scenario in a given studied region during its implementation process. The shaded yellow areas represent three different types of situations in which enterprises can be located. In the top part of the figure, there is the conceptual case where enterprises are not connected (1), but show both ICT adoption and absorption capabilities in higher to medium levels. Secondly, the right area of the picture shows enterprises with no absorption capabilities (2), but that are highly connected among them and with their institutional system, together with an important presence and use of ICTs. Finally, in the left area of the figure there are the enterprises with no ICTs (3) but with good levels of absorption capabilities and connectivity.

The shaded blue areas represent areas where the located enterprises just have good levels of one of the 3 axes. For enterprises located in these shaded areas a combination of policies is required as marked with the blue arrows in the figure.

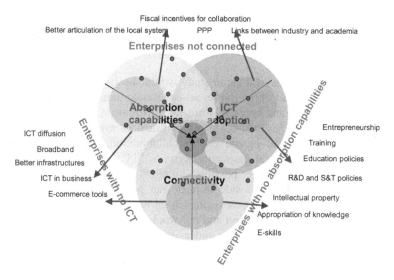

Source: Rivera León, Kataishi, 2009

Fig. 6. Policy mixes for enterprises and regions

It is important to underline the main steps that lead to the policymaking focus of this section. In Figure 5, the measurement proposal showed the importance of mapping the interactions in 3D (3 axes) in addition to a 2D representation (circles within axes). Following this methodology, the feasibility studies would try to map the condition in which the agent's actions take place for each location of a given enterprise, in order to identify behavioral-based clusters. It is expected that relatively homogenous characteristics among firms will be shown, together with some outlier cases that will have to be analysed in isolation. This will end in a similar representation to the one shown in Figure 6, showing firms positioned following the results of a mix of indicators described and explained extensively in Figure 4; but also positioned within three dimensions, enabling the diagnosis of the situation of a given region as a whole.

As it is well-know, the elaboration of policy mixes is not an easy task, especially in developing countries [10], since it implies several sets of restrictions that affect not only the elaboration of policies per-se but also this affects the effectiveness of interventions given the institutional and resources related limitations.

Despite this, this proposal is focused on the development of policies oriented towards specific issues that would be detected through the methodological process in itself, and *ceteris paribus* through the limitations and restrictions identified through previous research and in-depth analyses of the region.

Policymaking can be defined using multiple criteria. Two of them are of the interest of this paper. On one hand, the mechanisms that may lead policymaking processes could be based on the average position of firms shown in the 3D representation. On the other hand, policies could be oriented to the benefit of a given critical group, characterized by the lack of a number of attributes (i.e. a group of enterprises with no connectivity).

The area where the majority of firms would be positioned would determine the policy focus. As it is shown in Figure 6, the lack of some elements implies the relatively precise set of tools that would influence and modify the particular disposition of firms within the approach.

For example, if the results of a given study would show that most of the 'agents' are positioned in the top part of the figure, actions should then be oriented to simulate connectivity among firms and between firms and institutions. Having this acknowledgement, there would be several sets of policy practices available for governments to implement, such as fiscal incentives for collaboration, promotion of linkages among components of the local innovation system, PPP oriented policies, etc. The same logic can be applied to other possible situations derived from the Figure: concentration for firms on the right side implies that policies should be oriented to enforce firm's capabilities. If they were situated on the left side, the need for implementing efforts towards ICT adoption and use would be revealed. The possible policymaking actions are determined by the under- development and under-representation of the considered dimensions. It is also possible to find situations where interventions on more than one spheres are needed.

One last thing to consider is the relative importance of each considered dimension and their practical implications (absorption capabilities, connectivity and ICT adoption). Traditional interventions in developing countries are normally focused on the left side of Figure 6, relegating both, connectivity issues but specially absorption capabilities in enterprises.

In comparison, ICT related deficiencies are usually associated with resources driven polices, as the main goal is to 'connect' these technologies to users or firms, and the first step usually consists in the acquisition of the technology in itself, which can be done form example through financial assistance and support. Connectivity and absorption capabilities are more complex issues that would probably require much more than only financial resources, but rather medium and long-term strategic approaches to policymaking.

From this perspective, the key element that emerges from this complex system [6] appears to be the absorption capabilities of agents. Nevertheless, the modification of each parameter is very much related to generic development policy that involves education, Science and Technology and socio-economic transformation issues, among others.

The proposed methodological framework has the faculty of giving a macro-picture based on microeconomic data, and most importantly it makes able to drive qualitative conclusions based on quantitative data. It is also an ideal tool for measuring regional performance in the developing world, with the view of comparing this performance with developed countries without the logical bias related to the different stages of development. The implications in terms of policymaking are large and are to be exploited when introducing innovation policies.

6 Conclusions

This paper analyses the implementation process of Digital Ecosystems in multiple regions. The first section was oriented to summarize the state of the art of DEs definitions and goals.

Secondly, this work underlined the relevance of DEs as an approach that considers technology adoption as a social process, driven by social networking actions and collaboration. This implies at least three important elements. First, there has to be a form of collaboration among social actors (in other words, connectivity among agents is required for the implementation of DEs. Second, it is required that social actors have something to offer to other actors, otherwise the collaboration may not be possible and/or sustainable (it can be said that learning capabilities and knowledge storage of agents –absorption capabilities- have to be heterogeneous and close to the state of the art, and possibly determined by the local common practices). Third, it is supposed that ICTs can perform a "booster" role of both connectivity and absorption capabilities.

Having this into account, the paper described four spheres that represent the D.E. implementation scenarios: the social capital, the ICTs adoption level, the agent connectivity and absorptive capacities level. One of the main contributions of this work is the development of a measurement methodology capable of quantify the variables present in the four considered dimensions. In this sense, groups of indicators that are available in through European statistics and in some Latin American countries have been identified.

The second main contribution of the paper is the proposal of a representation of firms in a 3D graph that considers the dimensions previously underlined, in order to identify the deficiencies in a region and make an accurate diagnostic regarding the challenges that the implementation of DEs implies in a specific context. Finally, the proposed methodology allows to orient policymakers towards a set of actions to be taken into account during the implementation process.

References

[1] Moore, J.: The death of competition: leadership and strategy in the age of business eco-systems. Harper Business, New York (1996)
[2] Nachira, F., Nicolai, A., Dini, P.: Introduction – The Digital Business Ecosystems: Roots, Processes and Perspectives. In: Digital Business Ecosystems Book, European Commission (2007)
[3] Rivera León, L., Passani, A., Pennese, F.: D11.8: Preliminary study on methodologies for DE socioeconomic impact analysis. OPAALS Project (2009)
[4] Van Egeraat, C., O'Riain, S., Kerr, A., Sarkar, R., Chatterjee, J., Stanley, J.: A Research Agenda for bridging Digital Eosystems to regional development and innovation in the Knowledge Economy – Preliminary Report, WP11 OPAALS Project (2008)
[5] Camacho, H., Cámara, L., Cascante, R., Sainz, H.: El enfoque del marco lógico: 10 casos prácticos. Cuaderno para la identificación y diseño de proyectos de desarrollo. Fundación CIDEAL, Acciones de Desarrollo y Cooperación, Madrid (2001)
[6] Yoguel, G., Robert, V., Milesi, D., Erbes, A.: El desarrollo de las redes de conocimiento en Argentina. Programa Economía del Conocimiento en América Latina, IDRC-FLACSO, Mexico (2009)
[7] Van Egeraat, C., O'Riain, S., Kerr, A., Sarkar, R., Chatterjee, J., Stanley, J., Rivera León, L., Passani, A.: A Research Agenda for bridgin Digital Ecosstems to regional development and innovation in the Knowledge Economy – Preliminary Report, OPAALS Project WP11 D11.1 (2008)

[8] OECD: The well-being of nations: the role of human and social capital (2001)
[9] Bruno, N., Miedzinski, M., Reid, A., Ruiz Yaniz, M.: Socio-cultural determinants of innovation. Technopolis, Europe Innova Innovation Watch, Systematic Project (2008)
[10] Vessuri, H.: Science, politics, and democratic participation in policy-making: a Latin American view. Technology in Society 25(2), 263–273 (2003)

Appendix A – Set of Variables, Indicators and Data Sources for Measuring Collaboration Networks for Innovation through Digital Ecosystems in the EU

Category in new methodological framework	Indicator	Data source	Level of availability	Available years
Connectivity	Product innovation jointly with other enterprises or institutions	Question 2.2 (INPDTW) 'Who developed these product innovations'. Answer: 'Your enterprise together with other enterprises and institutions'. CIS IV, survey data	Only Spain and Italy/enterprise sizes and NACE sectors	2002 - 2004 data, gathered in 2005, forming the CIS IV 2004
	Process innovation jointly with other enterprises or institutions	Question 3.2 (INPCSW) 'Who developed these product innovations'. Answer: 'Your enterprise together with other enterprises and institutions'. CIS IV, survey data	Only Spain and Italy/enterprise sizes and NACE sectors	2003 - 2004 data, gathered in 2005, forming the CIS IV 2004
	Extramural R&D	Question 5.1 (RRDEX) 'Did your enterprise engage in the following innovation activities: Extramural R&D'. Answer: 'Yes'. CIS IV, survey data	Only Spain and Italy/enterprise sizes and NACE sectors	2003 - 2004 data, gathered in 2005, forming the CIS IV 2004
	Acquisition of external knowledge	Question 5.1 (ROEK) 'Did your enterprise engage in the following innovation activities: Acquisition of other external knowledge'. Answer: 'Yes'. CIS IV, survey data	Only Spain and Italy/enterprise sizes and NACE sectors	2003 - 2004 data, gathered in 2005, forming the CIS IV 2004
	Public financial support for innovation activities: local or regional authorities	Question 5.3 (FUNLOC) 'Did your enterprise receive any public financial support for innovation activities from: local or regions authorities'. Answer: 'Yes'. CIS IV, survey data	Only Spain and Italy/enterprise sizes and NACE sectors	2003 - 2004 data, gathered in 2005, forming the CIS IV 2004
	Public financial support for innovation activities: central government	Question 5.3 (FUNGMT) 'Did your enterprise receive any public financial support for innovation activities from: central government including central government agencies or ministries'. Answer: 'Yes'. CIS IV, survey data	Only Spain and Italy/enterprise sizes and NACE sectors	2004 - 2004 data, gathered in 2005, forming the CIS IV 2004
	Public financial support for innovation activities: the European Union	Question 5.3 (FUNEU) 'Did your enterprise receive any public financial support for innovation activities from: the European Union'. Answer: 'Yes'. CIS IV, survey data	Only Spain and Italy/enterprise sizes and NACE sectors	2005 - 2004 data, gathered in 2005, forming the CIS IV 2004
	Information and co-operation for innovation activities: suppliers	Question 6.1 (SSUP) 'How important to your enterprise's innovation activities were the following information sources: suppliers of equipment, materials, components, or software'. Answer: 'High'. CIS IV, survey data	Only Spain and Italy/enterprise sizes and NACE sectors	2005 - 2004 data, gathered in 2005, forming the CIS IV 2004
	Information and co-operation for innovation activities: clients	Question 6.1 (SCLI) 'How important to your enterprise's innovation activities were the following information sources: clients or customers'. Answer: 'High'. CIS IV, survey data	Only Spain and Italy/enterprise sizes and NACE sectors	2005 - 2004 data, gathered in 2005, forming the CIS IV 2004
	Information and co-operation for innovation activities: competitors	Question 6.1 (SCOM) 'How important to your enterprise's innovation activities were the following information sources: competitors or other enterprises in your sector'. Answer: 'High'. CIS IV, survey data	Only Spain and Italy/enterprise sizes and NACE sectors	2005 - 2004 data, gathered in 2005, forming the CIS IV 2004
	Information and co-operation for innovation activities: private researchers	Question 6.1 (SINS) 'How important to your enterprise's innovation activities were the following information sources: consultants, commercial labs, or private R&D institutes'. Answer: 'High'. CIS IV, survey data	Only Spain and Italy/enterprise sizes and NACE sectors	2005 - 2004 data, gathered in 2005, forming the CIS IV 2004
	Information and co-operation for innovation activities: universities	Question 6.1 (SUNI) 'How important to your enterprise's innovation activities were the following information sources: Universities or other higher education institutions'. Answer: 'High'. CIS IV, survey data	Only Spain and Italy/enterprise sizes and NACE sectors	2005 - 2004 data, gathered in 2005, forming the CIS IV 2004
	Information and co-operation for innovation activities: government	Question 6.1 (SGMT) 'How important to your enterprise's innovation activities were the following information sources: government or public research institutes'. Answer: 'High'. CIS IV, survey data	Only Spain and Italy/enterprise sizes and NACE sectors	2005 - 2004 data, gathered in 2005, forming the CIS IV 2004
	Co-operation for innovation activities	Question 6.2 (CO) 'Did your enterprise co-operate on any of your innovation activities with other enterprises or institutions'. Answer: 'Yes'. CIS IV, survey data	Only Spain and Italy/enterprise sizes and NACE sectors	2005 - 2004 data, gathered in 2005, forming the CIS IV 2004
	Enterpises using the Internet for interacting with public authorities	Percentage of enterprises using Internet for interaction with public authorities - for obtaining information	Countries	2003 - 2008
	E-government usage by enterprises	Percentage of enterprises which use the Internet for interaction with public authorities	Countries	2003 - 2008
Absorptive capabilities	Students on tertiary education levels as a percentage of total students. Levels 5-6 (ISCED 1997)	Eurostat Regional statistics/Regional Education Statistics	NUTS 2	1998 - 2006
	Employment in technology and knowledge-intensive sectors	Annual data on employment in technology and knowledge-intensive sectors at the regional level. Percentage of HTEC on Total (all NACE branches). Eurostat regional statistics/High-technology manufacturing and knowledge-intensive services sectors	NUTS 2	1996 - 2007
	Job-to-job mobility of employed HRST	Annual data on job-to-job mobility of HRST, employed, between 25 and 64 years, percentage	Countries	1996 - 2007 (no data for the EU25)
	R&D personnel in Business enterprise sector (BES)	Total R&D personnel (researchers/RSE and technicians-equivalent staff/TEC) as a percentage of active population. Eurostat regional statistics/Research and Development	NUTS 2	1997 - 2007
	Patent applications	Number of patent applications to the EPO per million of inhabitants. Eurostat regional statistics/Patent applications to the EPO (European Patent Office) by priority year.	NUTS 2	1994 - 2005
	Registration of industrial designs	Question 9.1 (PRODSG) 'Did your enterprise registered an industrial design'. Answer: 'Yes'. CIS IV, survey data	Only Spain and Italy/enterprise sizes and NACE sectors	2005 - 2004 data, gathered in 2005, forming the CIS IV 2004
	Trademarks	Question 9.1 (PROTM) 'Did your enterprise registered a trademark'. Answer: 'Yes'. CIS IV, survey data	Only Spain and Italy/enterprise sizes and NACE sectors	2005 - 2004 data, gathered in 2005, forming the CIS IV 2004
	Copyright	Question 9.1 (PROCP) 'Did your enterprise claimed a copyright'. Answer: 'Yes'. CIS IV, survey data	Only Spain and Italy/enterprise sizes and NACE sectors	2005 - 2004 data, gathered in 2005, forming the CIS IV 2004
	Life long learning	Participation of adults aged 25-64 in education and training. Thousands. Eurostat regional statistics/Regional socio-demographic labour force statistics - LFS adjusted series.	NUTS 2	1999 - 2008
	Intramural R&D expenditure (GERD)	Total intramural R&D expenditure (GERD) in the business enterprise sector (BES) as a percentage of total GERD. Euros per habitant. Eurostat regional statistics/Research and Development.	Countries	1997 - 2008
	Intramural (in-house R&D)	Question 5.1 (RRDIN) 'Did your enterprise engage in the following innovation activities: Intramural (in-hourse) R&D''. Answer: 'Yes'. CIS IV, survey data	Only Spain and Italy/enterprise sizes and NACE sectors	2002 - 2004 data, gathered in 2005, forming the CIS IV 2004
	Business investment	Business investment - Gross fixed capital formation by the private sector as a percentage of GDP	Countries	1997 - 2008
	Training	Question 5.1 (RTR) 'Did your enterprise engage in the following innovation activities: Training'. Answer: 'Yes'. CIS IV, survey data	Only Spain and Italy/enterprise sizes and NACE sectors	2002 - 2004 data, gathered in 2005, forming the CIS IV 2004
	Risk aversion/entreprenuership	Eurobarometer flash survey No. 160. Question 12 'One should not start a business if there is a risk it might fail' Answer: Agree. Page 102 report.	Countries	2004 (date of fieldwork)

	Access to Internet	Enterprises having access to the Internet, % of total	Countries/different size of enterprises	2003 - 2008
	Enterprises having purchased via computer mediated networks. Percentage of total enterprises.	Eurostat/Policy indicators	Countries/different size of enterprises	2008
	Enterprises having received orders via computer mediated networks. Percentage of total enterprises.	Eurostat/Policy indicators	Countries/different size of enterprises	2008
	Enterprises with broadband access. Percentage of total enterprises.	Eurostat/Policy indicators	Countries/different size of enterprises	2003 - 2008
	Enterprises having purchased on-line over the last calendar year (at least 1%). Percentage of total enterprises.	Eurostat/Policy indicators	Countries/different size of enterprises	2003 - 2008
	Enterprises having received order on-line over the last calendar year (at least 1%). Percentage of total enterprises.	Eurostat/Policy indicators	Countries/different size of enterprises	2003 - 2008
	Enterprises which have received orders via Internet over the last calendar year (excluding manually typed e-mails). Percentage of total enterprises.	Eurostat/Policy indicators	Countries/different size of enterprises	2003 - 2007
ICT adoption	Enterprises which have ordered via Internet over the last calendar year (excluding manually typed e-mails). Percentage of total enterprises.	Eurostat/Policy indicators	Countries/different size of enterprises	2004 - 2007
	Enterprises using LAN and Internet or extranet in reference year. Percentage of total enterprises.	Eurostat/Policy indicators	Countries/different size of enterprises	2003 - 2008
	Enterprises using open source operating systems. Percentage of total enterprises.	Eurostat/Policy indicators	Countries/different size of enterprises	2007 - 2008
	Enterprises selling on the internet and offering the capability of secure transactions. Percentage of total enterprises.	Eurostat/Policy indicators	Countries/different size of enterprises	2007 - 2008
	Enterprises's turnover from e-commerce	Share of enterprises' turnover on e-commerce. Enterprises' receipts from sales through electronic networks as a percentage from total turnover. I2010 benchmarking indicators	Countries/different size of enterprises	2003 - 2008
	Automated data exchange with customers or suppliers	Percentage of enterprises using automated data exchange with customers or suppliers	Countries	2008
	E-invoices	Enterprises sending and/or receiving e-invoices. Percentage of total enterprise with at least 10 persons employed. Survey data	Countries	2007 - 2008
	Use of software solutions for analysing clients	Enterprises using software solutions, like CRM (Customer Relation Management) to analyse information about clients for marketing purposes. Percentage of enterprises with at least 10 persons employed. Survey	Countries	2007 - 2008
	Availability of IT systems for employees working from home	Enterprises having remote employed persons who connect to the enterprise's IT systems from home, %	Countries/different size of enterprises	2006
	Interest in science and technology	Eurobarometer special survey No. 224. Question A1.5 'Interest in new inventions and technologies' Answer: Very interested. Page 169 report.	Countries	2005 (date of fieldwork)
	Attitude towards risk from new technologies	Eurobarometer special survey No. 224. Question A15b.6 'If a new technology poses a risk that is not fully understood, the development of this technology should be stopped even if it offers clear benefits' Answer: Disagree. Page 269 report.	Countries	2005 (date of fieldwork)
	Attitude towards science	Eurobarometer special survey No. 224. Question A12a.1 'Science and tecnology makes our lives healthier, easier and more comfortable' Answer: Agree. Page 226 report.	Countries	2005 (date of fieldwork)
	Attitude towards future	Eurobarometer special survey No. 225. Question B7.2 'The next generation will enjoy a better quality of life than we do now' Answer: Agree. Page 153 report.	Countries	2005 (date of fieldwork)
Social capital	Trust	Eurobarometer special survey No. 225. Question B6 'In general, would you say that you trust other people' Answer: Trust. Page 156 report.	Countries	2005 (date of fieldwork)
	Democracy	Eurobarometer special survey No. 223. Question D1.13 'Please tell me if you are satisfied with the way democracy works in your country' Answer: Satisfied. Page 7 report.	Countries	2004 (date of fieldwork)
	Quality of the educational system	Eurobarometer special survey No. 223. Question D2.4 'Please tell me if you are satisfied with the quality of the educational system' Answer: Satisfied. Page 16 report.	Countries	2004 (date of fieldwork)
	Quality of the health system	Eurobarometer special survey No. 223. Question D2.6 'Please tell me if you are satisfied with the quality of the health system' Answer: Satisfied. Page 19 report.	Countries	2004 (date of fieldwork)
	Tolerance and respect for other people	Eurobarometer special survey No. 225. Question B5.6 'Please indicate how important you consider it to be tolerance and respect for other people' Answer: Important. Page 148 report.	Countries	2005 (date of fieldwork)
	Life satisfaction	Eurobarometer special survey No. 225. Question B6 'How satisfied are you with the life you lead' Answer: Absolutely satisfied. Page 151 report.	Countries	2005 (date of fieldwork)

Exploring the Role of Value Networks for Software Innovation

Lorraine Morgan[1] and Kieran Conboy[2]

[1] Lero – the Irish Software Engineering Research Centre, University of Limerick
Limerick, Ireland
lorraine.morgan@.ie
[2] National University of Galway, Ireland, Galway, Ireland
Kieran.conboy@nuigalway.ie

Abstract. This paper describes a research-in-progress that aims to explore the applicability and implications of open innovation practices in two firms – one that employs agile development methods and another that utilizes open source software. The open innovation paradigm has a lot in common with open source and agile development methodologies. A particular strength of agile approaches is that they move away from 'introverted' development, involving only the development personnel, and intimately involves the customer in all areas of software creation, supposedly leading to the development of a more innovative and hence more valuable information system. Open source software (OSS) development also shares two key elements of the open innovation model, namely the collaborative development of the technology and shared rights to the use of the technology. However, one shortfall with agile development in particular is the narrow focus on a single customer representative. In response to this, we argue that current thinking regarding innovation needs to be extended to include multiple stakeholders both across and outside the organization. Additionally, for firms utilizing open source, it has been found that their position in a network of potential complementors determines the amount of superior value they create for their customers. Thus, this paper aims to get a better understanding of the applicability and implications of open innovation practices in firms that employ open source and agile development methodologies. In particular, a conceptual framework is derived for further testing.

Keywords: Networks, agile development, open source software, open Innovation.

1 Introduction and Research Motivation

Traditional theoretical and empirical studies of innovation concentrated on explanatory factors internal to firms such as investments in R&D and the production and acquisition of patents (Landry and Amara, 2001). The approach in which organisations generate, develop and commercialise their own ideas belong to the closed model of innovation (Fasnacht, 2009). According to Chesbrough (2003) closed

F.A. Basile Colugnati et al. (Eds.): OPAALS 2010, LNICST 67, pp. 20–30, 2010.

innovation is a view that successful innovation requires control and that firms need to be strongly self-reliant because of uncertainty with quality, availability and capability of others' ideas. Traditionally, new business development processes and the marketing of new products took place within the firm boundaries and exclusively with internal resources. Within the closed model, the innovation process is characterised by firms that invest in their own R&D, employing smart and talented people in order to outperform their competitors in new product and service development. In addition, after producing a stream of new ideas and inventions, firms must defend their intellectual property thoroughly against the competition (Dahlander and Gann, 2007).

However, more recent theories of innovation lay much emphasis on the importance of knowledge and networking. According to Nonaka et al. (2003) and Tidd et al. (2005), successful innovation is the result of combining different knowledge sets and such knowledge is frequently to be found outside the organization (Chesbrough, 2003; De Wit et al. 2007). As March and Simon (1958) suggest, most innovations come from borrowing from others rather than from inventing. In addition, changes in society and industry have led to an increased availability and mobility of knowledge workers and the development of new financial structures like venture capitalism. This has resulted in the boundaries of innovation processes to start breaking up (Chesbrough, 2003a) and the do-it-yourself mentality becoming outdated (Gassmann, 2006). It has been suggested that a paradigm shift is taking place in how companies commercialize knowledge, resulting in the boundaries of a firm eroding. This has been characterized as a move towards 'Open Innovation'. This concept has been defined as "the use of purposive inflows and outflows of knowledge to accelerate internal innovation, and expand the markets for external use of innovation, respectively" (Chesbrough, 2003a: XXIV). A general theme underling open innovation is that firms cannot continue to look inward in their innovation processes, isolating themselves from possible partners, collaborators and competitors. In other words, open innovation invites firms to open up their boundaries to achieve a flexible and agile environment. Thus, it is necessary that firms develop processes to ensure a flow of ideas across its boundaries while seeking input from network players such other companies, which include competitors, as well as customers, suppliers, third parties complementors etc. Conducting open innovation in the firm has various potential benefits, such as access to a wider knowledge pool (Chesbrough 2003), higher innovative performance (Laursen and Salter 2006) and, ultimately, higher market value (Alexy 2008). The open innovation paradigm shares some commonalities with both open source and agile development methodologies, both of which are described in more detail below. The remainder of paper focuses on a conceptual framework for future research followed by a conclusion.

1.1 Agile Development

Agile methods have emerged in recent years as a popular approach to software development. Some of the most popular include eXtreme Programming (XP) (Beck, 2000) and Scrum (Schwaber and Beedle, 2002). These methods have been well received by those in the system development community and there is strong anecdotal evidence to suggest that awareness and indeed use of these methods is highly prevalent across the community. Agile has been described as 'the business of innovation', relying on people and their creativity rather than on processes" (Highsmith and Cockburn, 2001). Highsmith (2002a) contends that "agile approaches are best employed to explore new ground and to power teams for which innovation and creativity are paramount". Indeed, agile methods stress individuals and interactions over processes and tools; working software over comprehensive documentation; customer collaboration over contract negotiation; and responding to change over following a plan (Highsmith and Cockburn, 2001). A particular strength of agile approaches is that they move away from 'introverted' development, where the team building the system are detached from the customer. Instead, agile approaches continually involve the customer in the development process, supposedly leading to the development of a more innovative and hence more valuable information system (Beck, 1999; Schwaber & Beedle 2002). Thus, agile methods, given their flexible and light-weight processes, place emphasis on close communication and collaboration in project teams (Beck, 2000; Schwaber and Beedle, 2002). Despite these claims, however, there is a lack of understanding of what constitutes innovation in software development in general and to which extent agile methods actually facilitate innovation. This is part of a much larger problem in terms of agile method research, where many benefits are claimed, but rigorous conclusive research to support these claims is lacking (Abrahamsson et al., 2009; Conboy, 2009).

Stakeholder involvement is considered imperative to the creative process, yet many with vested interest are never involved (Nonaka and Takeuchi 1995; Amabile 1996; Ekvall 1996; Mathisen and Einarsen 2004). In a truly creative environment, an organization's internal and external communication boundaries should be as porous as possible (Leonard-Barton 1995). While the customer plays an essential part in the agile process, this practice could be extended to include multiple stakeholders and even other organizations. We propose that it is useful to consider how the agile innovation process can benefit from becoming more 'open', e.g., by opening up the boundaries of a systems development entity to include other stakeholders besides the customer.

1.2 Open Source Development

Open source software has significantly transformed from its free software origins to a more mainstream, commercially viable form (Feller et al., 2008; Fitzgerald, 2006; Agerfalk et al., 2005). Indeed it demonstrates two key elements of the open innovation concept – namely the collaborative development of the technology and shared rights to the use of that technology (West and Gallagher, 2006). In other words, OSS allows more people to be involved in the process of software development besides the developers within the boundaries of a firm (Lee and Cole 2003). It has even been suggested that open source is the most prominent example of the revolutionizing of traditional innovation processes, the enabling factors of which include short design-build-test cycles, new releases with low transaction costs and a great number of ideas that are enabled by the number of programmers that are involved worldwide (Gassmann and Enkel, 2006). In its emergent form, OSS represented a community-based software development model where geographically dispersed programmers collaborated to produce software (West and O'Mahony, 2005). However, OSS has since transitioned into the realm of mainstream business and plays an important role in the business models for firms in high technology and other industries (Rajala, 2008; Fitzgerald, 2006; Overby et al., 2006).

The benefits of adopting OSS have been well documented in the existing literature (Morgan & Finnegan, 2007a, 2007b; Ven and Verelst, 2007). Reduced licensing fees, escape from vendor lock-in, increased quality and performance were just some of the benefits cited in this literature. Additionally, Morgan and Finnegan's (2007a) study of thirteen firms that had complete or partial adoption of OSS revealed that the increased collaboration and innovation allowed by OSS ranked two of the highest benefits. Further research by the same authors (Morgan and Finnegan, 2008) revealed that firms experienced many opportunities in collaborating with other companies, research institutes and OSS communities. Working as part of a value network enabled these firms to capture value in the form of competencies and tacit knowledge that in turn created superior value for the customer. However, while there are benefits for firms using OSS, in most cases OSS is basically treated as any other third-party software and typically only one-way interaction between the firm and the environment takes place, resulting in clear distinct boundaries between the two (Alexy and Henkel, 2009). Thus, investigating the role of open value networks for value creation and capture with OSS warrants further research.

2   Theoretical Framework

For our theoretical base, we propose a framework drawn from three central open innovation archetypes proposed by Gassmann and Enkel (2006). These include: (1) the outside-in process; (2) the inside-out process; and (3) the coupled process. This framework provides a useful lens to examine the implications and applicability of open innovation in organizations that employ both agile and open source methodologies. Open innovation can be analyzed at a number of levels, which include the intra-organizational and inter-organizational networking level (Chesbrough et al., 2006). Indeed, the implications that open innovation has *within* an organization and in particular the fact that it affects different parts of an organization differently are largely neglected in the current literature (Alexy and Henkel, 2009). While there exists much research about intra-organizational level networking in general to stimulate innovation (e.g. Lagerstrom and Andersson 2003; Foss and Pedersen, 2002; Tsai and Ghoshal, 1998), this type of networking has not been analyzed explicitly within the open innovation context (Vanhaverbeke, 2006). In particular, there is no research that we know of that addresses intra-organizational networking in an agile project environment. In order to address this, we have tailored Gassmann and Enkel's framework to include innovation that occurs outside the boundaries of a business unit and across the organization as well as outside the firm. In the context of OSS, Dahlander (2004) proposes that in addition to inter-organizational relations, it is relations with users and developers that constitute the

OSS community that are also important for the firm. The importance of competitors in a firm's value network has also been highlighted as these competitors often collaborate together to further develop or stimulate adoption of a shared technology, e.g. the Eclipse Foundation (West, 2007). Thus, for the purpose of this study, multiple stakeholders outside the boundaries of the firm will include the OSS communities, customers and competitors, in additional to inter-organizational relationships with other firms and research institutes.

2.1 The Outside-In Process

Companies that decide on an outside-in process as a core open innovation approach choose to cooperate with suppliers, customers third parties etc. and integrate the external knowledge gained (Gassmann and Enkel, 2006). This can be achieved by investing in global knowledge creation, applying innovation across industries, customer and supplier integration and purchasing intellectual property. According to Gassmann and Enkel (2006), if firms possess the necessary competencies and capabilities, they can successfully integrate internal company resources with the critical resources of other members such as customers, suppliers etc, by extending new product development across organizational boundaries. Companies such as HP and Sun have used an outside-in process by donating research and development to the Mozilla open source project while exploiting the pooled R&D and knowledge of all contributors (i.e., academics, user organizations, individual hobbyists, etc.) to facilitate the sale of related products. The result was that these firms maximized the returns of their innovation by concentrating on their own needs and then incorporating the shared integrated systems (West and Gallagher 2006). As the focus of this research in on both intra-organizational and inter-organizational value networks, an outside-in open innovation approach will refer to the integration of external knowledge and resources gained from multiple stakeholders outside the business unit and the boundaries of a firm.

2.2 The Inside-Out Process

This process focuses on the externalizing of company knowledge and innovation in order to bring ideas to market faster. This approach includes licensing IP or multiplying technology by transferring ideas to other companies. In addition, outsourcing can be used to channel knowledge and ideas to the external environment. The benefits of outsourcing include gaining access to new areas of complementary knowledge, managing capacity problems which allows for more flexibility, reduced time-to-market, sharing of costs and concentration of core competencies (Gassmann and Enkel, 2006). IBM for example have used an inside-out approach as part of its open source initiative that represented spinouts in the 1990s and, more recently, donated software patents to the OSS community (West and Gallagher, 2006). In the context of this study, an inside-out process refers to leveraging and transferring knowledge to multiple stakeholders outside the boundaries of both the business unit and firm and gaining certain advantages by letting ideas flow to the outside.

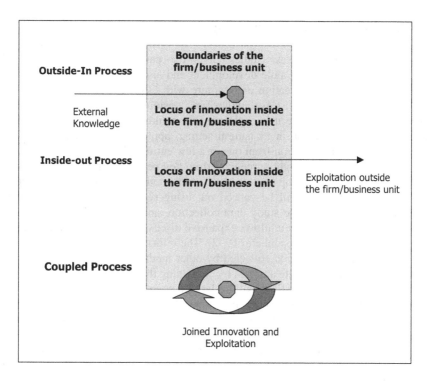

Fig. 1. Adaped Open Innovation Framework – Applying Open Innovation Principles in Firms that Employ Open Source and Agile Development Methodologies

2.3 The Coupled Process

This open innovation approach combines the outside-in (gaining external knowledge) with the inside-out process (to bring ideas to market). In order to accomplish both, these companies collaborate and cooperate with other companies (e.g. strategic alliances, joint ventures), suppliers and customers, as well as universities and research institutes. Indeed, companies like HP, Sun and IBM have also integrated elements of the coupled process by successfully cooperating with universities, research institutes etc., in both exploiting and sharing information and knowledge. To collaborate and cooperate successfully, a give and take of knowledge approach is crucial. Benefits of such an approach include an intensive exchange of knowledge and a mutual learning process. In this research, a coupled process will also refer to a combination of outside-in and inside-out as specified for this study. In particular, we will explore 1) how business units cooperate and interact with other business units in intra- organizational networks and 2) how firms cooperate and exchange knowledge with other firms, customers, communities, suppliers and competitors in value networks.

3 Proposed Research Methodology

The study will involve two case studies. Given the scarcity of empirical work in the area of open innovation and agile development and also the role of open innovation value networks for value creation and capture with OSS, a case study approach is considered most appropriate. Case studies are considered to be a suitable research approach for this study since it is exploratory in nature (Stake 2000; Yin 2003) and they explore a phenomenon in its natural setting, applying several methods of data collection to gather information from one or a few entities (Benbasat et al. 1987). We have already gained and sought agreement from key personnel in two firms – one of which employs agile development methods and another that utilizes open source software. Data collection will be carried out using semi-structured interviewing, a technique well suited to case study data collection, and particularly for exploratory research such as this because it allows expansive discussions which illuminate factors of importance (Yin 2003; Oppenheim 1992). The information gathered is likely to be more accurate than information collected by other methods since the interviewer can avoid inaccurate or incomplete answers by explaining the questions to the interviewee (Oppenheim 1992). The questions will be largely open- ended, allowing respondents freedom to convey their experiences and views of value networks and open innovation etc. (Yin 2003; Oppenheim 1992). The interviews will be conducted in a responsive (Rubin & Rubin 2005; Wengraf 2001), or reflexive (Trauth & O'Connor 1991) manner, allowing the researcher to follow up on insights uncovered mid-interview, and adjust the content and schedule of the interview accordingly. In order to aid analysis of the data after the interviews, all will be recorded with each interviewee's consent, and subsequently transcribed, proof-read and annotated by the researcher. In any cases of ambiguity, clarification will be sought from the corresponding interviewee, either via telephone or e-mail.

Data analysis will use Strauss & Corbin's (1998) open coding and axial coding techniques. Open coding is "the process of breaking down, examining, comparing, conceptualizing, and categorizing data" (Strauss & Corbin 1998). Glaser (1992) argues that codes and categories should emerge from the data, while with Strauss & Corbin's approach (1990) these are selected prior to analysis. The approach adopted in this study is more akin to the latter, where the interview questions and subsequent analysis will be based on the framework of open innovation presented earlier in this paper. This will provide a list of "intellectual bins" or "seed categories" structure the data collection and the open coding stage of data analysis. The next step will involve axial coding. Axial coding is defined by Strauss and Corbin (1998) as a set of procedures whereby data are put back together in new ways after open coding; whereas open coding fractures the data into categories, axial coding puts the data back together by making connections between the categories and sub-categories. As a list of codes begin to emerge, the analysis moves to a higher or more abstract level, looking for a relationship between the codes. Once a relationship has been determined, the focus returns to the data to question the validity of these relationships. Once the open innovation process (or lack thereof) is documented and analysed, the focus shifts to identifying the key benefits and challenges of open innovation and value networks in an open source and agile environment.

4 Future Work

This paper constitutes part of a research in progress aimed at exploring the applicability and implications of open innovation practices in firms that employ agile systems and open source development methodologies. At the moment we are in the process of collecting data from two companies that have agreed to participate in the study. Each case study has multiple embedded units of analysis corresponding to particular agile and OSS projects. Once data collection is complete, within-case analysis will be used to analyze the case-study data. Specifically, this paper argues for more open type of value networking that includes collaboration and reciprocal knowledge-sharing with other business units, customers, partners, communities and other relevant stakeholders pertinent to the business success of an organization, thus embracing open innovation principles. The conceptual framework proposed earlier raises some interesting research questions. In terms of 'outside-in' open innovation, how should firms choose which suppliers, customers, competitors and third parties to collaborate with? The 'inside-out' open innovation process also throws up some relevant questions. Deciding to change the locus of knowledge-sharing by transferring ideas to stakeholders outside the business unit or firm may prove challenging. Similar to the outside-in approach, a successful inside-out approach may be contingent on the firm's knowledge transfer capabilities and selection of appropriate stakeholders and their willingness and ability to engage and cooperate with each other. In relation to the 'coupled process' approach, how firms develop complementary internal and external value networks to create and gain external knowledge and ideas is a significant question. Additionally, there are more questions as to how the network is coordinated and maintained. Thus, it is crucial to understand how governance is shared across the network and how conflict is managed if it arises.

Acknowledgments

This work is supported by the European Commission through the FP6 project OPAALS (project no. 034824) and Science Foundation Ireland grant 03/C03/CE2/ 1303_1 to Lero – the Irish Software Engineering Research Centre, University of Limerick.

References

Abrahamsson, P., Conboy, K., Wang, X.: Lots Done, More to Do: The Current State of Agile Systems Development Research. European Journal of Information Systems 18, 281–284 (2009)

Agerfalk, P.J., Deverell, A., Fitzgerald, B., Morgan, L.: Assessing the Role of Open Source Sofware in the European Secondary Sotware Sector: A Voice from Industry. In: Proceedings of the First International Conference on Open Source Systems, Genova, Italy, July 11-15, pp. 82–87 (2005)

Alexy, O.: Putting a Value on Openness: The Effect of Product Source Code Releases on the Market Value of Firms. In: Solomon, G.T. (ed.) Proc. of the 67th Annual Meeting of the Academy of Management (2008)

Alexy, O., Henkel, J.: Promoting the Penguin? Intra-organizational Implications of Open Innovation (2009), http://ssrn.com/abstract=988363

Amabile, T.: Creativity in Context. Westview Press, Boulder (1996)

Beck, K.: Extreme Programming Explained. Addison-Wesley, Reading (1999)

Beck, K.: Extreme Programming Explained: Embrace Change. Addison-Wesley, Reading (2000)

Benbasat, I., Goldstein, D.K., Mead, M.: The Case Research Strategy in Studies of Information Systems. MIS Quarterly 11(3), 369–386 (1987)

Chesbrough, H.: Open Innovation: The New Imperative for Creating and Profiting from Technology. Harvard Business School Press, Boston (2003a)

Chesbrough, H.: Managing open innovation. Research & Technology Management 47(1), 23–26 (2004)

Chesbrough, H.: Open Business Models: How to Thrive in the New Innovation Landscape. Harvard Business School Press, Boston (2006)

Chesbrough, H., Swartz, K.: Innovating Business Models with Co-Development Partnerships. Research Technology Management 50(1), 55–59 (2007)

Chesbrough, H., Rosenbloom, R.S.: The role of the business model in capturing value from innovation: evidence from Xerox corporation's technology spin-off companies. Industrial and Corporate Change 11(3), 529–555 (2002)

Conboy, K.: Agility From First Principles: Reconstructing The Concept of Agility. Information Systems Development, Information Systems Research 20, 329–354 (2009)

Dahlander, L.: Appropriating returns from open innovation processes: A multiple case study of small firms in open source software, School of Technology Management and Economics, Chalmers University of Technology (2004),
http://opensource.mit.edu/papers/dahlander.pdf

Denzin, N.K., Lincoln, Y.S.: The Discipline and Practice of Qualitative Research. In: Denzin, N.K., Lincoln, Y.S. (eds.) Handbook of Qualitative Research, pp. 1–32. Sage Publications, Thousand Oaks (2000)

De Wit, J., Dankbaar, B., Vissers, G.: Open Innovation: the New Way of Knowledge Transfer. Journal of Business Chemistry 4(1), 11–19 (2007)

Dodgson, M., Gann, D., Salter, A.: The management of technological innovation: strategy and practice. Oxford University Press, Oxford (2008); Dubin, R.: Theory Building. Free Press, New York (1969)

Dyer, J.H., Singh, H.: The Relational View: Cooperative Strategy and Sources of Interorganizational Competitive Advantage. Academy of Management Journal 23(4), 660–679 (1998)

Eisenhardt, K.M., Santos, F.M.: Knowledge-based view: A new theory of strategy? In: Pettigrew, A., Thomas, H., Whittington, R. (eds.) Handbook of strategy and management, pp. 139–164. Sage, London (2002)

Ekvall, G.: Organisational Climate for Creativity and Innovation. European Journal of Work and Organisational Psychology 5, 105–123 (1996)

Feller, J., Finnegan, P., Fitzgerald, B., Hayes, J.: From peer production to productization: a study of socially-enabled business exchanges in open source service networks. Information Systems Research 19(4), 475–493 (2008)

Fitzgerald, B.: The Transformation of Open Source Software. MIS Quarterly 30(3), 587–598 (2006)

Gassmann, O.: Opening up the Open Innovation Process: Towards an Agenda. R&D Management 36(3), 223–228 (2006)

Gassmann, O., Enkel, E.: Constituents of Open Innovation: Three Core Process Archetypes. R&D Management (2006)

Ghoshal, S., Moran, P.: Bad for Practice: A Critique of the Transaction Cost Theory. The Academy of Management Review 21(1), 13–47 (1996)

Gomes-Casseres, B.: Competitive advantage in alliance constellations. Strategic organization 1(3), 327–335 (2003)

Guba, E.G.: The alternative paradigm. In: Guba, E.G. (ed.) The Paradigm Dialog, pp. 17–27. Saqe, Newbury Park (1990)

Highsmith, J., Cockburn, A.: Agile Software Development: The Business of Innovation. Software Management, 120–123 (2001)

Lagerstrom, K., Andersson, M.: Creating and Sharing Knowledge within a Transnational Team - The Development of a Global Business System. Journal of World Business 38(2), 84–95 (2003)

Landry, R., Amara, N.: Creatuvity, innovation and business practices in the matter of knowledge management. In: De La Mothe, J., Foray, D. (eds.) Knowledge Management in the Innovation Process: Business Practices and Technology Adoption, pp. 55–80. Kluwer Academic Publishers, Boston (2001)

Laursen, K., Salter, A.: Open for Innovation: The role of openness in explaining innovative performance among UK manufacturing firms. Strategic Management Journal 27(2), 131–150 (2006)

Lee, G.K., Cole, R.E.: From a firm-based to a community-based model of knowledge creation: The case of the Linux kernel development. Organization Science 14(6), 633–649 (2003)

Mathisen, G., Einarsen, S.: A Review of Instruments Assessing Creative and Innovative Environments Within Organisations. Creativity Research Journal 16(1), 119–140 (2004)

Morgan, L., Finnegan, P.: Benefits and Drawbacks of Open Source Software: An Exploratory Study of Secondary Software Firms. In: Feller, J., Fitzgerald, B., Scacchi, W., Sillitti, A. (eds.) Open Source Development, Adoption and Innovation. Springer, Heidelberg (2007b)

Morgan, L., Finnegan, P.: How Perceptions of Open Source Software Influence Adoption: An Exploratory Study. In: Proceedings of the 15th European Conference on Information Systems (2007a)

Morgan, L., Finnegan, P.: Deciding on Open Innovation: An Exploration of How Firms Create and Capture Value with Open Source Software. In: León, G., Bernardos, A., Casar, J., Kautz, K., DeGross, J. (eds.) IFIP 8.6 Proceedings. Open IT-Based Innovation: Moving Towards Cooperative IT Transfer and Knowledge Diffusion, vol. 287, pp. 229–246. Springer, Boston (2008)

Nonaka, I., Takeuchi, H.: The Knowledge-Creating Company. Oxford University Press, NY (1995)

Nonaka, I., Keigo, S., Ahmed, M.: Continuous Innovation: The Power of Tacit Knowledge. In: Shavinina, K. (ed.) International Handbook of Innovation. Elsevier, New York (2003)

Oppenheim, A.: Questionnaire Design, Interviewing and Attitude Measurement, Continuum, New York (1992)

Overby, E.M., Bharadwaj, A.S., Bharadwaj, S.G.: An Investigation of Firm- Level Open Source Adoption: Theoretical and Practical Implications. In: Jain, R.K. (ed.) Open Source Software in Business - Issues and Perspectives. ICFAI University Press, Hyderabad (2006)

Parise, S., Henderson, J.C.: Knowledge resource exchange in strategic alliances. IBM Systems Journal (40), 908–924 (2001)

Peppard, J., Rylander, A.: From value chain to value network: lessons for mobile operators. European Management Journal 24(2), 128–141 (2006)

Porter, M.E.: Competitive Advantage. The Free Press, New York (1985)

Powell, W.W., Koput, K.W., Smith-Doerr, L.: Interorganizational collaboration and the locus of innovation: Networks of learning in biotechnology. Administrative Science Quarterly 41(1), 116–145 (1996)

Rajala, R., Nissilä, J., Westerlund, M.: Revenue Models in the Open-Source Software Business. In: Amant, K.S., Still, B. (eds.) Handbook of Research on Open Source Software: Technological, Economic, and Social Perspectives, pp. 541–554. Information Science Reference, New York (2006)

Rubin, H., Rubin, I.: Qualitative Interviewing: The Art of Hearing Data. Sage, Thousand Oaks (2005)

Schwaber, K., Beedle, M.: Agile software development with SCRUM. Prentice-Hall, Englewood Cliffs (2002)

Straub, D., Gefen, D., Boudreau, M.-C.: The ISWorld Quantitative Positivist Research Methods Website. In: Galletta, D, ed. (2004),
http://www.dstraub.cis.gsu.edu:88/quant/

Strauss, A., Corbin, J.: Basics of Qualitative Research: Grounded Theory Procedures and Techniques, 2nd edn. Sage Publications, Newbury Park (1998)

Teece, D.J.: Profiting from technological innovation: implications for integration, collaboration, licensing and public policy. Research Policy 15(6), 285–305 (1986)

Teece, D.J., Pisano, D., Shuen, A.: Dynamic capabilities and strategic management. Strategic Management Journal (18), 509–533 (1997)

Tidd, J., Bessant, J., Pavitt, K.: Managing Innovation: Integrating Technological, Market and Organisational Change. Wiley, Chichester (2005)

Timmers, P.: Business Models for Electronic Markets. Electronic Markets 8(2), 3–8 (1998)

Trauth, E., O'Connor, M.: A study of the interaction between information technology and society: An illustration of combined qualitative research methods. In: Nissen, H.E., Klein, H.K., Hirschheim, R. (eds.) Information Systems Research: Contemporary Approaches and Emergent Traditions, pp. 131–144. North-Holland, Amsterdam (1991)

Vanhaverbeke, W.: The Interorganizational Context of Open Innovation. In: Vanhaverbeke, W., West, J., Chesbrough, H. (eds.) Open Innovation: Researching a New Paradigm, pp. 258–281. Oxford University Press, Oxford (2006)

Vanhaverbeke, W., Cloodt, M.: Open Innovation in Value Networks. In: Chesbrough, H., Vanhaverbeke, W., West, J. (eds.) Open Innovation: Researching a New Paradigm, pp. 258–281. Oxford University Press, Oxford (2006)

Wengraf, T.: Qualitative research interviewing: biographic narrative and semi- structured method. Sage Publications, London (2001)

West, J.: Value Capture and Value Networks in Open Source Vendor Strategies. In: Proceedings of the 40th Annual Hawaii International Conference on System Sciences (HICSS'07), Hawaii, pp. 176–186 (2007)

West, J., Gallagher, S.: Challenges of open innovation: the paradox of firm investment in open-source software. R&D Management 36(3), 319–331 (2006)

West, J., O'Mahony, S.: Contrasting Community Building in Sponsored and Community Founded Open Source Projects. In: Proceedings of the 38th Annual Hawaii International Conference on System Sciences, Waikoloa, Hawaii, January 2005, pp. 196–203 (2005)

Yin, R.K.: Case study research, design and methods, 3rd edn. Sage Publications, Newbury Park (2003)

Social Network Analysis of the Irish Biotech Industry: Implications for Digital Ecosystems

Chris van Egeraat and Declan Curran

National Institute for Regional and Spatial Analysis (NIRSA),
John Hume Building, NUI Maynooth, County Kildare, Ireland
{chris.vanegeraat,declan.g.curran}@nuim.ie

Abstract. This paper presents an analysis of the socio-spatial structures of innovation, collaboration and knowledge flow among SMEs in the Irish biotech sector. The study applies social network analysis to determine the structure of networks of company directors and inventors in the biotech sector. In addition, the article discusses the implications of the findings for the role and contours of a biotech digital ecosystem. To distil these lessons, the research team organised a seminar which was attended by representatives of biotech actors and experts.

Keywords: Digital Ecosystems; Social network analysis; Innovation; Biotech.

1 Introduction

The Digital Ecosystem (DE) put forward by the OPAALS Research Consortium is a self-organising digital infrastructure established with the aim of creating a digital environment for networked organizations which is capable of supporting co-operation, knowledge sharing, the development of open and adaptive technologies and evolutionary business models [1]. The Digital Ecosystem provides structures of communication and collaboration that can facilitate collective learning, knowledge flow and innovation across SMEs and other actors.

In order to understand sustainable digital ecosystems of SMEs and the contribution they could make to competitiveness of SMEs and regional development, we need to understand in depth the processes of knowledge flow and innovation. This paper sets out to address two main research questions. Firstly, what are the structural character-istics of knowledge and innovation networks in the Irish biotech industry and are these conducive to knowledge flow? Secondly, what does this mean for the roles and the contours of a biotech digital ecosystem? The first question is explored with social network analyses, providing insight into the structural characteristics of both formal and informal networks. The second question was partly addressed through consulta-tion of biotech actors and experts.

Section two of this paper presents the concepts and themes on which the study focuses. This is followed by the research design and methodology in section three. Section four introduces the biotech sector in Ireland. Next, section five presents the findings of the social network analysis. The paper ends with conclusions and a discussion of the implications of the findings for digital ecosystems.

F.A. Basile Colugnati et al. (Eds.): OPAALS 2010, LNICST 67, pp. 31–43, 2010.

2 Digital Ecosystems and Networks

Recent studies of innovation emphasize the collective, collaborative processes that underlie innovation. The situation of slowly changing networks of organisations will be replaced by more fluid, amorphous and transitory structures based on alliances, partnerships and collaborations. These trends have been characterised as a transition towards 'open innovation' [2] and 'distributed knowledge networks' [3].

Knowledge economies can be thought of as ecosystems. Economic ecosystems are assemblages of interdependent institutions in which the welfare of the component organisms is dependent on the interactions between them. They tend to evolve towards an optimum state due to gradual adaptation. The evolution is accelerated by the promotion of higher and more efficient levels of knowledge flow/sharing. Towards this, *digital* ecosystems seek to exploit the benefits of new ICTs in terms of enhanced information and knowledge flow.

Economic ecosystems tend to be organised on a territorial basis as expressed in related concepts such as clusters [4] and regional systems of innovation [5]. Most territorial economic development concepts recognise that networks are an important aspect of innovation and clustering processes [6] Network theory and analysis can therefore lead to a better understanding of innovation and clustering processes [7].

The roots of the network concept and network theory go back to the end of the 19th century [8]. In sociology, anthropology, and psychology, network analysis was initially employed in a range of empirical context. For a long time surprisingly little attention was devoted to the role of networks in economic activity but this has changed drastically in more recent times. Since the early 1990s an increasing body of economists, economic sociologists and economic geographers have been focussing on the role of networks in economic activity, innovation and regional development. In this paper we focus on business/innovation networks. In broad terms a network can be defined as a set of actors linked through a specific type of connections [8].

A range of network forms and types can be identified. For the current research project we made a basic distinction between formal and informal networks (facilitating formal and informal knowledge exchange). Formal networks are configured as inter-organisational alliances while informal networks are based on inter-personal ties. In our view, formal networks include both the longer-term strategic networks based on strategic alliances and joint ventures, as well as the shorter-term project networks distinguished by [8]. In formal networks firms or institutions are linked in their totality, via, for example, joint research projects or buyer-supplier agreements.

In informal networks, the connected persons principally represent themselves. Because the persons are employed by firms and institutions, the links between these persons indirectly also link the institutions, providing a pipeline for (informal) information flow between these institutions. A large variety of informal networks exist including networks of former students, professional networks, networks of friends, members of sport clubs, networks of corporate board members, and so forth. Informal networks can develop on the back of formal business activity, as is the case with networks of former colleagues or former business relations that have developed a friendship. However, the characteristic of such informal networks is that the network is no longer based on these (former) formal relations. Informal networks have different levels of organisation or institutionalisation. Some professional networks (informal from the firms' point of view) can be strongly institutionalised while other networks, for example those based on friendship are virtually unorganised.

In this paper formal networks are seen as pipelines for formal knowledge exchange while informal networks are linked to informal knowledge exchange.

Rather than treating regional networks as a distinct type of network [8] we work from the perspective that all (types of) networks have a spatiality. Thus all, formal and informal, networks have a spatiality that may include local, regional, national and global aspects. During the 1990s, the interest in networks became strongly focused on regional networks. The cluster literature paid a great amount of attention to space of flows and the positive role of networks in regional clustering processes. However, it was assumed that the space of flows and the space of place showed a great deal of overlap [7]. The global aspects of networks tended to be ignored. Regions were treated as isolated islands of innovation.

Although remaining highly influential, these ideas became increasingly challenged by empirical studies that showed that firms in even the most developed clusters are often highly depended on non-local relations and networks for their knowledge. In fact, the non-local relations often play a crucial role in providing new (from the perspective of the region or cluster) knowledge. In the context of the biotech industry these ideas were supported by [9] [10] [11]. Recent contributions to the knowledge - based theory of spatial clustering specifically incorporate the idea that firms in clusters are connected to both local and non-local networks and depend on local and non-local knowledge flows through 'local buzz' and 'global pipelines' [12] [13]. Clusters are understood as nodes of multiple and multi-scalar knowledge connections [8].

This is not to say that the spatiality of the networks is irrelevant for the competitiveness of firms and regions. Firstly, from a neoclassical perspective one can point to the fact that proximity between actors in a network increases the efficiency of knowledge flow. Secondly, more important is the fact that the scale of some networks is strongly regional or national in character by nature. The membership of most regional/national professional organisations, chambers of commerce, industrial organisations etc, is nearly entirely regional/national. Many social networks, such as networks of former school-friends, are starting to include an increasing amount of globally dispersed members, but retain a strong national character. In particular, many informal networks tend to have a strong regional/national character, although some informal networks tend to have a significant international membership, e.g. epistemic communities.

Disagreement exists as to the salience or importance of the informal knowledge exchange both for the innovation capacity and competitiveness of firms and for regional clustering processes [14]. Some contributions argue that informal networks are important channels for knowledge exchange and that individuals in different firms and institutions informally provide each other with technical and market-related knowledge that can be of great value to the firm. Others are of the view that, although informal knowledge exchange does occur, the knowledge generally has limited commercial or strategic value. Individuals will only exchange general knowledge that is of relatively low value to the firm, for example information about new job openings. In addition, the knowledge may not flow freely throughout the local network but, instead, circulate in smaller (sub-) communities.

One of the aims of this paper is to increase the insight into the quality of informal and formal networks in the biotech industry, notably whether the structure is conducive for knowledge exchange.

3 Methodology and Data Sources

This paper sets out to address two main research questions:

1) What are the structural characteristics of knowledge and innovation networks in the Irish biotech industry and are these conducive to knowledge flow?
2) What does this mean for the roles and the contours of a biotech digital ecosystem?

The first research question was addressed through social network analysis. Social network analysis, one of the dominant traditions in network theory [8], is based on the assumption of the importance of relationships among interacting units or actors and that units don't act independently but influence each other. Relational ties between actors are viewed as channels for transfer or flow of resources [15]. The social network analysis tradition has developed a range of conceptual devices that can facilitate an analysis of regional business ecosystems, including structural equivalence, structural holes, strong and weak ties and small worlds. This paper focuses on the small world concept.

Networks of relationships between social actors, be they individuals, organizations, or nations, have been used extensively over the last three decades as a means of representing social metrics such as status, power, and diffusion of innovation and knowledge [16] [17]. Social network analysis has yielded measures both of individual significance, such as centrality [18], and of network efficiency or optimal structure [19]. Analysis of network structures becomes important when one is interested in how fragile or durable observed networks are. For example, what do network characteristics such as sparseness or clustering imply for the stability of the network structure? One established framework for analysing network structure is that of "small world" network analysis. *Small world analysis is concerned with the density and reach of ties.* A small world is a network in which many dense clusters of actors are linked by relationships that act as conduits of control and information [20] [21]. In keeping with the age-old exclamation "it's a small world!", this type of network allows any two actors to be connected through a relatively small series of steps or links – despite the fact that the overall network may be quite sparse and actors may be embedded in distinct clusters. As a result, actors in the network may in reality be "closer" to each other than initially perceived.

These small world networks, with high clustering and short global separation, have been shown by Watts [16] to be a general feature of sparse, decentralized networks that are neither completely ordered nor completely random. Small world network analysis offers us a means by which we can gain insights into network structures and the role of these structures in facilitating (or hindering) the flow of innovation and knowledge throughout the entire network. Watts [16] and Kogut and Walker [17] advocate comparing an observed network with a randomised network (i.e. a random graph) that has the same number of actors (nodes) and same number of relationships (links) per actor as the observed. Simulations by Watts [16] show that the structural stability of small worlds is retained even when a substantial number of relationships are replaced with randomly generated links. The network becomes more globally connected rapidly but the dense clusters are slow to dissolve. Thus, actors in the network can strategise and, rather than being disrupted, the small world structure is still replicated. In this way, networks that appear sparse can in fact contain a surprising degree of structure.

Small world analysis has been productively applied in the context of biotech clusters [22] and has important application in the context of regional biotech digital ecosystems. Knowledge will flow most efficiently in biotech ecosystems with small world characteristics. Where small world characteristics are absent, these can be created by adding a relatively small number of remote links to the network where the level of local clustering is already high [8].

The formal description of small world networks presented here is as per Watts [16], with the networks represented as connected graphs, consisting of undifferentiated vertices (actors) and unweighted, undirected edges (relationships). All graphs must satisfy sparseness conditions. The small world network analysis that follows in Section five is characterized in terms of two statistics:

- *Characteristics path length (L)*: the average number of edges that must be traversed in the shortest path between any two pairs of vertices in the graph. L is a measure of the global structure of the graph, as determining the shortest path length between any two vertices requires information about the entire graph.
- *Clustering Coefficient (C)*: if a vertex has k_v immediate neighbours, then this neighbourhood defines a subgraph in which at most $k_v(k_v-1)/2$ edges can exist (if the neighbourhood is fully connected). C_v is then the fraction of this maximum that is realised in v's actual neighbourhood, and C is this fraction averaged over all vertices in the graph. In this way, C is a measure of the local graph structure.

In order to determine what is "small" and "large" in this analysis, Watts [16] determines the following ranges over which L and C can vary:

1. The population size (n) is fixed.
2. The average degree k of vertices is also fixed such that the graph is sparse ($k<<n$) but sufficiently dense to have a wide range of possible structures ($k>>1$).
3. The graph must be *connected* in the sense that any vertex can be reached from any other vertex by traversing an infinite number of edges.

Fixing n and k enable valid comparisons to be made between many different graph structures. This also ensures that the minimum value for C is 0, while the maximum value for C is 1. The sparseness condition ensures that, while the network is sufficiently well connected to allow for a rich structure, each element operates in a local environment which comprises of only a tiny fraction of the entire system. Finally, the requirement that the graph is connected guarantees that L is a truly global statistic.

Data collection started with an inventorisation of biotech companies in Ireland (see the next section on the Irish biotech industry). Following this, two separate datasets were compiled for our social network analysis of the Irish biotech industry. In order to compile the first dataset, a rigorous internet search of official company websites and media sources has been conducted. In this way, it can be ascertained whether a director of a given Irish biotech company also holds a directorship on another Irish biotech company. Joint directorships are then taken to represent a conduit of informal knowledge flow between the respective companies. This dataset also contains information on the founders of each company; serial entrepreneurs, who form numerous companies; and spin-off companies. The database also identifies whether these spin-off companies emerged from existing private companies or universities. The date of establishment of all spin-offs and existing companies is also included in the dataset, allowing us to undertake an analysis of the evolution of the Irish biotech industry over time. We have endeavoured to verify the database through consultation with industry experts.

The second dataset has been compiled from patent data available from the Irish Patent Office (http://www.patentsoffice.ie/), US Patent and Trademark Office (http://www.uspto.gov/), and *Esp@cenet*, the European Patent Office (http://ep.espacenet.com/). For each Irish biotech company that that has registered patents, we can establish the researchers who worked on each patent; their employer at the time, and whether they were foreign-based or located in Ireland. We take this formal research collaboration to represent formal knowledge flow between Irish biotech companies.

The second research question deals with the meaning of the results of the social network analysis for the roles and contours of a biotech digital ecosystem. To distil these lessons, the research team organised a seminar. This seminar was attended by 14 representatives of biotech companies, industrial promotion agencies, third-level colleges, venture capital companies, software companies, the OPAALS community and other industry experts.

4 The Irish Biotech Industry

The OECD [23] defines biotech as the application of science and technology to living organisms, as well as parts, products and models thereof, to alter living or nonliving materials for the production of knowledge, goods and services. In order to narrow the definition to 'modern' biotech the OECD employs a list based definition that includes various techniques and activities: synthesis, manipulation or sequencing of DNA, RNA or protein; cell and tissue culture and engineering; vaccines and immune stimulants; embryo manipulation; fermentation; using plants for cleanup of toxic wastes; gene therapy; bioinformatics, including the construction of databases; and nanobiotech.

Partly due to the lack of official statistics and partly due to the ambiguous nature of the definition it is difficult to determine the size of the Irish biotech industry. Our 'universe' of firms in the modern biotech industry in Ireland was based on existing survey material [24], the list of firms included on the 'Biotechnology Ireland' website (hosted by Enterprise Ireland), information from interviews with industry experts and internet search. The final list included 80 biotech firms. Fifty two of these companies are Irish-owned. All but two of these indigenous companies are small or medium sized. It is estimated that the majority of indigenous companies in the list are micro-enterprises, employing less than 10 staff - often start-up companies or campus companies. The majority of the other indigenous companies are small enterprises, employing less than 50 staff.

5 Results of Social Network Analysis of the Irish Biotech Sector

Figure 1 presents a sociogram of the network connections in the Irish biotech industry, using data on directorships. Some directors are director of more than one company, providing links or ties in the network which can support information flow and diffusion of the digital ecosystem concept. Figure 2 presents a sociogram of the network connections in the Irish biotech industry, but now using patent data. On the face of it the sociograms would suggest that the networks have a low density.

However as discussed in the methodology section, the structure of the networks may be such that despite low overall density, short path length and high clustering may still be features of the network. This would suggest that rather than being a sparse network unsuited for swift flows of knowledge, there may actually be potential for rapid diffusion of knowledge (and adoption of a biotech digital ecosystem) through the network if the right actors are targeted. Using both datasets, both informal knowledge flows and formal knowledge can be analysed and their resulting network characteristics compared. The results of the small world network analysis are now presented.

Table 1 presents the results from the Irish biotech network of directors and companies, analysing the directors and companies separately (i.e. deconstructing a 2-node network into its constituent 1-node networks). While directors may be connected to each other by virtue of being on the board of the same company, this type of intra- company link is avoided by analysing the company-only 1-node network. Thus, presenting the results of both 1-node network analyses serves as a useful robustness check. In keeping with the formal description of small world networks presented in the methodology section, two central findings can be gleaned from Table 1. First, it is clear that both directors and companies are highly clustered ($C = 0.948$ and 0.669, respectively). This is particularly evident when compared to the low degree of clustering generated by a random network with the same number of nodes and ties as the highly structured observed networks ($C = 0.039$ and 0.062, respectively). Second, though the director and company networks are highly clustered, they are not characterised by long path lengths. This is in keeping with Watts' [16] findings that even as a network moves from a structured to a random graph, the path length decreases rapidly but the clustering is persistent. For the purposes of our Irish biotech study, this highly clustered/short path length characteristic of the directors network and the network of companies connected via directors has practical implications for the diffusion of informal knowledge and tacit knowledge throughout the entire network. It indicates that while knowledge is capable of travelling rapidly through the entire network, the challenge is get the knowledge to flow between the distinct clusters. It is exactly this challenge that a digital ecosystem can help overcome.

Comparable results emanating from the network of Irish biotech researchers and the network of Irish biotech companies via patents are presented in Table 2. While the findings outlined above can be interpreted as capturing informal knowledge flows, the results of Table 2 are based on patent data and therefore refer to formal knowledge flows. Once again, the salient findings are those of high clustering and short path lengths for both the researcher and company networks. However, in this instance the company network is noticeably less clustered via patents than it was through directors. This suggests that formal knowledge flows through the network in a different, slower, manner than informal knowledge. This may also have important practical implications both for understanding the process of knowledge diffusion in the Irish biotech industry and for ensuring optimal design and operation of a digital ecosystem in such a setting.

Finally, Table 3 presents a comparison of small world networks identified in a range of existing studies and allows us to assess "how small" the networks in the Irish biotech industry are. The small world network statistics of the Irish biotech industry are compared with comparable statistics from a study of networks of German firm owners [17] and a study that reported on three types of networks [25]: a network of film actors connected by participation in films; a power grid network representing links between generators, transformers, and substations in the western United States; and *C. Elgans*, which is the completely mapped neural network of a worm. Comparison across the networks illustrates once again the strong small world characteristics of the director network and the network of companies connected via directors, as well as the lesser degree of clustering in the small world network of Irish biotech researchers and the network of Irish biotech companies via patents.

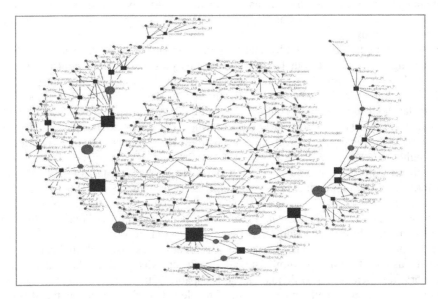

Fig. 1. Network of Irish Biotech Directors and Companies, based on directorship data

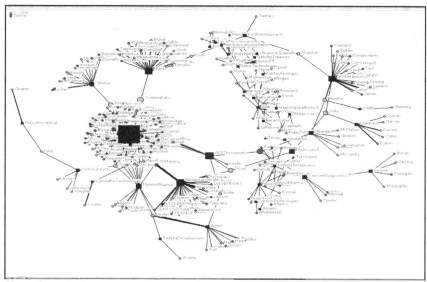

Note: Green indicates researchers based in Ireland: red indicates researcher based abroad.

Fig. 2. Network of Irish Biotech Researchers and Companies, based on patent data

Table 1. Irish Biotech Industry Directors and Companies (via Directorships) Network Statistics

Variable	Directors	Companies
Density		
Density (for all directors/firms)	0.018	0.016
Total no. of ties	1,622	118
Average no. of ties	5.5	2.7
(between those connected)		
Clustering		
Cluster coefficient	0.948	0.669
Random Cluster coefficient	0.039	0.062
Path Length		
Average Path length among those connected	3.538	2.912
Random Average Path Length	3.127	4.111

Note: No. of directors: 302; No. of firms: 86; no. of connected firms: 43.

Table 2. Irish Biotech Industry Researcher and Companies (via patents) Network Statistics

Variable	Researchers	Companies
Density		
Density (for all researchers/firms)	0.163	0.041
Total no. of ties	16,110	64
Average no. of ties (between those connected)	52.5	2.78
Clustering		
Cluster coefficient	0.975	0.439
Random Cluster coefficient	0.570	0.099
Path Length		
Average Path length among those connected	2.091	2.256
Random Average Path Length	2.013	3.264

Note: No. of researcher: 315; connected researchers: 307; No. of firms (that have registered patents): 40; connected firms: 23.

Table 3. A Comparison of Small World Network Statistics

	Path Length		Clustering		Actual-to-Random Ratio for:	
Network	Actual	Random	Actual	Random	Length	Clustering
Irish Biotech Directors	3.538	3.127	0.948	0.039	1.131	24.31
Irish Biotech Companies (via directors)	2.912	4.111	0.669	0.062	0.708	10.79
Irish Biotech Researchers	2.091	2.013	0.975	0.570	1.039	1.711
Irish Biotech Companies (via patents)	2.256	3.264	0.439	0.099	0.691	4.434
German Firms[1]	5.64	3.01	.84	.022	1.87	38.18
German Owners[1]	6.09	5.16	.83	.008	1.18	118.57
Film Actors network[2]	3.65	2.99	.79	.001	1.22	2,925.93
Power Grid network[2]	18.70	12.40	.08	.005	1.51	16.00
C. Elegans network[2]	2.65	2.25	.28	.05	1.18	5.60

[1] Kogut and Walker [17]; [2] Watts and Strogatz [25].

6   Conclusions and Implications for the Biotech Digital Ecosystem

Previous case study based research on innovation processes in the Irish biotech industry [26] showed that individual innovation projects involved little collaboration and informal information flow between regional actors. From this, one might have anticipated low density, sparse and weakly clustered networks. However, the social network analysis shows that networks do exist in the Irish biotech industry and that both the formal networks, connected through patents, and the informal networks, connected through directorships, have small world characteristics. This means that the network structures are conducive to knowledge flow. However the formal network is noticeably less clustered than the informal network, which suggests that the informal networks are far more conducive to knowledge flow than the formal networks. Knowledge in the formal network will flow and diffuse in a different, slower manner. The results also suggest that in both types of networks there remains scope for improving the structural characteristics of the network by creating links between distinct clusters in the network.

The social network analysis has provided new insight into network structures of the Irish biotech industry. At the same time one must not lose sight of the fact that the results are in most cases only suggestive of efficient knowledge flow. It remains unclear how much knowledge flows through the links and how far the knowledge travels through the network [14]. In addition, some knowledge is more strategic than other. It is therefore important that future research investigates what actually flows across the links [8]. This is, of course partly dependent on the type of actors in the network.

The findings, and the discussion of these findings with industry actors and experts, suggest important implications for the role and structure of a digital ecosystem in the Irish biotech sector. In the Irish biotech industry, a digital ecosystem is unlikely to play a significant role in promoting regional development by facilitating efficient and secure communication and knowledge flow between regional actors (partners), collaborating in a specific innovation project (i.e. as a project management tool). The actual numbers of collaborations is simply too small for a digital ecosystem to have a significant impact on regional development in this way.

In the Irish context, a digital ecosystem is more likely to stimulate regional development by acting as a more general communication tool and knowledge resource, connecting all regional players in the biotech industry (irrespective of whether or not these actors are partners in a specific innovation project). It could provide a more efficient medium for existing networks of individuals and firms to exchange informal knowledge, thereby better exploiting these existing networks.

The digital business ecosystem in the biotech industry should involve the entire social world of the firms, linked to the specific inter-firm networks that firms have and, more importantly, also to the loose web of ties that people within innovation projects share with others in the industry. Innovation seems to be driven strongly by engagement with public spaces and 'communities' where information sharing is relatively open. Innovation remains rooted in an engagement with a community that involves accessing diverse sources of knowledge through decentralized networks, loosely defined ties, and quasi-public spaces. Public spaces are crucial to innovation. A digital ecosystem could play the role of a new type of 'public space' [27]. The digital ecosystem environment can also actively be employed to stimulate or create new links between distinct clusters in a network.

A biotech digital ecosystem in Ireland should include strong assistance/support functionality. Companies and individual actors provide information about their knowledge assets and requirements. One of the central questions becomes "what knowledge that could be of value to me do you have, and are you willing to share?" This may be particularly beneficial, to young companies and new actors, but not exclusively so.

The digital ecosystem should provide a multi-level data/communication structure. Some levels are shared by all firms and individual actors while others are only accessible to smaller groups. The different levels mediate knowledge and information with different levels of sensitivity, requiring different levels social proximity and trust.

Given the important knowledge generating role of the universities, one of the most valuable roles of a digital ecosystem in the biotech industry is to facilitate knowledge transfer from these universities and research institutions. Universities and their lead scientist would therefore be the most important players and potential catalysts in a digital ecosystem organised on a regional basis.

On the basis of the proceeding of the seminar with industry actors and experts, we suggest that the following digital ecosystem applications have the greatest potential in the Irish biotech industry:

- A forum for regional actors (in universities; research institutions and private enterprise) to consult each other on a reciprocal basis about the location of (regional and extra-regional) actors and sources of knowledge.
- A regionally-based science forum for biotech scientists and technicians. Here biotech scientists and technicians in companies and universities can ask for advice about, and interactively discuss, scientific and technical problems.
- A biotech sector dedicated electronic interactive labour exchange, matching skilled people to jobs.
- A directory tool, providing information about regional actors, and promote
- Ireland as a biotech region.

Acknowledgments. The authors wish to thank Sarah Maloney for her excellent research assistance.

References

1. Nachira, F., Nicolai, A., Dini, P., Le Louarn, M., Rivera, L.: Digital Business Eco-systems. Office of the Official Publications of the European Communities, Luxembourg (2007)
2. Chesbrough: Open Innovation. Harvard Business School Press, Boston (2003)
3. Asheim, B., Boschma, R., Cooke, P.: Constructing regional advantage: platform policy based on related variety and differentiated knowledge bases. Utrecht University Working Paper (2007)
4. Porter, M.: The Competitive Advantage of Nations. Macmillan, London (1990)
5. Cooke, P.: Regional Innovation Systems, Clusters and The Knowledge Economy. Industrial and Corporate Change 10(4), 945–974 (2001)
6. Ingstrup, M., Freytag, P., Damgaard, T.: Cluster Initiation and Development: A critical view from network Perspective. Paper presented at the Conference Euromed Management, France (2009)

7. Ter Wal, A., Boschma, R.: Applying Social Network Analysis in Economic Geography: Theoretical and Methodological Issues. Annals of Regional Science 43(3), 739–756 (2008)
8. Grabher: Trading routes, bypasses and risky intersections: mapping the travels of networks, progress human geography. Progress in human Geography 30(2), 163–189 (2006)
9. Owen Smith, J., Powel, W.: Knowledge Networks as Channels and Conduits: the Effects of Spillovers in the Boston Biotechnology Community. Organisation Science 15(1), 5–21 (2006)
10. Coenen, L., Moodysson, J., Ryan, C., Asheim, B., Phillips, P.: Comparing a Pharmaceutical and an Agro-food Bioregion: On the Importance of Knowledge bases for Socio-spatial Patterns of Innovation. Industry and Innovation 13(4), 393–414 (2006)
11. Moodysson, J., Jonsson, O.: Knowledge Collaboration and Proximity: The Spatial Organization of Biotech Innovation Projects. European Urban and Regional Studies 14(2), 115–131 (2007)
12. Bathelt, H., Malmberg, A., Maskell, P.: Cluster and Knowledge: Local Buzz, Global Pipelines and the Process of Knowledge Creation. Progress in Human Geography 28(1), 31–56 (2004)
13. Gertler, Wolfe: Spaces of Knowledge Flow: Clusters in a Global Context. In: Asheim, B., Cooke, P., Martin, R. (eds.) Clusters and Regional Development: Critical Reflections and Explorations, pp. 218–236. Routledge, London (2006)
14. Dahl, M., Pedersen, C.: Knowledge flow through informal contacts in industrial clusters: Myth or Reality. Research Policy 33, 1673–1686 (2004)
15. Wasserman, S., Faust, K.: Social Network Analysis: Methods and Applications. Cambridge University Press, New York (1994)
16. Watts, D.: Networks, Dynamics, and the Small-World Phenomenon. American Journal of Sociology 105, 493–527 (1999)
17. Kogut, B., Walker, G.: The Small World of Germany and the Durability of National Networks. American Sociological Review 66(3), 317–335 (2001)
18. Freeman, L.C.: Centrality in Social Networks: Conceptual Clarification. Social Networks 1, 215–239 (1979)
19. Yamaguchi, K.: The Flow of Information through Social Networks Diagonal- Free Measures of Inefficiency and the Structural Determinants of Inefficiency. Social Networks 16, 57–86 (1994)
20. Milgram, S.: The Small World Problem. Psychology Today 2, 60–67 (1967)
21. White, H.: Search Parameters for the Small World Problem. Social Forces 49, 259–264 (1970)
22. Casper, S., Murray, F.: Careers and Clusters: Analyzing the Career Network Dynamics of Biotechnology Clusters. Journal of Engineering Technology Management 22, 51–74 (2005)
23. OECD (2006)
24. InterTradeIreland: Mapping the Bio-Island. InterTradeIreland, Newry (2003)
25. Watts, D., Strogatz, S.: Collective Dynamics of 'Small World' Dynamics. Nature 393, 440–442 (1998); White, H.: Search Parameters for the Small World Problem. Social Forces 49, 259–264 (1970)
26. Van Egeraat, C., O'Riain, S., Kerr, A.: Social and Spatial Structures of Collaboration and Innovation in the Knowledge Economy. Deliverable 11.2 for EU FP6 OPAALS Research Project (2009), http://wiki.opaals.org/DeliverableAbstracts
27. Lester, R., Piore, M.: Innovation. Harvard University Press, Cambridge (2006)

Digital Ecosystem for Knowledge, Learning and Exchange: Exploring Socio-technical Concepts and Adoption

Amritesh and Jayanta Chatterjee

Industrial and Management Engineering, IIT Kanpur, India
{amritesh,jayanta}@iitk.ac.in

Abstract. Knowledge is an indispensable element that ensures healthy functioning of any socio-technical system. Despite a terminological ambiguity, it is discussed by many researchers. Learning is a process to seek and recreate knowledge within socio-technical systems. In this paper we attempt to explicate the terminological ambiguities of knowledge and explore knowledge processing and creation cycles as relevant for socio-technical systems. Further we present insights about theories of learning discussed by different scholars. We extend the paper towards new models of knowledge exchange inspired by and flow inspired by digital ecosystem concepts.

Keywords: Digital Ecosystems, Knowledge, Learning, Socio-Technical Systems, Agriculture Innovation in India.

1 Introduction

Digital ecosystem (DE) is described as a socio-technical infrastructure and processes bridging three different disciplines: Social Science, Computer Science, and Natural Science [1]. The social science here refers to the behavioral aspects and addresses the instances of communities, culture, practices. Computer Science refers to the technological aspects of information and communication and addresses the issues of infrastructure, application environment and services to support the behavioral dimension. Natural Science act as a metaphor to observe, learn and apply the self-sustainability features such as 'self generation', 'self catalysis' etc. to the behavioral dynamics. In context of knowledge transfer and learning, a DE can be viewed as a socio-technical system which supports knowledge creation, recreation, diffusion, absorption and exchanges that support the dynamics of social innovation.

The socio-technical theory is one of the approaches initially conceptualized in the domain of organizational design where the main focus was given to people, technology and work environment. One of the seminal articles [2] defines the socio-technical systems as:

a work system is made up of two jointly independent, but correlative interacting systems - the social and the technical. The technical system is concerned with the processes, tasks, and technology needed to transform inputs to outputs. The social system is concerned with the attributes of people (e.g., attitudes, skills and values), the relationships among people, reward systems, and authority structures.

F.A. Basile Colugnati et al. (Eds.): OPAALS 2010, LNICST 67, pp. 44–61, 2010.

According to this conceptualization, both social and technical systems jointly interact to produce organizational output. Social system holds its own requirements during adoption of the technology, while technical system may have their own constraints. These two systems when considered separately can be contradictory or complementary to each other. Earlier authors [2],[3] argue for the principle of 'joint optimization' of both of these two systems for effective functioning of any organization. In socio-technical context, knowledge can be viewed as one of the organizational outputs and learning as the process to internalize knowledge. In this we will elaborate upon the social and technical elements necessary for knowledge exchange and learning.

Knowledge is the central element which interacts with the individual and organization, gets embedded into them and drives it in a progressive manner. The human mind observes and interprets the surrounding in terms of concepts, theories, values, and beliefs, which form knowledge embodied in the mind. Knowledge is also recognized as a mental construct which resides within every individual and acts as a framework to understand, evaluate and acquire new knowledge. In this respect, Knowledge is highly personal to the knower who gathers facts by his/her objective observation of nature. In the book 'Personal Knowledge-Towards a Post Critical Philosophy', [4] highlights the personal component 'Intellectual Passion' as an individual urge that drives the exploration of patterns existing around the nature. Knowledge is created in the mind through mental processes such as cognition, comprehension, and through exchange in a social interaction. Knowledge once acquired, is reflected through actions, words, behavior, habit and attitude of the individual. It is also expressed into written or symbolic forms for dissemination, storage, protection, sharing, and future usage. Knowledge creation is understood as organizational phenomena and elaborated by the SECI model [5]. Another model of knowledge creation is presented by Boisot [6] in his 'social learning cycle'. Both of these knowledge processing theories can be related to the socio-technical theory. Knowledge creation happens by the contribution of both social as well as technological factors. Social components include behavioral phenomena such as learning (e.g. social learning and organizational learning) and knowledge sharing (which also includes knowledge co-creation). Technological components deal with the information and communication tools that help in codification, storage, dissemination and transfer of knowledge. Moreover, advanced information, communication and media technologies are also helping social interaction and experience sharing among individuals and communities.

2 Knowing about Knowledge

It is widely accepted that a valued part of the knowledge remains in the tacit form in the minds of the individual. The iceberg model of knowledge is widely accepted to give a general distinction between tacit and explicit spaces of knowledge. It is commonly understood that we can write less than what we can speak, and we can speak far less than what we know. There is always a large repertoire of unexpressed knowledge present within every individual. It is well evident that knowledge requires language and words for elaboration and codification. Nonaka argues upon the proposition that tacit knowledge can be (difficult to) externalized and codified into documents and physical artifacts external to the mind. An alternative argument exists against the codification strategy of knowledge, which says that all the tacit knowledge can never be captured and converted into explicit form [7]. This idea against codification also complies primarily with the technical know-how dimension of tacit knowledge. For example, riding a bicycle, swimming, convincing the customers, interpersonal skills etc. are unique experiential knowledge with respect to the individual and can never be transformed completely into the hard form, but only learned through practice. Innovation in a large socio-technical system needs dynamic exchange of tacit knowledge often in ad hoc transient networks like farmer's fairs or through 'problem based learning'.

From epistemological perspective, any knowledge is valid only within the context and settings in which it is acquired and learnt. Before applying the previously acquired knowledge (as available in books and standard procedures), it is essential to critically compare both the contexts, i.e., context of knowledge acquisition and context of knowledge application. Therefore, knowledge of context is itself is a separate kind of knowledge which is critical during the process of learning and practicing. According to Nonaka, tacit knowledge resides in the humans as justified beliefs. Justified beliefs are the knowledge which are tested and evaluated by the current state of reasoning ability by the knowledge holder and liable to continuous modifications through learning. This notion can be related with another perspective to categorize knowledge, which is put forth by Firestone & McElroy [8] under 'unifying theory'. The unifying theory assumes that knowledge is produced by complex adaptive systems[1]. Authors propose three types of knowledge which are assumed to be tested and evaluated.

i.   Tested, evaluated, and surviving structures of information in physical systems that may allow them to adapt to their environment (e.g. genetic and synaptic knowledge).
ii. Tested, evaluated, and surviving beliefs (in minds such as mental models) about the world that is subjective and non-sharable?
iii. Tested, evaluated, and surviving, sharable (objective), linguistic formulations about the world (i.e. speech- or artifact-based or cultural knowledge).

These three types of knowledge are called 'Biological Knowledge', 'Mental Knowledge', and 'Cultural Knowledge' respectively. Biological knowledge is highly personal in nature and is about becoming conscious of internal mechanisms of the body and mind which influences day to day activities. Mental knowledge can be understood as a latent knowledge that complies with the tacit dimension discussed earlier. Cultural knowledge reflects in the oral and behavioral expressions of individuals during work and social interactions. The brings out the latent knowledge of the mind to an observable form, which can be further investigated, analyzed, learned and adopted by the others.

---

[1] Examples of a complex adaptive system can be the ecosystem, where organisms interact with other species in the ecosystem and learn to survive and grow [9]; or it may be human brain whose neurons are connected by synapses to a complex network [10].

3 Knowledge Processing and Creation

3.1 Information Perspective: DIKW Model

Theories of information management attempted to situate 'knowledge' at a relative position in a pyramidal structure known as 'Wisdom Hierarchy' or 'DIKW Hierarchy' which has four successive states: data, information, knowledge, and wisdom [11]. Data is placed at the bottom of the pyramid followed by information, knowledge and wisdom staying at the top. Data are symbols representing observations and raw facts (A state of Know Nothing) which is external to our mind and can be interpreted in many ways. Information is processed and inferred data (Know what) with respect to some specific context of the recipient or interpreter. Knowledge is information combined with understanding, context, interpretations and reflections (Know-How which becomes richer than the information. Wisdom is the evaluated understanding (Know-Why) that helps to make strategic decisions, such as why, where, and when to apply the knowledge. The level of complexity, context dependency, and integrity gradually increases from bottom to top of the pyramid.

3.2 Cognitive Perspective: E2E Model

The four hierarchical constructs of DIKW pyramidal model is further extended by the arguments of cognitive system of knowledge, which presents a complexity based view. This view restructures the earlier conception of DIKW hierarchy with two additional constructs: 'existence' and 'enlightenment' and opens up the closed boundaries of the pyramidal knowledge system. This chain of six successive constructs (Existence-Data-Information-Knowledge-Wisdom-Enlightenment) is known as existence to enlightenment (E2E) model [12]. This proposition further says that data, information, knowledge and wisdom, all are constructs coming out as a result of abstraction of the existence at different levels. Such abstractions exist in a continuum and the highest level of abstraction is the state of enlightenment, which is hypothesized to be the most intelligent form of understanding. The relationship between these constructs is considered to be non-pyramidal and non-linear in nature.

3.3 Knowledge Processing Cycle

The cognitive perspective (E2E model) of knowledge is basically related with the process of learning through observation and abstraction of the different parts of reality existing in the world. In E2E model, Knowledge is processed by the mind by combining the context and conditions. Knowledge flow is directed towards the mind which usually happens during the process of learning or understanding. The nature of knowledge processing in DIKW model has alomost similar direction of flow. In this case, knowledge is created by interpreting the data and information with their respective contexts. Tuomi [22] challenges this pyramidal structure and proposed a reverse hierarchy of Knowledge, Information and Data. He contends that knowledge must exist to formulate information, and data emerge only after availability of information.

The abstract of these arguments relates to the top-down and bottom-up processing of knowledge and creates a cycle of 'learning' and 'elaboration' which leads to the creation of implicit and explicit knowledge respectively (Figure 1). Any knowledge artifact can be created by some agent (by his previously acquired knowledge) and represented as data of information. This data or information is observed, interpreted (through analysis and synthesis) and internalized by another agent and form a part of his knowledge repertoire. Elaboration leads to externalization of knowledge into knowledge artifacts, and learning leads to internalization of knowledge artifacts in the mind of the recipient. The nature of learning and elaboration are highly dependent upon the ability of the agent and his/her environment or interface.

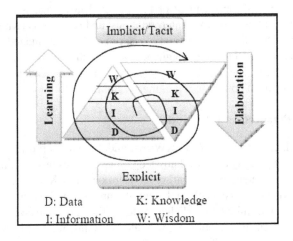

Fig. 1. Knowledge Processing Spiral

From above discussions it can be inferred that knowledge is a dynamic object which can be created in the mind in implicit (and tacit) form and can be explicated by expressing into different forms. The explicated knowledge when detached from the source and context, then it becomes an information (and/or data). When information is observed by the recipient and combined to his prepossessed context, then as a process of learning it becomes a new form of knowledge and stays within the recipients mind. Knowledge keeps on combining and updating with experience over the time. But this process is individualistic and cannot be easily controlled by centralized 'top down' system. In a goal oriented socio-technical innovation system we are however concerned with time bound (at least progressive) results. The preceding discussion elaborates the problems that can arise when one want to 'manage' and control a process which is inherently 'uncontrolled'. Our experimental Knowledge Exchange Network (KEN) system therefore draws upon the duality of figure 1, and conceptualizes an interactive network.

3.4 Knowledge Creation and Transformation

Knowledge can be processed with the help of mind and computing devises. But according to the earlier discussions in this paper, processed knowledge unless situated inside mind can't be termed as useful 'knowledge'. Every time any knowledge is processed, it changes its form and learning state of the knowledge processing entity. Here in this case, knowledge processing entities are essentially individuals. The theories of knowledge creation explain the phenomena of learning and elaboration happening among individuals, groups and organizations. Knowledge creation can be regarded as creation of both implicit and explicit knowledge as a result of cognition, social interaction, comprehension, and elaboration.

4 Theories of Learning

Learning can be understood as an integral part of the knowledge creation phenomena. As an organizational phenomenon, learning can be described from two different angles: Adaptive Learning and Generative learning [13]. The two types of learning are the part of macro level strategic debate between organizational learning [23] and learning organization [14]. These two learning concepts are elaborated in the following subsection.

4.1 Learning at Institutional and Individual Levels

Generally both adaptive and generative learning practices are followed by the organization in different proportions. Learning strategy is partially pushed by the top level management (e.g., radical technological changes) to match the organizational vision, and partially it is pulled by the individual and groups towards operations involving organizational innovations and incremental improvements.

Literature on Organizational Learning also discusses learning as a problem solving activity which involves error detection and corrections and to make decisions. Problems can occur either during the people trying to math their ability with the fundamental elements of the organization such as its structure, goals and objectives. In this context, learning is classified into two types: single loop learning, and double loop learning [15]. These two types of learning happen in two separate problem space: one is the space where organizational tasks are performed, and other is the fundamental principles, goals and objectives at which organizational tasks are defined. When people are working on a well defined job under formally set organizational policies and principles, then this gives rise to the phenomena of single loop learning. On the other hand, when people questions upon the job definition, organizational policies and principles, then it is the case of double loop learning. Single loop and double loop learning practically correlates with generative and adaptive learning respectively. Learning during problem solving is directly influenced by the way problem is understood and disseminated. In one of the popular works on Organizational Learning, [16] refers to 'bounded rationality' and discusses about the limited human ability to adap to the new and complex environments in finit time. The author discusses about the organizational learning in the domain of adopting organizational culture and transfer of innovation among the employees.

Apart from the strategic perspective, learning can also be seen from a process perspective at individual level. Learning takes place by social interaction, individual cognition and comprehension depending upon the nature and forms of knowledge. Jenson and others [17] propose two different modes of learning: STI (science, technology and innovation) mode and DUI (doing, using and interacting) mode respectively - which assist in knowledge creation and innovation.

i.   STI Mode: This mode can be assumed close to theoretical learning (i.e. learning with codified documents/literatures etc.), which is associated with the 'Know what' and 'Know why' kind of 'explicit knowledge', and primarily deals with 'Science, Technology and Innovation'. The important aspects of know-what and know-why may be obtained through reading books, attending lectures and accessing data bases. This mode of learning-innovation refers to procedures where a firm develops a scientific and technological understanding in R&D laboratories (e.g. 'just in case' learning). The STI mode of learning start with a local problem make use of global knowledge and end up in creating 'potentially global knowledge' in abstract forms which can be applied in many different situations.

ii. DUI Mode: This mode of is associated with tacit and experiential knowledge ('know how' and 'know who'), which is embodied in the organizational structure and relationships. It is rooted in practical field experiences of the individual and group. This kind of learning is generally and often localized. Communities of Practice (CoPs) [26] are one of the examples of this mode of situated or localized learning. This kind of learning happens in real time (or 'just in time') on the job employees who keep confronting new work challenges. This results into enhanced skill of employees and extends their repertoires.

4.2 Boisot's Social Learning Cycle

Boisot [6] described the knowledge transition in a three dimensional space which he named I-Space (or Information Space) where learning takes place through six different phases: Scanning, Problem Solving, Abstraction, Diffusion, Absorption, and Impacting.

i.   Scanning: This phase involves gaining insights by searching potentially fruitful pattern of (diffused) data which is less codified and abstract. This data may be subjected to multiple interpretations. Scanning is quick when data is abstract and codified and slow otherwise.

ii. Problem Solving: This is a process of giving structure and coherence to the insights. This involves exploring and extracting novel patterns from the scanned data and giving it a defined structure by applying imagination and independent thought. The new insights are partially codified.

iii. Abstraction: The partially codified insights are generalized to diverse range of situations. This involves reduction of the abstract insights to specific features particular to different situations.

iv. Diffusion: Insights are shared with the target population. Higher degree of abstraction and codification of insights increase the diffusion of data at distributed locations by electronic medium.

v. Absorption: The codified insights are applied to different situations in a learning-by-doing or learning-by-using fashion which produce new learning experiences and behavioral changes during practice. New uncodified/tacit knowledge is created as a result of absorption.

vi. Impacting: The abstract knowledge is embedded into concrete practices e.g., artifacts, technical or organizational rules, or behavioral practices. Tacit knowledge is reflected in formal work practices through which standards are formulated.

5 Summarizing the Key Concepts

Following table enlists all the concepts of knowledge and learning explored in the previous sections.

Table 1. List of Concepts Explored

	Author	Year	Key Concepts
Knowledge	Polanyi	**1958**	Personal Knowledge, Tacit Knowledge
	Nonaka	1994	SECI Model
	Nonaka & Konno	1998	The Concept of 'Ba'
	Wilson	2002	Distinguishing Tacit and Implicit Knowledge
	Firestone & McElroy	2005	'Biological', 'Mental' and 'Cultural' Knowledge
	Rowley	2007	DIKW Hierarchy
	Faucher, Everett, & Lawson	2008	E2E Model of Knowledge
	Tuomi	1999	Reverse Hierarchy of Knowledge
Learning	Argyris	1982	Single Loop Learning and Double Loop Learning
	Senge	1990	Learning Organizations;
	Simon	1991	Bounded Rationality
	McGill, Slocum, & Lei	1992	Adaptive (Instrumental) Learning and Generative Learning
	Senge & Fulmer	1993	Anticipatory Learning and Systems Thinking
	Boisot	1998	Social Learning Cycle
	Jensen, Johnson, Lorenz, & Lundvall,	2007	STI and DUI modes of learning.

6 Socio-technical Approach

We attempt to look at the digital ecosystem from a socio-technical vantage point for observing the static and dynamic states of knowledge and learning. Firstly we try to elaborate upon the basic components of this system.

6.1 Socio-technical Elements

i.   Actors: Since knowledge resides in the mind of the people, therefore they are the central actors of knowledge creation. Knowledge creation and learning can happen at individual level, group level, organizational level and social level at large. Mind and body are the two participating element with the actor. Apart from those, culture, ideals and skills are the inherent attributes of the people which influence the system. Organizational values, social norms, self esteem, incentives, peer to peer network etc. are the environment factors which influences the knowledge and learning of the social actors.

ii. Technology: In the context of knowledge creation, there can be three different segments of technology:

   a) Information Technology (codification/compression/ user interface design etc.)
   b) Communication Technology (telecommunication/ internet/ broadband wireless network, collaborative interaction etc.)
   c) Media technology (video, animation, audio, image/design, graph/chart, text/ number etc.)

Information technology provides electronic spaces and facilitates representation, storage, and transfer of both semantic and episodic knowledge of the individual and organization. Additionally it helps in organizational knowledge processing and support semantic and episodic learning. Communication technology provides network and connectivity among the knowledge actors for their multilevel interactions and exchange of knowledge. Media technology provides richness or viscosity[2] to the knowledge flow and facilitates dynamic dissemination of knowledge objects. It provides highly interactive and dynamic interfaces which influence larger parts of human senses and helps not only in semantic learning, but also in episodic learning.

iii. Knowledge: Knowledge is not of any use unless it is processed, personalized and transferred to the application spaces (in our case towards social innovation).

iv. Learning: Learning happens through cognition which involves comprehension as well as apprehension. Problem representation (definition) and solving are the two fundamental aspects of learning. Due to cognitive limitations, problems are conceived by different individuals in different perspectives. Technology helps to integrate various aspects of a problem and overcome the phenomena of 'bounded rationality'. Apart from that, technology helps to create a better reflection of organizational image and overview by various dissemination methods. 'Scanning' and 'systems thinking' [24] are the fundamental aspects of problem representation and solving. Better representation of problems and

---

[2] Viscosity refers to richness of knowledge and concerned with the gap between value of transmitted and absorbed knowledge. Long and continued mentoring exemplifies the phenomena of highly viscous knowledge transfer.

organizational processes facilitates not only single loop learning, but also in double loop learning. Technology can assist in both STI and DUI modes of learning which can build scanning ability and systems thinking approach within the individuals. Highly interactive design spaces encourage people to express their creativity and learn by DUI mode. Well design dissemination and communication spaces helps in STI mode of learning.

In the social context (for example in Indian agriculture), learning can also be understood seen as knowledge transfer between different social strata having varying accessibility to knowledge. One of the most valuable thought is given by Freire [25], who elucidated the phenomena of learning in a particular socio-political environment.

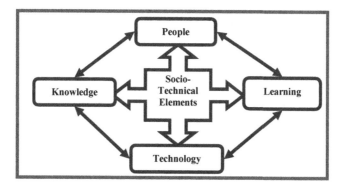

Fig. 2. Fundamental Socio-Technical Elements

6.2 Socio-technical Environment

Technology is a key determinant to create a characteristic environment where knowledge creation takes place. Environment has a major influence upon the behavioral patterns of the knowledge creating agents. In context of knowledge creation we limit our view on technology to information, communication, and media technology.

Extending the SECI model of knowledge creation Nonaka and Konno [18] highlights upon the four knowledge creation spaces: 'originating Ba', 'interacting Ba', 'cyber Ba' and 'exercising Ba' which corresponds to the four respective phases of SECI model. 'Ba' can be any physical space, mental space, or virtual space where knowledge is created in the organization. In the authors' conceptualization, Cyber Ba refers to the virtual spaces of interaction what is limited only to facilitate combination mode of knowledge creation. Cyber ba is highly influenced by drastic growth in information, communication and media technology. The technological trend has opened the conceptual boundary of 'Cyber Ba' to overlap and partially accommodate all other three modes of knowledge creation which includes socialization, externalization, and internalization (Figure 3). The extension of virtual world (or Cyber Ba) to other domains helps in the following ways:

i. Experience sharing through storytelling, empathizing and sympathizing (socialization) in voice based transactions. The virtual spaces that support socialization are: virtual chat rooms, audio blogs, video sharing virtual spaces etc.

ii. Interacting with the group and make a shared understanding over a subject (externalization). Examples of supporting spaces are Yahoo group, Google group, Online Forums etc.

iii. Analyzing and Synthesizing codified knowledge from different sources in order to innovate (combination). Examples include document sharing websites, social networks having personal and sharable spaces, collaborative web spaces (Wikipedia) etc.

iv. Learning from the interactive virtual spaces (personalization) such as online design environment, animations, and games.

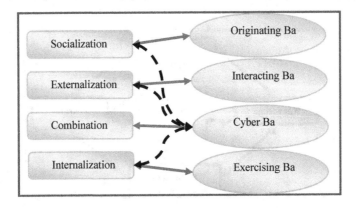

Fig. 3. Extension of Cyber Ba to all four modes of knowledge creation

6.3 Moving towards Social Innovation

We will now explore the role of knowledge exchange and learning among various constructs that lead to social innovation. In the article 'Knowing Life', Polanyi [4] discusses about the limits of observational capacities through which knowledge is acquired and how new knowledge further increases the same observational capacities through which knowledge was acquired. It is difficult to identify the initial point where there was no knowledge. Here we assume that some insights are always possessed by everyone at any point of time which comes from past experience, cognition and experimentations. Such insights help in the process of natural selection and creates urge to explore more knowledge of particular type of relevance to the respective agents. Personal insights are added with multiple modules of information coming from different domains. It is further enriched with collaborative interactions happening in social groups such as Communities of Practices, Knowledge Networks. Phenomena like 'produsage' [19], 'co-experience' [20] etc. illustrate the domain of the collaborative interactions in more detail and one can easily situate these concepts in the context of socio-technical innovation.

Knowledge construction is driven primarily by such collaborative interactions, while previously acquired insights act as a measure to believe and justify the trueness of the constructed knowledge. The trueness of knowledge however changes over time and space with change in social value propositions, purpose and utility. Therefore continuous learning and finding newer applications of knowledge becomes essential to keep the knowledge recent and useful. This stage of knowledge evolution unravels the possible roots of innovation that can occur in various social processes. The expansion of 'cyber ba' increases the possibilities of collaborative interactions and continuous learning, which paves the way for social innovation and strengthens the economic competency among the socio-economic actors. Figure 4 represents a framework to understand the relationships among different concepts that foster framework to understand the relationships among different innovation.

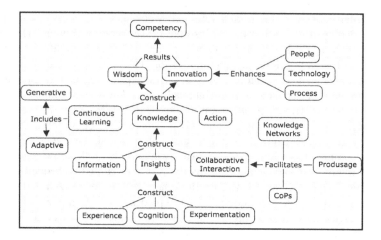

Fig. 4. From Cognition to Competency

7 Knowledge Driven Agriculture in India

Indian agriculture can be considered as one of the oldest socio-technical systems which engage directly or indirectly over 500 million people. The social part of this socio-technical system is composed of agriculture community which includes agriculture scientists, researchers, farmers, and service providers. Technology is embedded at every level in agriculture: right from the production of seeds, fertilizers, pesticides etc to tractors, tillers, and other farm equipments or in post harvest practices.

The Indian agriculture sector is expected to meet the growing need of food grains for a rapidly growing nation, new strategies are needed to enhance the agricultural growth rate to at least four percent from the current level of 1.8 to 2 percent. This level is essential to sustain a double digit growth rate for Indian GDP while retaining inflation at a manageable level. In order to pursue this target, knowledge and innovation are at the core of the strategy around which agricultural innovation can happen.

Several generalizations of the concepts of Knowledge are elaborated in earlier sections of this paper. For this particular domain of agriculture, we assume here that knowledge is socially constructed and adds to the personal insights. Online environment of collaborative interaction and community based learning or other ICT based learning opportunities enhances the potential of innovation and hence can introduce economic competency in this sector.

7.1 Review of Previous Experiences

Thus ICT implementation efforts across Indian villages attempt to connect agriculture community at all levels of practice and research. One of the efforts in this direction is DEAL (digital ecosystem for rural livelihood; website: www.dealindia.org) project which was deployed in 2006 at Indian Institute of Technology Kanpur. This project aimed to form a digital knowledge network among the Indian agriculture community to facilitate diffusion of knowledge and innovation. DEAL was structured around the concept of 'self-catalysis' (autopoiesis) [21] which proposes that if circularity in conversation can be created and maintained, then knowledge objects keep on reproducing and evolving. The structural design of DEAL has two different natures of knowledge spaces: 'Gana Gyan' and 'Gyan Dhara' focused to serve the knowledge needs of the organized, explicit and unorganized, emergent or transient agriculture communities respectively. 'Gana Gyan' implies folk knowledge existing in people's mind. This knowledge is primarily unstructured and embedded into work practices and insights. Gyan Dhara', on the other hand is formalized and validated knowledge which emerge from commonly accepted knowledge models of scientific community. This knowledge is highly structured in nature and can be disseminated, and circulated through internet. But paradoxically this digital distribution can result into ground level implementation only through social networks in the real world. These two streams of knowledge generation were targeted for different segments of agriculture communities. 'Gyan Dhara' was targeted for the people who constitute the knowledge storing and producing community. These include: agriculture scientists, researchers, field extension experts. 'Gana Gyan' on the other side is meant primarily for the communities who are knowledge users. They are lead farmers, small farmers, agriculture traders, student-researchers, Kisan (farmer) call-centers, agri-clinic etc. Scientific knowledge could thus be complimented and experimented with the field experiences, and the field experiences could be validated and generalized by the scientific knowledge. DEAL served for acquisition and visualization of knowledge and extended its services to the village level agriculture consultancy institutions known as KVKs (Village Knowledge Centers). The underlying assumption was, 'by infusing knowledge connectivity to human agencies and creating socio-technical feedback loops' critical success conditions can be created to energize rural economic infrastructure in India.

7.2 Challenges Encountered

Indian agriculture has been facing many technological challenges such as: unavailability of proper harvest management information infrastructure. Indian agriculture extension system relied upon the model of vertical diffusion for knowledge transfer between KVK scientists and farmers (Figure 5).

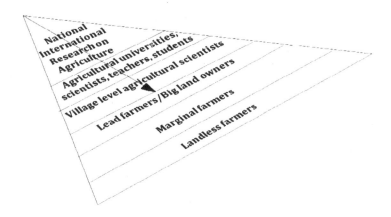

Fig. 5. Vertical Diffusion Model

Structure of communication was influenced by institutional hierarchy which impeded velocity and viscosity of knowledge transfer. Geographic distance was a critical barrier for the knowledge from one location to reach another in limited time. That gave birth to the phenomena of 'bounded rationality' among the field practitioners while taking critical decisions particularly during adoption of new techniques and methods. Same phenomena influenced the scientists whose research conclusions often lacked enough empirical instances.

In the beginning, the prime focus of DEAL was targeted to create digital knowledge repository and providing its accessibility to all the stakeholders of agriculture community in a networked environment. The network infrastructure established by DEAL to facilitate knowledge flow between lab and land possessed potential to overcome skepticism that farmers usually have while adopting new variety of seeds, fertilizers, pesticides, and farm machineries.

Despite having numerous promises to benefit Indian agriculture, DEAL has so far encountered many challenges which seek further attention. DEAL's knowledge based services are composed of two dimensions of knowledge: 1) tacit and experiential (Gana Gyan); and 2) validated and codified (Gyan Dhara). The level of knowledge diffusion to respective segments of agriculture community, and progress gaps are presented in the Figure 6. The upper quadrants of this figure represents the top end of the pyramid which include agriculture scientists and researchers from KVKs, State Agriculture Universities (SAUs), Indian Council of Agriculture Research (ICAR), and other similar Institutions. DEAL has been able to deliver in both of its knowledge dimensions to this section. For example, tacit and experiential component of knowledge flow is marked by participation in online spaces for Questions and Answers, Communities of Practices etc.

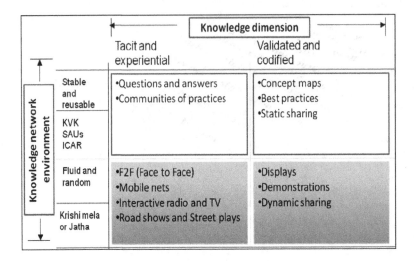

Fig. 6. Knowledge Dimension Vs Network Environment

Validated and structured dimension of knowledge flow is observed in the form of creation of concept maps, best practices etc. However this kind of knowledge contribution lacks direct p2p network based collaboration which is essential for social innovation.

Knowledge is acquired primarily through the contribution of commissioned scientists and researchers whose work is to structure and validate expert knowledge. Voluntary sharing of knowledge among the scientists did not reflect much interest which is a major cause of concern. The acquired knowledge remained in the silos of the repository and no significant interdisciplinary flow is observed. 'Self-catalysis' as a phenomenon is yet not self evident.

The lower quadrants of figure 6 reflect agriculture community who are primarily farmers and field practitioners. In this section, the tacit dimension of knowledge flow relies on face to face communication, mobile networks, radio, televisions, road shows and street plays etc. Validated and structured knowledge flow happens through displays, demonstrations, and dynamic sharing among the community members. The DEAL portal has not been able to contribute enough to this section. Farmers having rich field experience are yet to take much interest in this present form of digital media.

7.3 The Ripple Model

The initial flow of knowledge in India agriculture extension system followed a pattern of vertical diffusion. DEAL attempted to flatten this vertical hierarchic structure to some extent and reduced the time to get validated knowledge. However, it is unable to encourage collaborative activities, voluntary knowledge sharing, learning and innovation among the members of agriculture community. So a new approach is framed upon the 'Ripple Model' (Figure 7) which promises to produce ripple effect of knowledge flow through collaboration and interactions happening at multiple levels between and within the closely formed communities. This model emphasizes community formation among the people who are at immediate position in the vertical knowledge flow hierarchy. For example, at the highest level, the people from national and international research (level 1) can more likely to associate with scientists, teachers and students at Agriculture Universities (level 2). This section has the highest potential to engage in collaborative knowledge creation when connected through a social network. The network can be considered as 'Collaborative Innovation Network' (COIN). This network can also extend partially to village level scientists (level 3) and diffuse innovation oriented community behavior towards this end. At next level network can be formed by close association among people at level 2 and village level scientists (level 3) which can further extend up to big farmers and landlords (level 4). At this level of collaboration, expert knowledge flows from advanced university scientists to village level scientists. The network formed at this level can be considered predominantly 'Collaborative Learning Network' (CLN). Both the networks COIN and CLN despite having different nature of collaboration, doesn't exist in isolation, rather remain connected and keeps influencing each other. This also allows transition of members from CLN to COIN and so on. At the farmers end (level 5 and level 6), network creation can happen by common interest groups. This network can be called 'Collaborative Interest Network' (CIN). To reach out to the small or landless farmers (economically the most deprived) a new pedagogy through new forms of knowledge mediators will have to be created. It is assumed that this new pedagogy will complement India's current ambitions NREGA (National Rural Employment Guarantee Act) and will be created by the users' with p2p collaboration with local knowledge producers.

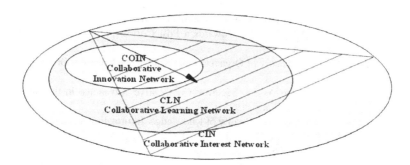

Fig. 7. Ripple Model

The 'produsage' and 'co-development' field experiment will provide new insights to understand the outermost ripples. 'Produsage' gives theoretical insights about knowledge co-creation by multiple agents who are collaborating and providing their individual inputs to create a common knowledge artifact. For example, many agriculture scientists having different domains of specialization can contribute their expert insights to create crop specific knowledge model. This kind of collaboration also results into co-experience mixed with social emotions. This co-experience can further create stronger ties among the collaborating entities and creates more potential for social interaction in future. This possibility has a strong affinity to visualize the phenomena of 'self catalysis' among the participating agents.

8 Conclusion

In this paper we have explored the socio-technical theories in relation to the DE framework to examine the phenomena of knowledge and learning in socio-technical systems. We extended the concept of Ba to the enlarged scope of 'Cyber Ba' and its possibility of transformation into social media. We have elaborated the case on Indian agriculture and presented a review upon previous experiences of DEAL implementation. The unmet gaps of the DEAL experience will be addressed through a new knowledge and innovation approach 'Ripple Model'. We propose a new design for transforming structural features of DEAL to shape it in the form of a new media for social innovation. This will need further research and investigation to track its performance over the next period.

References

[1] Dini, P., Iqani, M., Revera-Leon, L., Passani, A., Chatterjee, J., Pattnayak, D., et al.: D12.10: Foundations of the Theory of Associative Autopoietic Digital Ecosystems: Part 3. In: WP12: Socio-Economic Models for Digital Ecosystems, OPAALS (2009)

[2] Bostrom, R.P., Heinen, J.S.: MIS Problems and Failures: A Socio-Technical Perspective. Part I: The Causes. MIS Quarterly 1(3), 17–32 (1977)

[3] Trist, E.: The Evolution of Socio-Technical Systems: A Conceptual Framework and Action Research Program. Ontario Ministry of Labour (1981)

[4] Polanyi, M.: Personal Knowledge- Towards a Post Critical Philosophy. Routledge and Kegal Paul Ltd. London (1958)

[5] Nonaka, I.: A Dynamic Theory of Organizational Knowledge Creation. Organization Science 5(1), 14–37 (1994)

[6] Boisot, M.H.: Knowledge Assets: Securing Competitive Advantage in the Information Economy, Oxford, United States (1998)

[7] Hildreth, P.M., Kimble, C.: The Duality of Knowledge. Information Research 8(1) (2002)

[8] Firestone, J.M., McElroy, M.W.: Doing Knowledge Management. The Learning Organization 12(2), 189–212 (2005)

[9] Levin, S.A.: Ecosystems and Biosphere as Complex Adaptive Systems. Ecosystems, 431–436 (1998)

[10] Schuster, H.G.: Complex Adaptive Systems. In: Radons, G., Just, W., Haussler, P. (eds.) Collective Dynamics of Nonlinear and Disordered Systems, pp. 359–369. Springer, Heidelberg (2001)

[11] Rowley, J.: The Wisdom Hierarchy: Representations of the DIKW hierarchy. Journal of Information Science 33(2), 163–180 (2007)

[12] Faucher, J.L., Everett, A.M., Lawson, R.: Reconstituting Knowledge Management. Journal of Knowledge Management 12(3), 3–16 (2008)

[13] McGill, M.E., Slocum, J.W., Lei, D.: Management Practices in Learning Organizations. Organizational Dynamics 21(1), 5–13 (1992)

[14] Ortenblad, A.: On Differences between Organizational Learning and Learning Organization. The Learning Organization 8(3), 125–133 (2001)

[15] Argyris, C.: Organizational Learning and Management Information System. Data Base (1982)

[16] Simon, H.A.: Bounded Rationality and Organizational Learning. Organization Science 2(1), 125–134 (1991)

[17] Jensen, M.B., Johnson, B., Lorenz, E., Lundvall, B.A.: Forms of Knowledge and Modes of Innovation. Research Policy 36, 680–693 (2007)

[18] Nonaka, I., Konno, N.: The Concept of Ba-Building A Foundation for Knowledge Creation. California Management Review 40(3) (1998)

[19] Bruns, A.: Produsage, Generation C, and Their Effects on the Democratic Process. In: Media in Transition, Boston, vol. 5 (2007)

[20] Katja, B.: Defining co-experience. In: Proceedings of the 2003 international conference on Designing pleasurable products and interfaces, pp. 109–113. ACM, Pittsburg (2003)

[21] Luhmann, N., Bendnarz, J., Baecker, D.: Social Systems (Writing Science). Standford University Press (1996)

[22] Tuomi, I.: Data is more than knowledge: implications of the reversed knowledge hierarchy for knowledge management and organizational memory. Journal of Management Information Systems 16(3), 103–117 (1999)

[23] Senge, P.M.: The Fifth Discipline: The Art and Practice of the Learning Organizations. Doubleday Currency, New York (1990)

[24] Senge, P., Fulmer, R.M.: Simulations, Systems Thinking and Anticipatory Learning. Journal of Management Development 12(6), 21–33 (1993)

[25] Freire, P.: Pedagogy of the oppressed. Continuum, New York (2006)

[26] Wenger, E.: Communities of practice-Learning as a Social System. Systems Thinker (1998)

The Diffusion of Social Media and Knowledge Management – Towards an Integrative Typology

Frauke Zeller[1], Jayanta Chatterjee[2], Marco Bräuer[1,3],
Ingmar Steinicke[3], and Oxana Lapteva[3]

[1] Ilmenau University of Technology, Institute for Media and Communication Studies,
Ehrenbergstr. 29, 98693 Ilmenau, Germany
frauke.zeller@tu-ilmenau.de
[2] Indian Institute of Technology Kanpur, Kanpur-208016, India
jayanta@iitk.ac.in
[3] Kassel University, Department for Computational Philology,
Georg-Forster-Str. 3, 34127 Kassel, Germany
{braeuer,ingmar.steinicke,oxana.lapteva}@uni-kassel.de

Abstract. This paper introduces a first outline of a typology of distributed knowledge co-creation in virtual communities based on Porter's typology of virtual communities. The typology is based on empirical results from the analyses of social media, and a discussion of case study results from India proves the adaptability as well as usefulness of the typology. At the same time, the case study serves as an example to depict a socio-economic perspective on social media and knowledge management in Digital Ecosystems.

Keywords: Social media; web 2.0; typology; knowledge management; knowledge co-creation; produsage; presumption.

1 Introduction

Over the past years, researchers from the media and communications field as well as researchers from adjoining fields have observed that the traditional terminological and conceptual differentiation between mass media production on the one hand and their reception on the other hand is becoming increasingly blurred. The rapid diffusion of Internet technologies and the emergence of online communities--particularly online knowledge communities--have lead to the need for a theoretical reconceptualisation. New terms such as 'produsage' and 'prosumption' [1] [2] describe emerging patterns of media production and consumption that blur the traditional role-division of producer and user, of sender and recipient. This change has far reaching consequences for the fields of economy and politics, and also for the traditional mass media sector.

So far marketing strategists as well as media and communication scholars often use the same terminology such as social media, web 2.0, social software, and online/virtual communities. However, communication scholars refer to broader trends of parallelised, decentralised content creation and distributed creativity regarding so- called *social*

F.A. Basile Colugnati et al. (Eds.): OPAALS 2010, LNICST 67, pp. 62–75, 2010.

media products (e.g. YouTube, Facebook) that depict a post-industrial mode of socio-technical innovation. These innovations are built on iterative, evolutionary models and fluid, ad hoc processes that are not necessarily nor preliminary controlled by the hierarchies and laws of classical economics. The terminological aberrations that often describe similar, however not identical aspects of those processes (e.g. social media, social software, web 2.0, etc.) complicate the systematic study of respective innovations. In this respect, typologies are needed that allow the systematic classification of the various innovations observed.

Moreover, in the field of Digital Ecosystems research new paradigms such as Dynamic Service Composition etc. also focus on a flexible and interchangeable approach to software adoption and development, where the end-users are provided with the opportunity of easily adopting as well as adapting software (in a broad sense) for their own businesses, certain forms of produsage and prosumption can also be found. Given the importance of communities as a core factor in Digital Ecosystems, where knowledge (co-)creation and knowledge management represent the vital driving force as well as catalysts regarding a sustainable DE community, we argue that insights from the social sciences can contribute significantly to a structured approach and understanding of the social and technological (i.e. software applications) mechanisms as well as the connection of both in virtual knowledge communities.

In this paper we will therefore explore the modes of self-catalysed content co- creation as socio-economic innovation (produsage and prosumption) and propose a first outline of a typology on distributed knowledge co-creation.

2 Theoretical Background

Due to the increased number and complexity of products and services that compete with each other in a global knowledge economy [3] the importance of knowledge and its production as well as management has increased immensely. Accordingly, *knowledge management* (KM) has become a core issue in organisations in the last decades. The basic function of KM is the transformation of its employees' tacit knowledge into explicit knowledge and vice versa which is a circular process of socialisation, externalisation, combination, and internalisation [4]. The functions of KM can be summarised as (a) making existing knowledge transparent, (b) controlling of recently or prospectively needed knowledge, (c) supporting the exchange of knowledge, and (d) the interlinking and retrieval of information [5].

Recently KM has surpassed the organisational context towards more community-oriented ('gemeinschaftlich', see [6]) ways of knowledge production. In this respect the term of *communities of practice* (CoP) as introduced by Wenger seems to depict this shift best: According to Wenger "communities of practice are groups of people who share a concern or a passion for something they do and learn how to do it better as they interact regularly" [7]. Hence, communities of practice are groups whose members participate in joint sharing and learning processes based on common interests. Members engage in discussions, help each other and share information [8].

According to Wenger the domain, community and practice together build a community of practice. The practice is developed through a variety of activities such as problem solving, requests for information, seeking experience, reusing assets, coordination and synergy, discussing developments, the documentation of projects, visits, and the mapping of knowledge and identification of gaps [7].

In recent years, ICT has facilitated the development of communities of practice whose members are locally dispersed and CoPs are being recognised as valuable organisational assets to "overcome the inherent problems of a slow-moving traditional hierarchy in a fast-moving virtual economy" [8].

It is possible to regard CoPs as a representation of the so-called *Mode 2 of knowledge production.* The traditional mode of knowledge production that used to be organised according to disciplinary boundaries is fundamentally changing [9]. More than 15 years ago, a new mode of knowledge production coined as 'Mode 2' has emerged. Not only the character of knowledge production but also knowledge validation and dissemination are transforming. Communities of practice produce their knowledge in the context of applications, across disciplinary borders, and the communities are composed of members with a huge variety of skills, professions, and scientific domains. Along these characteristics, flat hierarchies and transient organisational structures are preferred. Since the expertise of any community member is regarded as a potential asset, communities of practice need to ensure that contributions from any member can be taken up by the community--and this is the point where 'social' media come into play.

Certain technologies such as weblogs, folksonomies, and social networking sites have in common that communities make use of them to manage their own knowledge. Such media are usually coined as *social media* and are part of O'Reilly's notion of a *web 2.0* [10]. Lietsala & Sirkkunen [11] suggest to use the notion of social media as an umbrella term under which one "can find various and very different cultural practices related to the online content and people who are involved with that content" [11]. According to the authors, web 2.0 is not a synonym for social media as it represents an even looser concept referring to online services and technologies. What is important to keep in mind is that social media put an emphasis on the *content* whereas social software refers to the code, the technology, and software used for social media applications [11].

The most prominent example of a social medium is Wikipedia. Regarding Wikipedia, the differences between a classical encyclopaedia project and this new form of knowledge production are salient: In principle any web user can become a 'co-'author and add any entry to this encyclopaedia or is allowed to amend entries as editor. The same user can contribute with her knowledge and benefit from the knowledge of other users. "These produsers engage not in a traditional form of content production, but are instead involved in produsage - the collaborative and continuous building and extending of existing content in pursuit of further improvement." [1].

Another example are weblogs, which "could only be created by people who already knew how to make a website first. A weblog editor had either taught herself to code HTML for fun, or, after working all day creating commercial websites, spent several off-work hours every day surfing the web and posting to her site." [12]. However, since open source weblog software is not only available for free but also easy to use, the number of weblogs has increased enormously. Knowing HTML is no longer an obstacle which needs to be dealt with when sharing one's knowledge in the World Wide Web. Moreover, by means of using blogging services it is not even necessary anymore to own personal webspace. And given that many Web 2.0 applications had been and still are collaboratively developed in open source software communities, it is not only the *content* of weblogs and wikis which are 'prosumed', it is also the *software* itself. This again refers to the underlying paradigms and ideas of the Digital Ecosystems 'movement': Software applications are meant to be not only adoptable but also adaptable to the specific needs and services of the different DE community members (i.e. SMEs). Hence, with the dynamic contents that are to be managed by means of the software, the software itself is also changed (adapted).

So far, other formats than Wikipedia (such as Lycos IQ or Yahoo! Answers) have not found much attention from communication scholars albeit representing both knowledge-management systems and social media as the users generate the knowledge themselves. What constitutes most distinctively the usage of social media is that users generate their own content and that they are no longer relying on the content created in markets under economic and mass media logics. In this context KM is not only an issue of organisational success that huge corporations can afford. Resource weak actors are given new tools for managing their knowledge, individuals can profit from the participation in knowledge communities, be it for political participation, hobbyist activities, self-help, or other topics. In addition, SMEs may improve their position in the competition with corporations.

Nevertheless, the term social media remains somewhat vague describing phenomena like user-generated-content on the Internet in communities with the help of specific technologies and software. As media are used for communication which is a social process, media are also an integrative part of this process. This means that media are social phenomena and hence have immanent social characteristics. With regard to the various practices on the Internet, almost every website covers features usually described as social media. Even so called web 1.0 practices like personal homepages, IRC, and newsgroups could be considered as social media as they involve social processes as well. Accordingly the attribute 'social' is not specific enough, and the term *participatory media* would describe the concrete phenomenon more precisely [13].

However, we would argue that a typology helps to better understand and systemise the various phenomena of joint knowledge creation that were introduced above and that are usually also coined as social media. When it comes to the characterisation of social media we argue that it is appropriate to draw on dimensions that had already been used to categorise the phenomenon of online communities, a line of research that gained academic interest before the emergence of the terms 'social media' or 'web 2.0'. One example was devised by Porter [14], who differentiated between the dimensions purpose, place, platform, population, and profit model when creating a typology of online communities.

3 Porter's Dimensions of Virtual Communities

According to Porter, "a virtual community is defined as an aggregation of individuals or business partners who interact around a shared interest, where the interaction is at least partially supported and/or mediated by technology and guided by some protocols or norms" [14]. As Porter notes, one advantage of her definition is the inclusion of business partners, acknowledging that communities can also operate in an economic system. Furthermore, Porter speaks of 'technology' in an abstract manner instead of focussing exclusively on Internet-based technologies and on the 'virtual' aspect of those communities. Community life can also take place in the 'real' world and not only in the virtual setting.

Porter's proposed typology differentiates two first-level categories. The first category deals with the establishment of virtual communities. They can be member-initiated or organisation-sponsored (for-profit or non-profit sector). The second category acknowledges the relationship-orientation: Member initiated communities can be social or professional. Organisation sponsored organisations can foster commercial, non-profit or governmental relationships.

The next step is to define the attributes of virtual communities. Porter proposes the following five attributes: *purpose* (content of interaction), *place* (extent of technology-mediation of interaction), *platform* (the technical design of interaction in the virtual community, where designs enable synchronous communication, asynchronous communication or both), *population interaction structure* (the pattern of interaction among community members as described by group and type of social ties), and *profit model* (return on interaction) [14]. With the help of those attributes, researchers should be provided with a consistent set to describe virtual communities. However, does this typology also cover all aspects of virtual knowledge communities?

4 Porter's Typology Assessed

On the basis of different social media (question-answer-communities*, Wikipedia*, blogs, etc.), a qualitative content analysis and structure analysis were conducted. The guiding research question was: How does Porter's typology of virtual communities fit to knowledge communities and social media? Is it sufficient or does it need adaptation according to empiric phenomena?

The results revealed that knowledge management can be regarded as the main *purpose* of the researched communities like Wikipedia, Yahoo! Answers or social media in general. The *profit models* can vary – Yahoo! makes revenues through

---

* If available, German editions were used for analysis.

advertisement, Wikipedia in turn is exclusively financed by donations. The *place* is almost exclusively virtual for all analysed communities even though local face-to-face meetings can be observed for Wikipedia as well. The platform is basically based on asynchronous interaction that can be explained through the complexity of the transformation from tacit to explicit knowledge and vice versa. The dimension of population resp. pattern of interaction needed further specification and hence should be described as the *role diversity* of community members. Generally it can be stated that any user can be both producer and recipient of content. But with respect to the functions of KM, when it comes to the controlling of recently or prospectively needed knowledge, each community has own rules and practices regarding the inclusion and exclusion of relevant or irrelevant content. Therefore Wikipedia developed a democratic hierarchy between 'conventional' users and elected administrators who for example have the authority to delete articles (Fig. 1).

In Yahoo! Answers the administrators (Fig. 2) are not elected by other users but nominated by Yahoo! instead. However the nomination procedure in Yahoo! Answers was not transparent and the moderators act on an anonymous basis. It is not clear whether a 'conventional' user can receive administrator privileges. However, it is evident that an administrator could also be a 'conventional' user by means of a second account. Even though the work of the administrators in both communities is dependent on input by other users who can indicate faulty or poor content, the power of the roles within these communities is not equally distributed. There are users with less and users with more power. The rules as to how to divide this power can either be transparent or vague.

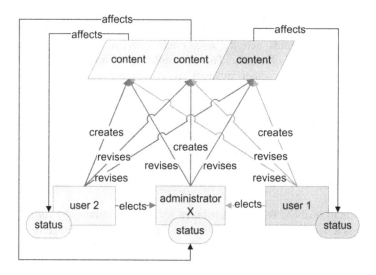

Fig. 1. Interaction pattern in Wikipedia

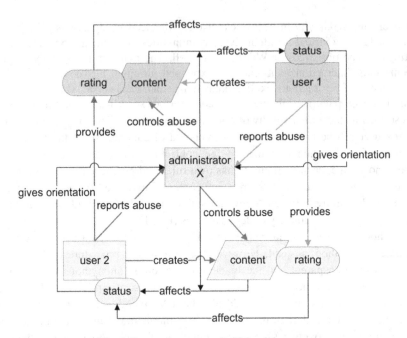

Fig. 2. Interaction pattern in Yahoo! Answers

In contrast to social media like YouTube, where spontaneous entertainment is a core purpose of the users, knowledge managing communities follow a long term approach. So it is of interest to ensure that the stored information can be transformed into knowledge through perception in future. Therefore, knowledge managing communities developed several *sustainability mechanisms*. These mechanisms are for example reputation systems, activity based credit systems, or rankings: users may be able to rate the content of others. The accumulation of all ratings of certain content can be compared to other evaluations of content and hence these ratings can be ranked. Additionally to simple statistics (e.g. number of content contributions, number of logins, and other activities), rating based statistics can be used for creating the reputation of each user in the community. On the one hand these mechanisms are meant for the motivation of the users to contribute regularly and on the other hand to provide indicators for the quality of the content.

The results of our research have lead to the integration of two further attributes to Porter's typology of virtual communities: *role diversity* and *sustainability mechanisms*.

With the help of these two new attributes we hope to initiate the further development towards a new typology for social media and knowledge management in Digital Ecosystems, allowing a distinct and clear categorisation of the various social and technological aspects. Whereas the notion of 'role diversity' refers to the concepts of produsage and prosumption, representing a central aim of Digital Ecosystems, 'sustainability mechanisms' are of equal importance to Digital Ecosystems' communities. The example of reputation and credit systems can also be integrated in a discussion of trust regarding DE. Since the concept of 'trust' represents a major field in both computer science and social science oriented research of Digital Ecosystems, not only theoretical discussions but also practical applications and interpolations are needed. These practical applications can in turn be instilled by already existing virtual communities' practices, that cannot be applied as is, but by means of a structured approach in order to adapt the underlying mechanisms to the Digital Ecosystems context.

Table 1 summarises therefore Porter's features of virtual communities together with our proposed additional features regarding social media in Digital Ecosystems: role diversity and sustainability mechanisms. It goes without saying that these two new features also represent crucial attributes for communities that do not populate the Digital Ecosystems dimension, however particularly in this dimension they appear to be pivotal.

Table 1. Porter's Features of Virtual communities and additional features of social media

Porter's Teatures of Virtual Communities	Additional features of social media in Digital Ecosystems
Purpose	Role diversity
Place	Sustainability mechanisms
Platform	
Population interaction structure	
Profit model	

5   Social Media and Knowledge Management and a DEAL's Perspective

So far this paper has focussed on analysing the features of virtual communities with a clear theoretical focus. This section will provide insights drawn from a series of Digital Ecosystems pilot-projects in India in order to discuss the potential impact of the produsage and prosumption concepts on socio-economic growth.

The Digital Ecosystem pilot projects (DEAL-www.dealindia.org) for Indian Agricultural Extension Services (IAES) were analysed from a 'produsage' and 'prosumption' perspective. According to these analyses 'produsers' generate social capital that contributes towards 'self sustenance' of the ecosystem. According to our proposed typology from section 4, this analysis mainly focuses on the issues of role- diversity and sustainability mechanisms in a Digital Ecosystem.

IAES operates through KVKs--Krishi Vigyan Kndra (Agri-Science Centres), directed by the State Agricultural Universities (SAU), Indian Council of Agricultural Research (ICAR) Institutes or by the NGOs engaged in rural development (Non- Governmental Organizations). The projects focus on Northern India and selected five KVKs for the longitudinal research. The projects attempted to create digital knowledge bases for fostering the process of creation and dissemination of new information and innovative knowledge among farmers (the tacit knowledge holder), KVK experts (the knowledge moderator) and scientists of the state agriculture universities and other institutions (the explicit knowledge holder).

In the pre DEAL framework, there were established and clearly defined formal relationships between different layers of actors—academic (Agricultural institutes, IITK), administrative (ICAR, ZCU-the Zonal Control Unit—the implementing and monitoring authority) and functional field units (KVKs), but very few ties among the members at the same layers. For example, ZCU in Kanpur had direct link with all five KVKs and farmers had close link with their respective local area KVKs but there were almost no direct links among the five KVKs or the farmers of different villages with each other before DEAL.

The OPAALS authors at IITK have incorporated some network diagrams which are drawn by using NetDraw, to give a clear idea about the effect of DEAL in the network structure. Their analysis shows that DEAL is able to boost the information flow by increasing the number of ties (from 77 to 183), and group reciprocity (from 0.3585 to 0.7745) of the existing network where all the existing nodes are present and IITK is the only new actor introduced.

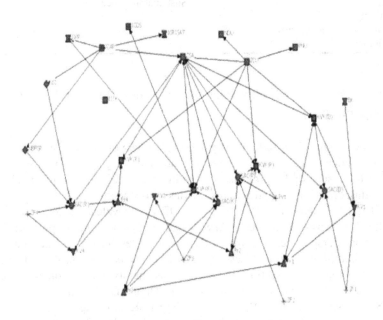

Fig. 3. Network ties before DEAL

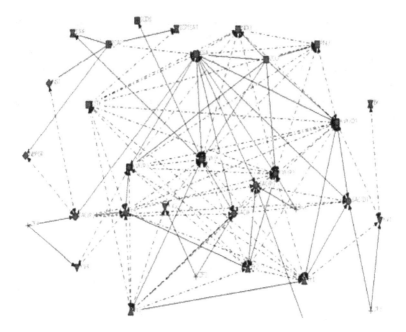

Fig. 4. Network ties after DEAL

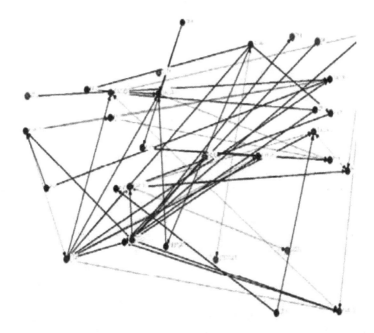

Fig. 5. Reciprocal ties before DEAL

The dotted lines, in figure 4, are the newly created links, where as the solid lines show the pre existing relationships among the actors. These ties are formed in the process of knowledge prosumption by the mutual engagement of the network members, facilitated by DEAL. Empirically, DEAL is able to generate weak ties within and between different groups of farmers, KVKs and different research institutes. These ties help to access the resources (information) associated with different nodes by reducing the network distance. So it helps in increasing the resource mobilisation in the network.

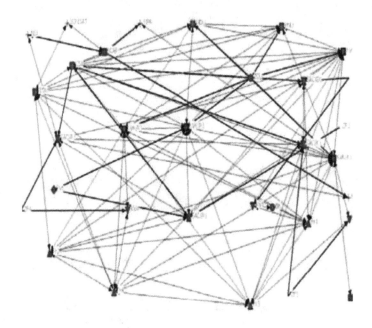

Fig. 6. Reciprocal ties after DEAL

Before DEAL, the reciprocity among ties was also low. Initially better reciprocity was observed only in the informal ties among the actors. But most of the ties within (e.g. between two academic institutions) and between (e.g. between a KVK and ZCU) layers were formal and there was low reciprocity. So the whole system was like a top- down push system. The implementation of DEAL helped in reducing the gap and increasing the group reciprocity which is an indicator of increased interactions among the members in the network. In figures 5 and 6, the light lines denote the reciprocal ties and the dark lines denote non-reciprocal ties. Increase in reciprocal ties has a positive impact on the collaborative knowledge co-creation process and increased 'prosumption' among members enhance the reciprocity. Thus there is a cyclical process in the network between content prosumption and increased interaction which also corresponds to the feature role diversity in our typology.

While responding to the question of 'self sustainability' of the community, the researchers proposed that the currency of reputation and trust generated through produsage and prosumption motivates further collaboration and community participation. Therefore, DEAL is able to enhance the social capital by increasing the mobilisation of resources in the network and also by enhancing the voluntary collaboration in the content creation process.

The DEAL example shows that this project can be described well with the help of Porter's features of virtual communities and the proposed additional features for social media in Digital Ecosystems (role diversity and sustainability mechanisms). An analysis focusing on the issue of role diversity and sustainability mechanisms seems to be capable of detecting important and potentially critical aspects of virtual knowledge communities. The results of such analyses may help to re-design such communities in order to facilitate their success.

6 Summary: Social Media and Knowledge Management and a Digital Ecosystem's Perspective

We have used approaches from the field of media and communication together with socio-economic discussions regarding the shift from traditional divisions between producer and recipient in virtual knowledge communities that use social media as interaction tools and environments. By means of this, we tried to:

- Develop a structured approach to the fuzzy term 'web 2.0' in order to arrive at an analytic environment for the deployment of social media in virtual communities. Contribute to the field of Digital Ecosystems by focussing on a driving factor of such environments, that is virtual knowledge communities.
- Using the concept of 'communities of practice' in order to introduce shifts in knowledge production and management (produsage and prosumption).
- Arrive at a typology of distributed knowledge co-creation (in Digital Ecosystems) by exploring the modes of self-catalysed content creation as socio-economic innovation (produsage and prosumption).

Regarding a Digital Ecosystem's perspective, the findings depicted in this article provide a small albeit significant field of work. Given the broad topic of knowledge production and management, together with discussions and theoretical explorations of the 'medium' (in our case 'social media'), we focused on the development an integrative typology of knowledge management in/with social media that should be used as a platform for further research (empirical and theoretical). The DEAL case shows that such research could provide promising results that can inform the designers and facilitators for knowledge communities and Digital Ecosystems. At the same time, we draw upon prior work [15, 16, 17, 18] conducted in the course of OPAALS and other work [19], in order to work with a sound theoretical and empirical basis.

Finally, based on our findings, we propose three main paths for further research that connect to existing research fields and results in Digital Ecosystems:

- *Modes of social media deployment in Digital Ecosystems*, including questions of accessibility and quality issues; social networks and applications; case studies and applications.
- *The integration of trust in Digital Ecosystems' knowledge communities*, including questions regarding knowledge verification; return of investment; identity; governance structures.
- *Knowledge modelling in Peer-to-Peer networks*, including questions as to the formalisation of knowledge and business modelling; knowledge visualisation; distributed knowledge production and management.

References

1. Bruns, A.: Blogs, Wikipedia, Second Life, and Beyond: From Production to Produsage. Peter Lang, New York (2008)
2. Humphreys, A., Grayson, K.: The Intersecting Roles of Consumer and Producer: Contemporary Criticisms and New Analytic Directions. Sociology Compass 2, 1–18 (2008)
3. Dunning, J.H. (ed.): Regions, Globalization, and the Knowledge-Based Economy. Oxford University Press, Oxford (2002)
4. Takeuchi, H., Nonaka, I.: Hitotsubashi on Knowledge Management. Wiley, Singapore (2004)
5. Blumauer, A., Pellegrini, T.: Semantic Web und semantische Technologien: Zentrale Begriffe und Unterscheidungen. In: Pellegrini, T., Blumauer, A. (eds.) Semantic Web: Wege zur vernetzten Wissensgesellschaft, pp. 9–25. Springer, Heidelberg (2004)
6. Tönnies, F.: Gemeinschaft und Gesellschaft. Wissenschaftliche Buchgesellschaft, Darmstadt (2005); 1st edn. (1887)
7. Wenger, E.: Communities of Practice: A Brief Introduction,
   http://www.ewenger.com/theory/index.htm (retrieved November 22, 2009)
8. Lesser, E.L., Storck, J.: Communities of Practice and Organizational Performance. IBM Systems Journal 40(4), 831–841 (2001)
9. Gibbons, M.C., Limoges, H., Nowotny, S., Schwartzman, P.S., Trow, M.: The New Production of Knowledge. Sage, London (1994)
10. O'Reilly, T.: What Is Web 2.0? Design Patterns and Business Models for the Next Generation of Software,
    http://oreilly.com/web2/archive/what-is-web-20.html
11. Lietsala, K., Sirrkunen, E.: Social Media: Introduction to the Tools and Processes of Participatory Economy. Tampere University Press, Tampere (2008)
12. Blood, R.: Weblogs: A History and Perspective,
    http://www.rebeccablood.net/essays/weblog_history.html
13. Rheingold, H.: Using Participatory Media and Public Voice to Encourage Civic Engagement. In: Lance Bennett, W. (ed.) The John D. and Catherine T. MacArthur Foundation Series on Digital Media and Learning, pp. 97–118. The MIT Press, Cambridge (2008)
14. Porter, C.E.: A Typology of Virtual Communities: A Multi-Disciplinary Foundation for Future Research. Journal of Computer-Mediated Communication 10(1) (2004)
15. Bräuer, M., Steinicke, I., Zeller, F.: Self-Reflection of Community Building, Communication and Collaboration Processes in the NoE. Deliverable for the European Commission, RP6 Information-Society-Technology IST-2005-034824 (2009)

16. Bräuer, M., Crone, A., Dürrenberg, C., Lapteva, O., Zeller, F.: Appropriateness of Communication and Collaboration Tools in an International Virtual Research Community. In: Proceedings of the 2nd International OPAALS Conference on Digital Ecosystems, Tampere, Finland, pp. 20–27 (2008)
17. Zeller, F., Lapteva, O., Crone, A.: Studies/Papers on Discourse Organisation of Epistemic Cultures: Theoretical and Methodological Analyses and Practical Interpolation. Deliverable for the European Commission, RP6 Information-Society-Technology IST-2005-034824 (2007)
18. Bräuer, M., Dini, P., Dory, B., English, A., Iquani, M., Zeller, F.: Principles, Models, and Processes for the Development of the Open Knowledge Space. Deliverable for the European Commission, RP6 Information-Society-Technology IST-2005-034824 (2007)
19. Steinicke, I.: Wissen und Wissensmanagement in online Gemeinschaften: eine qualitative Inhaltsanalyse von Mechanismen zur Sicherung der Validität von Wissen in wissensverwaltenden online Gemeinschaften, Master Thesis (2008)

Digital Ecosystems Adoption at Local Level: A Preliminary Comparative Analysis

Francesco Botto[1], Antonella Passani[2], and Yedugundla Venkata Kiran[1]

[1] Create-Net, Italy
[2] T6, Italy
{francesco.botto,vkirany}@create-net.org, a.passani@t-6.it

Abstract. This article focuses on the process of Digital Ecosystem adoption at territorial level. We present our views on the European approach to Digital Ecosystem. We try to define the process of DE territorial adoption as it has been modeled in last years. We carried out a preliminary analysis of ongoing comparative research about DE adoption and described the methodology used in the research process. With reference to the research preliminary data, we project how and to what extend the theoretical model of DE adoption elaborated in DBE and OPAALS projects and has been adapted and used at local level. The territorial process described here is introducing DE in different industrial sectors and originate from very diverse socio-economic situations. For this reason, interviewers are introducing DE at local level by adapting the DBE model to local needs. A preliminary analysis of these strategies and of actors involved in the process is presented. A first gaze on the technological side of DE adoption in term of infrastructure used and service developed is also provided. From the analysis, we indicate steps for future research and delineate those open questions that deserve a deeper analysis in the near future.

Keywords: Digital Ecosystems, Social Science, DE Territorial adoption, Regions, Comparative analysis.

1 Introduction

Digital Ecosystems (DEs) is a novel, not yet stabilized research field: its research objects are emerging and only a limited number of initial implementations exist. One of the main approaches is known as the "European" DE approach that emerged from the European Commission FP6 DBE projects cluster [1]. The European approach to DE is based upon an open framework with key ingredients including flexible community leadership, use of a P2P, distributed and decentralized infrastructure and Open Source Software. In this paper we apply this approach together with the socio- technical theory.

We believe that a comparative analysis of DE adoption in local/regional settings could provide useful inputs and motivations for future research. Therefore, the objective of this paper is to give preliminary results of such a comparative analysis we are conducting on experience collected from several Digital Ecosystem adoptions at local level.

F.A. Basile Colugnati et al. (Eds.): OPAALS 2010, LNICST 67, pp. 76–91, 2010.

The paper is organized as follows. In Section 2, we start by describing digital ecosystems as it was defined within the framework of the DBE project and then how it has been further developed within the OPAALS project. Then we outline the process that can lead a region or a community to the use of the digital ecosystem approach in a self- sustaining way by briefly describing how these processes have been modeled recently. We present our research methodology in Section 3 which is based on an initial questionnaire, interviews for deeper investigation and a workshop. In Section 4 we present and analyze preliminary data about how this model has been used, adopted and developed in different local contexts. We conclude the paper in Section 5 by highlighting some issues for the further work of our comparative analysis.

2 DE Territorial Adoption[1]

We consider the DE [1] adoption, or deployment at local level, as a socio-technical process [2], [3]. This translates a process for technological environment development and knowledge creation and sharing in different local contexts, and maximizes its potentialities in term of economic development, social capital improvement, ICT diffusion and knowledge diffusion democratization. In socio-technical systems, society and technology construct and reconstruct each other in a complex process. Moreover, to understand DEs we used the metaphor of the socio-technical infrastructures [4], [5]: DEs are artifacts emerging from practice, directly connected to human activities and material structures that should be jointly analyzed with the technological and social frameworks (see: [6]). Consequently, DE adoption is a long- term investment that implies also a process of network-building, participation and the activation of multiple collaboration and involvement of diversified stakeholders (universities, intermediate actors, SMEs, police makers and knowledge hubs) (see [7]).

In the DBE projects cluster two understandings of territorial adoption have been evolved. We will therefore present the models stemming from the DBE project and from the OPAALS one, and then we will argue for the need of a comparative adoption analysis.

2.1 The DBE Model

At the end of DBE project [7] the process of local adoption of DEs has been defined as influenced by different variables, and as a process that needs to be adapted to local needs, user behaviors, and specific historical/economic junctures (see Figure 1).

Although the DE adoption process, as DE technology, needs to be planned with the aim of being adaptable to specific local needs, we synthesized the process in the following steps:

1. DE concept dissemination and awareness building.
2. Socio-economic regional analysis: Regional/territorial Maturity Grade and/or DEII.

---

[1] An important disclaimer is needed here: these process models are intended to be used by local stakeholders in a phase in which (as in the current one) the technology is not yet stable and the user community is still limited. Once the technology will be fully developed (both at the infrastructure level and in terms of basic services) the process will need to be fine-tuned.

3. Regional Catalyst definition and engagement.
4. Industry/sector/community definition.
5. Users' definition (Cluster and SMEs identification and selection, or research community identification, etc).
6. Development of a shared road-map for the development of the first habitat.
7. Training.
8. Service development and ecosystem population.
9. Pilot action evaluation (with DEII) and planning of systemic deployment of DEs.
10. Steps from 3 to 8 can be replicated for different habitats adapting the activities to the specific needs of each industry/sector/community.

This first model, that has the positive characteristic of being easily understood by local stakeholders, is based on a useful, but problematic simplification: it is mainly based on a linear process.

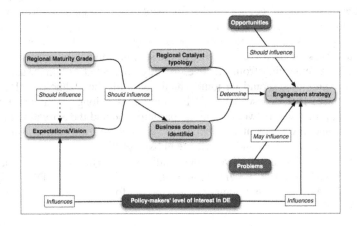

Fig. 1. DBE model of local implementation process for DEs

2.2   The OPAALS Model

An addition to the DBE model has been proposed [8], [9]. In this complementary model, the concrete actions to be taken at local level are analysed in depth and a participatory methodology is suggested. Thus, in this second approach the top-down nature of the first model is mitigated, and the local community (future users) gains a bigger role in the DE definition process. The top-down element of DE adoption, however, cannot be eliminated; due to the fact that DE deployment at local level is undoubtedly a political action that needs to involve policy-makers and needs to be connected to the institutional process of innovation policy development (this is particularly true at the present stage of digital ecosystem technology development). [8], [9] brought forth the following points to the attention of research community for consideration:

- Start from concrete local needs
- Work with people at various levels: In addition to the policy makers, innovation should involve both the management and the lower levels of business organizations and communities
- Work on what makes sense for participants, not only on the digital ecosystem idea: instead of 'implementing' a digital ecosystem (or a digital community ecosystem) as the core objective, focus on developing meaningful innovations for the community and use the digital ecosystem idea as a tool
- Avoid using the term "digital ecosystem" or "ecosystem": the result of this innovation should be something meaningful for local communities also in its label. It is improbable – but not impossible – that they will adopt their own vocabulary.

In this model (Figure 2), therefore, the top-down approach meets the bottom-up processes that always exist at local level, and the Regional Catalyst becomes the principal actor responsible for the area in the middle, in which the institutional level needs to meet the necessities and aspirations of the community level.

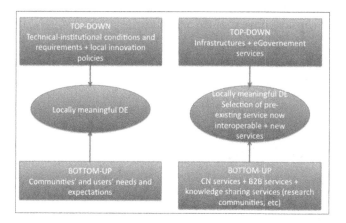

Fig. 2. A second model of DE adoption, Passani (2009) abstraction and further development from Botto and Szabo (2008)

The role of Regional Catalyst[2], in this model, is that of translating the digital ecosystem research into the 'local language' and of facilitating the process of bottom-up introduction of DE. It is consistent with the hermeneutic approach of the OPAALS projects and can be seen as an application of the participatory design and action research ideas to a complex local innovation process [10], [11].

---

[2] It is important to notice that the role of Regional Catalyst can be played in a collaborative way by more then one local actor. For example a local development agency can act as RC in the first steps of the process positively engaging the institutional level. Then a research centre or an innovative software enterprise can take action as Regional Catalyst when dealing with participatory service design. This issue has been already covered during the DBE project, see D31.6 (Passani, 2007).

The combination of the two DE adoption models accounts for Foucault's discursive approach by taking in consideration the definition that local actors give to a DE. In addition, it also acknowledges issues of power and democratic processes by introducing participatory decision-making processes.

In this new approach to DE adoption, partially already tested in Trento and in Lazio regions, we can see a better balance between the top-down and the bottom-up approaches and the role of the Regional catalyst evolved significantly. In this model the knowledge generated by the OPAALS project (and other projects in the DBE cluster) reaches out beyond the local level because of the emphasis on the complexity of the language layer of innovation. The local DE adoption can be seen as a process in which knowledge is provided by the research community to local stakeholders that then can use it for creating new knowledge, accordingly to their concrete needs.

Power asymmetries (between the institutional and the user layer on one hand and between knowledge providers and users on the other) are also acknowledged and mitigated thanks to participatory decision-making process (participatory development of a DE adoption roadmap) and participatory service development. The discourse layer is also taken into account enabling and facilitating users to develop their own definition of what a digital ecosystem can be for them. Figure 3 shows a global summary of the participatory process of DE adoption.

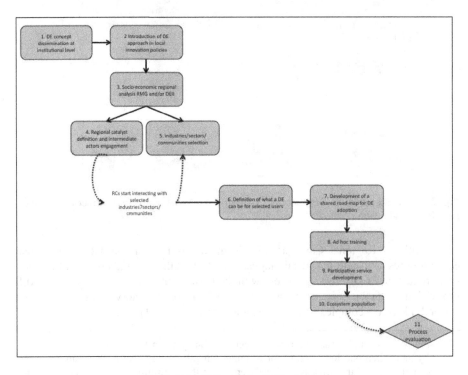

Fig. 3. OPAALS model of DE adoption

2.3 The Need for a Comparative Analysis

This paper is not the first report regarding the DE adoption cases in the DBE cluster projects. In the literature, a description of five regional cases is presented: Aragon (Spain), West Midlands (UK), India, Ireland and Brazil (see: [1]). Since the presentations of adoption cases are performed by the catalyst persons, they have the value of the first information coming from the field of real life adoption of technologies and services. Therefore the presentations are relevant because of their self-narration of adoption phases, problems encountered, expected impacts and first lessons learned. Besides this, DBE projects produced several deliverables about territorial experiences from the point of view of the regional catalysts [12], [13] and from that of the SMEs engaged [14]. The Peardrop project produced a dissemination analysis of the Aragon (Spain) and Welles (UK) cases designed especially for policy- makers.

Conversely, previously descriptions lack of (1) a common analysis framework, and (2) a critic, ex-post evaluation approach that could evidence what is working or not and why. A serious and clear analysis of successes and failures is commonly considered as a warranty of farsighted research. For this reason we started working on a comparative DE adoption analysis.

3   Methodology

As we will describe, this preliminary analysis takes into consideration the OPAALS regional cases community and some other experiences related to DE cluster projects (‚Eurokleis, SEAMLESS, Pearldrop and ONE) and local experiences carried out by OPAALS colleagues (DBE Lazio project).

The research strategy we delineated is articulated as follow:

1. Preliminary *questionnaire* to start investigating the core dimensions.
2. *Interviews* for further investigation of the cases.
3. *Workshop*: participative work analyzing and fine-tuning research outputs.

The analysis that follows is based only on the data gathered at the first level of the research. In this first round we took in consideration 10 cases of DE deployment, carried out by 8 organizations. Table. 1. contains the list of the organizations (both institutions and companies) to which we sent the questionnaires: all of them responded by providing the required information.

Table 1. Institutions that filled out the questionnaire

No	Institution/company	Nation
1	T6 Ecosystems srl	Italy
2	CREATE-NET	Italy
3	University of Modena and Reggio Emilia	Italy
4	Indian Inst. of Technology Kanpur (IITK)	India
5	Cambridge University	UK
6	NUI Maynooth	Ireland
7	CM International Ireland	UK
9	Instituto Tecnoloogico Aragon (ITA)	Spain

The aforementioned organizations have been or are being actually engaged in projects of DE deployment, not necessarily in the region in which they are legally located.

The map in Fig. 4 visualizes the regions in which DE deployment projects have been or are being currently activated. This first round of questionnaires took into consideration only some of the projects of DE deployment, and we cannot consider this selection exhaustive of the DE community universe. In fact, in the map, we visualized (with numbers in black) regions that are working on DE deployment and that we interviewed in this first round survey (see also Section.4); with numbers in red we indicate those regions and territories that have shown interest in the DE concept and may develop a local DE in the future.

3: Europe
6: India
21: Buenos Aires area, Argentina

Fig. 4. Regions working on the DE deployment[3]

Most of the regions marked with numbers in red belong to Den4Dek project: a network of regions working together with the specific aim of delineating a regional DE deployment plan. Beside the territories marked in the map, we have to acknowledge possibly many other territories that encountered the DE approach in the past (as was the case of West Midlands and Tampere during the DBE project) or in other projects with whom we would need to enter in touch in the next phases of this research. In the next section we will describe why we started our analysis from this group of organizations and how we are going to possible enlarge the sample.

---

[3] Region marked with black number are reported at the beginning of Section 4. The list of the regions marked in red is as follows: 11= Basque country (Spain); 12= Valencia (Spain); 13= Umbria (Italy); 14= Extremadura (Spain); 15= Galicia (Spain): 16: Madeira (Portugal); 17= East-Slovak region (Slovak); 18=State Vorarlberg (Austria); 19= Southern Great Plain region (Hungary); 20= Helsinki region (Finland); 21= Buenos Aires area (Argentina).

3.1 Cases Selection

It is important to underline that this is a qualitative survey with a yet not well identified universe. In fact, we are not aware of a complete list of territories and/or organizations working on DE concept and deployment. Due to this situation, we decided to use our social capital, our social and scientific networks, as first instrument for selecting the cases. We started this survey with organizations and territories participating in the OPAALS Network of Excellence. Thanks to the participation in this project and in others (ONE, DEN4DEK, DBE) we contacted researchers that are still working on DEs, excluding experiences that can be considered closed (such as the cases of West Midlands and Tampere from the DBE project). We are using a snow ball effect in order to enlarge the case sample; the criterion we are using during this process is to focus on cases that are actually deploying DE or that have at least a clear plan about how to implement it and that, possibly, have already contacted policy-makers and other territorial stakeholders.

4 Analysis

As we mentioned this first round of questionnaires took into consideration only some regional experiences, more specifically this analysis gives account of the projects undertaken/being undertaken in the following territories:

1. Lazio region (Italy)
2. Trentino (Italy)
3. European area, sector of European Cooperatives
4. Emilia Romagna (Italy)
5. Aragon (Spain)
6. India
7. Dublin (Ireland)
8. Wales (UK)
9. East Midlands (UK)
10. Cambridge and Peterborough (UK)

Beside regional experiences we can see also a different approach (Case 3 above), developed by CREATE-NET, which applied the DE approach not to a specific territory but to a commercial sector: in this case the one of European Cooperatives. In this case, DE concept is going to be spread at European level, possibly impacting many and different territories, but - as we will see in the following paragraphs – the strategy of deployment is not regional based/focused.

All the people interviewed have been or are engaged in DE cluster projects, in some cases in more than one; consequently each project of DE deployment here analyzed is linked with one or more EU projects. We can speak of a DE community of researchers and practitioners that - thanks to different European projects - developed and applied the concept of DE in different territories. As the Table 2 makes evident, some projects are more represented in this survey that others, but all DE cluster projects are represented at least through one of the person interviewed.

Table 2. Link between regional experiences and EU projects

Regional experience and EU projects	
Project	*Respondents that participated to the project*
DBE	● ●
OPAALS	● ● ● ● ● ● ●
EFFORT	●
DEN4DEK	● ● ●
PEARDROP	● ●
ONE	●
SEAMLESS	●
EUROKLEIS	●

European projects represent also the first contact with the concept of DEs for the great majority of the interviewed. We asked them, *"When and how did you have the first knowledge on DEs?"* and with only one exception (that found out about DE concept thought EU community web) all the others came across the concept because of personal contacts with researchers or EU representatives engaged in one of the above mentioned projects.

We then asked what pre-existing situation the interviewees intended to change thanks to the introduction of DE at local level. We can interpret the outputs of this questions as an answer to a core question: "why to introduce DE?". The Table 3 presents the gathered data.

Table 3. Motivations for DE introduction

Why to introduce DE	*Score*
Fostering SMEs networking	● ● ● ● ●
Provide to SMEs and clusters new and more interoperable ICT solutions	● ● ● ● ● ●
Foster collaboration and information flow within Public Administration (PA) and between PA and enterprises	● ●
Support and further develop local innovation strategies	● ●

Table 4. Sectorwise DE adoption progress

Sector in which DE adoption has been or is going to be introduced	
Sector	*Score*
Tourism	● ● ●
ICT-based regional innovation system for the public sector	●
Longitudinal to different productive sectors	●
Textile	●
Logistic	●
Agriculture	●
Biotech/biomedical	● ● ●
Environmental tech	●

The aim to provide locally new and more interoperable ICT solutions (score 6) and the aim to foster SMEs networking (score 5) motivated the catalysts more than fostering collaboration with and within the PA (score 2) and develop local innovation strategies (score 2).

SMEs are at the center of many DE adoption actions, with reference to the industrial sector or belonging we can see that researchers and practitioners are working in many fields, adapting DE to different industrial fabric (Table 4). The touristic and the biotech/biomedical sectors are considered suitable for DE adoption.

Table 5. Innovation strategies and methodologies

Innovation strategy/methodology		
Case	*Initial*	*Changes*
1	DBE strategy	Privilege under-developed provinces; less involvement of the Regional Authority; carry on only a pilot project; reconsider the technology.
2	DBE strategy	• Start from concrete local needs • Work with people at the many levels (Managers, workers, citizens…) • Work on what make sense for participants • Avoid terms like DEs or Ecosystems outside the political arena
3	Identify an economic segment and companies that could take up the technology developed during and after the project lifetime	No
4	Facilitate collaboration between small companies of different sectors	No
5	Common sense	No
6	ICT led Networking and Communication	More role of social media in community creation
7	(no answer)	Strategy constantly changing to suit the very rapidly changing economic and business profile of regional development
8	Promotion of business ecosystems on the basis of advanced IT	No
9	Look at enhanced models for clustering & networking to achieve regional competitiveness	No
10	Look at enhanced models for clustering & networking to achieve regional competitiveness	No

Table 6. DE introduction strategy

DE introduction strategy		
Case	*Initial*	*Changes*
1	DBE and OPAALS projects strategy: introduce the DBE SW.	Found a Catalyst other than the Regional Authority.
2	DBE strategy: introduce the DBE SW and social network.	We facilitate the emergence of: • shared and formalized interest from the stakeholders • creation/governance rules from the Local Government. We renounced the DBE SW and considered specific services like ONE.
3	Tackle the DE adoption from a productivity point of view.	Improved the connections with the EU cooperatives.
4	By creating a network of interoperable intermediaries able to increase substantially the ICT adoption.	We are adding special purpose ICT services to solve the lack of suitable ICT tools for internal operation.
5	DBE strategy: move influencers, introduce the DBE SW in SW SMEs.	We initially planned to engage directly the users but we realized it was better to engage the SW SMEs and then they will engage the users.
6	Top-down push strategy: integrate the DE strategy with the existing development plans.	We shifted to study the online social interactions through various modes of social media platforms.
7	awareness raising; demonstrator to show DE capabilities; meetings of key players; dissemination	It is constantly under review
8	Project management tool	Rather than a project management tool the DE will be introduced as a more general knowledge sharing tool
9	...	No
10	Initially we intended to support cluster managers.	The project moved towards a support tool for collaboration between biomedical companies, clinicians and research departments.

Regarding the strategy and/or methodology of their specific innovation case (Table 5), initially the Trento and Lazio cases (1, 2) adopted the DBE strategy described in Section 2, but later reconsidered it substantially. Also the India and Ireland cases (6, 7) describe a change in their innovation strategy. The other cases do not mention the DBE strategy model and did not change their strategy after during the adoption process.

About the DE technologies introduction strategy (table 6) explicitly or implicitly most of the cases aimed to adopt the DBE project software tools at the beginning. The technology introduction strategy changes regard probably the movement through more reliable tools. The case 3 is different because it is connected to a specific service development project, working on a technological framework other than the other cases: from the beginning the ONE platform provided a web service.

Table 7. Actors involved

Actors and roles				
Case	*Public & influencers*	*Catalysts*	*Research centers*	*Enterprises, associations & users*
1	3	3	1	16
2	1	2	9	30
3				6
4	8		12	70
5	6	1		140
6	?	?	?	
7	2	1 (?)	6	33
8			5	80
9	5	3	2	15
10	9	1	3	29

Actors and roles involved in the 10 cases of adoption are extremely heterogeneous (Table 7). Unfortunately the data are not comparable since in some cases (i.e. case 3) the few users are in fact associations of users, therefore numbers are not so much reliable.

The relationships between the catalysts and the regional authority are the following:

- 1 case: part of the regional authority;
- 2 cases: independent from the regional authority, with a long and stable subcontracting activity;
- 3 cases: independent from the regional authority, with several project financed by the region;
- 3 cases: collaborates with the regional authority only for the DE related activities;
- 1 case: no relationship with the regional authority.

The technologies that have been chosen to support the DE are various (Table 8). As explained before the DBE project did not provide reliably working applications and actually only tree cases are considering this option. Only four cases are explicitly moving through the adoption of Open Source Software (OSS) and five cases are considering a Peer-to-Peer (P2P) architecture. Considering that the DBE project cluster was built upon the P2P and OSS columns, this is a very informative data. Most of the cases (eight) are considering specific software applications to implement the DE, and four are using social network devices.

Table 8. Technologies considered

Technologies							
Case	WEB portal	Soc Net	Specific services apps	DBE or OPAALS tools	P2P	OSS	Mob
1	●	●		●		●	
2	●	●	●				
3			●		●		
4			●		●		
5	●	●	●	●	●	●	
6	●		●		●		●
7			●			●	
8		●			●	●	
9			●	●			
10			●				

In regions there is a huge variety of services that currently are or will be available in the DE (Table 9): social networking – four cases choice -, business to consumer, back-off management, Open Negotiation Management (ONE), transportation optimization, tourism, knowledge management and labour exchange. The ONE platform provides a service that is composed by: social networking, sales and negotiation management, contract definition, accounting and consulting. The project has been lead by CREATE-NET, therefore the two cases related to this research center (2 and 3) are actually considering this service for DE adoptions.

Table 9. Services considered

Services									
Case	Soc Net	B2 C	Back office man.	O N E	Transport optimiz.	Tour ism	Know. man.	Labour exch.	State of art
1	●	●	●						--
2	●			●					-
3	●			●					-o-
4			●		●				+
5						●			-
6							●		++
7									(?)
8	●						●	●	(?)
9									--
10									--

The state of the art of services adoption is described in the last column and says that only in the Indian case the services are currently under adoption:

- The service is a prototype, but the adoption is an idea (-o-): 1 case
- The services will be defined (--): 3 cases
- The services are being defined (-): 2 cases
- The services are under construction (+): 1 case
- Under adoption (++): 1 case

Finally we investigated the governance of the DEs (Table 10). The Governance system is not defined yet in three cases (Lazio, Aragon and Wales). Most of the cases are considering the creation of or involvement a foundation or a company or a board to govern the system. Regarding the service management, the distributed self-management suggested by the DBE cluster project is actually under consideration in only three cases (Aragon, East Midland, Cambridge and Peterborough). A more centralized solution is in operation or will be implemented in the other cases.

Table 10. Governance in DE adoptions / development

Case	Governance	
	Governance system	*Service manager*
1	Not yet defined.	Not yet defined.
2	Main rules: Local Government; Monitored: Board (Local Gov, Res Centers, Catalyst).	Catalyst will manage centralized services. Board will supervise.
3	Open ecosystem and public infrastructure. A foundation will coordinate the developers.	Initially a startup.
4	Based on equal rights and policies. Promoted by trusted intermediaries.	Intermediaries.
5	Not yet defined	Easily managed by the business owner.
6	Partially moderated and self-governed	IITK manages the platform.
7	(no answer)	(no answer)
8	Not yet defined.	The industrial promotion agency or the university.
9	Community Interest Company will provide a legal structure.	Project partnership.
10	Collaboration agreement initially, later more formal structure.	Project partnership.

5 Conclusions

As we saw in the previous paragraph, regional experiences of DE adoption are very different from many points of view such as: DE introduction strategy, reasons behinds

DE adoption, industrial sectors involved and so on. Another important difference deserves to be mentioned in this conclusive section.

We described the OPAALS model of DE adoption in Section IIB. It is composed of 11 steps; as obvious regional experience we considered in this article are at different steps of DE adoption. We have one case in stage two: "Introduction of DE approach in local innovation policies"; one case in stage 3 "Socio-economic regional analysis", one case in stage 5 "Industries/sectors/communities definition", one case in stage 6 "Definition of what a DE can be for selected users", one more case between this stage and the followers "Development of a shared read-map for DE adoption". Only two cases are in the last phases, one on "Participatory service development" and another in "Ecosystem population". As evident, we are speaking - in most cases - about ongoing processes that are open to modification and are still sensible to various risks.

We will dedicate next phases of our research to the investigation of such risks, by analyzing more specifically what are the main difficulties that DE adoption process may imply. We are also really interested in describing the problems, issues, possible conflicts that occurred during the process among engaged actors.

Moreover, if in this article we focused our attention on similarities among cases; in the next step we will prioritize on the differences among them. We will go deeper on each case with the help of face to face, semi-structured interviews and we proceed with a more biographic approach in order to achieve better understanding of the actors, roles and responsibilities of different players at local level. We are really interested in understanding in depth: who is working at local level, with whom, following which specific goal and building what kind of social networks. Similarly, we will further investigate the technological aspects of DE adoption; in this respect we recognized once more the necessity of undergoing face to face interviews, in order to develop a shared language and restrict the multi-semantic nature of many terms.

In our next work, we plan to reconsider the variables used during this first stage and analyze interdependencies, moving toward a multiple-variable approach. Finally, the concrete role that DE is going to play in territorial policies will deserve a special attention. It is crucial to understand if DE can become a framework for sustainable regional development or will obtain only a minor role, due to the limited investment and impacts.

References

[1] Nachira, F., Nicolai, A., Dini, P., Le Louarn, M., Rivera Leon, L. (eds.): Digital Business Ecosystems, Europen Commission, Bruxelles (2007),
http://www.digital-ecosystems.org/book/debook2007html

[2] Bijker, W.E., Hughes, T.P., Pinch, T.J.: The Social Construction of Technological Systems. MIT Press, Cambridge (1987)

[3] Bijker, W.E., Law, J.: Shaping Technology / Building Society. MIT Press, Cambridge (1992)

[4] Star, S.L., Griesemer, J.G.: Institutional Ecology, 'Translations' and Boundary Objects: Amateurs and Professionals in Berkeley's Museum of Vertebrate Zoology, 1907-39. Social Studies of Science, vol. 19, pp. 287–420 (1989)

[5] Star, S.L., Ruhleder, K.: Steps Toward an Ecology of Infrastructures: Design and Access for Large Information Spaces. Information Systems Research 7(1), 111–134 (1996)

[6] Botto, F., Passani, A.: Del 7.1 – The relationship between Community Networks and Digital Ecosystems, OPAALS project Deliverable (2007),
http://files.opaals.org/OPAALS/Year_1_Deliverables/WP07/OPAALS_D7.1_final.pdf

[7] Passani, A.: The Territorial Prospective of Digital Business Ecosystems. In: Nachira, F., Nicoai, A., Dini, P., Le Louarn, M., Rivera Leon, L. (2007)

[8] Botto, F., Szabo, C.: Del 12.7 – Sustainability and Business Models for Digital Community Ecosystems, OPAALS project Deliverable (2008),
http://files.opaals.org/OPAALS/Year_3_Deliverables/WP12/D12.7.pdf

[9] Botto, F., Szabo, C.: Digital Community Ecosystems: Through a new and Sustainable Local DE Development Strategy. In: Proceedings of the IEEE-DEST 2009 Conference, Istanbul, May 31- June 3, pp. 91–96 (2009)

[10] Greenbaum, J., Kyng, M.: Design at Work: Cooperative Design of Computer Systems. Lawrence Erlbaum Associates, Hillsdale (1991)

[11] Reason, P., Bradbury, H.: Handbook of Action Research. Sage, London (2006)

[12] Censis: SME Engagement Analysis and Collaborative Problem Solving, DBE Project Deliverable (2006), http://files.opaals.org/DBE/deliverables/Del_27.3_DBE_SME_engagement_and_collaborative_problem-solving.pdf

[13] Passani, A.: Regional Case Studies, DBE Project Deliverable (2007),
http://files.opaals.org/DBE/deliverables/Del_31.6_DBE_Regional_Case_Studies.pdf

[14] Darking, M., Whitley, E.: Studying SME Engagement Practices, DBE Project Deliverable (2005),
http://files.opaals.org/DBE/deliverables/Del_27.2_DBE_Studying_SME_Engagement_Practices.pdf

The Role of the Region in Knowledge Transfer for Economic Development

Associate, London School of Economics
Houghton Street, London, WC2A 2AE
neil.rathbone@daventryhouse.com

Abstract. This paper outlines the increasingly important role of regions in
Knowledge Transfer. Drawing on experiences from two ICT projects involving
Digital Ecosystems, it explores why the regions are the best catalyst for
knowledge transfer and suggests methodologies for the development of
effective ICT strategies for regional economic development.

Keywords: ICT, regions, regional economic development, SMEs, knowledge
transfer.

1 Introduction

The increasing globalisation of economic and cultural activity, underpinned by the
Internet and other media that transcend geography, could lead one to believe that
regions, the traditional geographical units of government and community, are
outdated and no longer necessary.

However, as we shall see in this paper, the opposite is true. Regions have an
increasingly important role to play and should be encouraged to develop the specific
qualities that make them effective in the adoption of new and emerging technologies.
Indeed, what we call a region may well change over time; from a unit that is defined
by historical or geographical convenience, to one that is dynamically defined by
economic potential and knowledge transfer activity.

More specifically, using knowledge gained from two international ICT projects
that developed a particular model of regional participation, we will see how the next
generation of Internet technologies, Digital Ecosystems, has a strong basis in
devolution and democracy, which naturally lends itself to the direct involvement of
regions and regional actors.

We will also look at specific techniques that provide a practical model of how
regions can approach technology development and adoption, and how individual
regions have leveraged their diversity to find their own ways to rise to the challenge
presented by the leading edge of ICT.

2 Regionalisation as a Reaction to Global Change

Before delving into the specifics of knowledge transfer, it is worth looking at what
may be on the wider horizon for regions.

F.A. Basile Colugnati et al. (Eds.): OPAALS 2010, LNICST 67, pp. 92–99, 2010.
© Institute for Computer Sciences, Social Informatics and Telecommunications Engineering 2010

In his book 'Jihad v McWorld'[i], Benjamin Barber points out the irony of Jihadists carrying out their anti-Western bombings wearing Nike branded trainers. He argues that the two extremes of globalisation and tribalisation are in reality a counterpoint to each other and will probably always co-exist: global consumerism producing an increasingly sterile and homogenised 'McWorld' against which stronger localisation satisfies the human need for identity, roots, history, and a sense of belonging.

As well as the psychological need for local communities, there is an increasing imperative due to global issues such as food shortages and climate change. The UN Food and Agriculture Organisation in its 2009 November summit[ii] called for more food to be produced closer to those who need it, and warned of a threat to security of supply for those dependent on imports, especially from countries that may be adversely affected by climate change. Transporting food great distances is also becoming recognised in the public conscience as an unnecessary contributor to fossil fuel consumption and the personal carbon footprint. This means that we can anticipate regions becoming more self-sustaining and operating across a wider range of activities.

Finally, within all but the smallest countries, many organisations and associations find it advantageous to have a local presence via 'branches' and 'groups'. This recognises that, while representation has to be carried out at national and supranational level, localisation of relations with members is better than trying to serve heterogeneous end-users with the homogeneous output of a centralised facility. The motivation that comes from being able to make a visible difference within your immediate environment means that organisations with a regional network are able to tap into volunteer energy and a wider range of skills and personalities.

So, despite globalisation and digital media, there are a number of general current trends that indicate that the region is far from dead. Let's now look at the specific qualities that regions bring to knowledge transfer.

3 Value in Diversity

Across the world, regions are so different in their size, structure, authority, power, resources, culture, history, and behaviours that one region might hardly recognise another as its counterpart. However, this lack of uniformity is what underpins the value of the region as a conduit between global technological progress and local practical application. The 'one size fits all' solutions that are likely when thinking on a national or international scale fail to recognize the diversity and the complexity of adapting initiatives and programmes to it.

One might suppose that the European Union, which has often been called a 'Europe of the Regions' would have some consistency, but in fact there is huge contrast between, say, the powerful federalist states of the German Lander, the subjugated, purely administrative role of the UK regions, and the smaller nations like Finland that behave as if they are simply one region.

As an example, I would like to contrast two regions that participated in Digital Ecosystems. Both are located in developed countries within the EU, yet due to their fundamentally different environments they adopted different approaches to being partners in Digital Ecosystems research.

4   Contrasting Two EU Regions: UK's West Midlands v Spain's Aragon

Despite being research projects, DBE and OPAALS, of which you will hear more elsewhere in this conference, both managed to successfully involve regions and their SMEs. Before we look at the generic principles that we established from this participation, let us take a moment to contrast two of the participating regions, both of which are from well-established EU Member States.

West Midlands

The UK's ancient history is one of feuding regional kingdoms and successive invasions to produce a densely-packed mongrel nation within a small island. Much of its subsequent political history has been about uniting England, Wales, Scotland, and Ireland. Although it is now being reversed to a modest extent by the creation of 'regional assemblies', much of the UK's modern history has been about unity at national level, and regions have been purely an administrative convenience: a way of dividing the country up into manageable chunks, reflected in invented regional names like 'West Midlands' which have no geographical, historical or cultural sense of identity.

Under Prime Minister Margaret Thatcher, The UK had a long period of anti-interventionist government during the 1970s and 1980s, during which many state functions were moved to the private or non-profit sector. This produced a more lean and fit organisational landscape but also produced duplication and competition among agencies, and confusion among businesses.

Thus when the UK West Midlands wanted to participate in the Digital Business Ecosystem project the participation was led by one of the region's universities, who in turn had to get funding for their participation from the Regional Development Agency and other sources. These funding sources each had their own agendas, and so participation was a matter of appealing to those agencies and accommodating their various individual interests. Interestingly, there was no restriction on working outside their region, and so they also covered the neighbouring East Midlands region and were even able to bid (though unsuccessfully) to set up and operate a national Open Source centre.

The university did not have day to day contact with SMEs and so had to engage other agencies, specifically a kind of industry IT club, to help to attract SMEs to events and identify target adopters. Thus the core role of the university was to bring together the various funders and SME agencies in a complex landscape, and it took on two new staff specifically contracted for the project.

Aragon

Spain also had a turbulent past, finding itself, since pre-historic times, at a crossroads between Northern Europe, the Middle East, and North Africa. General Franco governed Spain as a Fascist dictatorship for almost 40 years from 1939 to 1978, when the transition to democracy brought a sharp reversion to former regional kingdoms, with substantial devolution of power and autonomy. For economic development, such autonomous regional governments tend to operate though their own tied agencies, although these may be constituted as independent not-for-profit companies.

The region of Aragon has an ICT agency, ITA, which decided to become involved in the Digital Business Ecosystems project. Although constituted as an independent not-for-profit company, ITA is in effect owned by the regional government and enjoys a monopoly status on ICT development. IT formulates the government's regional ICT plan, delivers government funding to SMEs, as well as providing advice and involving SMEs in projects. As such they enjoy massive long term 'social capital' with SMEs and an established working relationship with many of them.

ITA was able to directly contact suitable SMEs and to command their attention and respect, based on its past relationships. The other side of this position is that they have been anxious to preserve this 'social capital', especially when research has failed to meet its time scales or objectives and the SMEs were in danger of becoming disillusioned. ITA has also been able to mobilise complementary government funding for implementation, and so, largely as a result of its stable, central position in regional ICT, has been able to take a long-term and holistic approach.

5 Matching Initiatives to Environments

Both West Midlands and Aragon achieved considerable success in bringing SMEs into the research and exploitation. The main point is that they did it in entirely different ways, adapting to their local situation.

A key ingredient for success is to understand the interplay of action and environment and how to modify and adapt for different local situations. Throughout Europe, many knowledge transfer initiatives have been demonstrated to be effective. The methodologies have been published and others have been keen to adopt this 'best practice'. However, the transplanted initiatives have often failed. The root cause of such failure is normally that the environment is different. This is where the region has an invaluable role to play, providing that it understands it's own environment and how it differs from others.

There is a good analogy with farming: the crop that thrives in the soils of one region may quickly wither and die in those of another. As any farmer knows, it is a question of knowing your own environment and of putting the right crops in the right conditions, although with intelligence, knowledge, and some investment, there is a limited degree to which one can change the environment to suit the crop. Thus if one region achieves success in a particular environment it may be that other regions can construct a similar environment and use it equally successfully, or change the nature of the initiative in order to adapt it to their own environment. To continue the analogy, this would be the equivalent of selecting a different variety of a crop to suit the local soil and weather. The main point here is to study and fully understand why and how an initiative or methodology works. This involves researching both the initiative and its interaction with its environment. Even a scheme that has never before enjoyed success may become a star if it is transplanted into a sympathetic environment.

During the DBE and OPAALS projects, much work was done to produce analyses of the regional partner's environments, focusing in particular on human and organisational networks. This knowledge is available and should inform regions that are developing ICT strategies, or for that matter any knowledge transfer strategy. During this analysis, working with these and other regions, some important generic concepts were developed and employed project-wide, which we will look at now.

6   Important Generic Concepts

Regional Catalysts

While each of the DBE and OPAALS regions rightly approached their participation in different ways, a common concept was that each had a single organisation that was able to act as a 'catalyst', in the sense that the organisation itself remained unchanged, but it facilitated the desired interactions between the project and local actors. This concept became know as the 'Regional Catalyst' and has provided an enduring model of regional engagement that can be readily transferred to any size or type of region, and does not only apply to sub-national level. Indeed, IPTI, the conference organiser, is our very successful Regional Catalyst for Brazil.

Translating activities into local culture is also an important role for the Regional Catalysts. From simple things, like language translation, to more subtle translation of research messages into business language. In the reverse direction, the Regional Catalysts have an invaluable role in articulating the needs of business and translating them into research language, such as by developing use cases, and providing interpreted feedback. These inter-language and inter-cultural barriers between research and industry cannot be over emphasised. They can cause a complete breakdown if ignored.

It is best that Regional Catalysts are existing organisations. Most regions have some existing infrastructure that can be leveraged. New organizations are a last resort, as they can suffer from many disadvantages and risks:

- Lack of 'social capital' within the region – even fear of the unknown
- Cost, time and uncertainty of establishment
- Perceived temporary nature leading to uncertainty in staff and partners
- Lack of experience, mature systems, and infrastructure
- Inheritance or imposition of inappropriate rules and regulations

Likewise it may be tempting to form groups, partnerships, or committees implement projects. These can suffer disadvantages too:

- Infighting
- Multiple agendas
- Knowledge fragmented between members of the group
- Bureaucracy, slowness, and cost
- Poor internal and external communication

Government itself is rarely the ideal agent to directly carry out activities that require characteristics such as close contact with industry, building relationships with international partners from different cultures, making judgments about participants, and forcing things to happen in creative ways.

Industry and end-user groups or associations are unlikely to make good Catalysts as they tend to have a very short-term view and fail to bridge the gap between research and industry. Instead they may be valuable in identifying and supporting 'Drivers'.

Driver SMEs

It is notoriously difficult to involve SMEs in research. Research requires investment and contains many risks, such as technical or market failure, or simply being outstripped by others who come along later and stand on your shoulders. To be successful SMEs in general have to be very focused on their day-to-day business, reducing risk and seeking short-term and certain profit. Digital Ecosystems technology is transparent and unintelligible to the end user, a situation that is likely to become a more widespread problem as technology becomes ever more sophisticated. Why should any SME invest time and money in a technology that they can neither understand nor reap profit from in the near future? No amount of 'blue skies' arguments and 'greater good' sentiments make any business sense in terms of the typical SME. Another approach is needed.

Drivers are those few successful and respected SMEs that are already innovative on the basis of personal interest. They have both the mindset and the capability to adopt the technology. They make excellent early adopters, and provide information, case studies, and leadership, in support of more widespread adoption by the mass of 'follower SMEs'. The Drivers in our case were software developers who could develop end-user applications. Often they were already active in Free/Open Source Software, so were receptive to this aspect of Digital Ecosystems and enthusiastic about the possibilities.

Such Drivers can in turn bring their end-user customers into the project. This technique uses the social capital of the supply chain. In some cases, both in our project and in other examples of collaborative R&D involving intermediaries, the SMEs that are the ultimate end-user target are not even aware that they are participating. They trust their suppliers to ensure that they get what they need and don't need to know that they are 'guinea pigs'. It is the Driver who takes on board the risk that failure will damage their social capital and manages that risk.

It was found to be worth supporting such Driver SMEs intensively. We did this in terms of training and practical support, plus financially, with small-scale standardised funding packages delivered via the Regional Catalysts. This meant that our selected Driver SMEs did not have to make a business case to justify their participation in research, which at best will only have long-term gains, against competing business objectives with more certain short-term returns. The participation was cost neutral.

7 How to Prepare a Region as a Knowledge Transfer Agent

We have seen from the West Midlands versus Aragon comparison, that there is no rigid formula for success. Success comes from the interplay of actors and actions within an environment. However, from our formal self-analysis as a project we can identify several factors that will help ensure success:

- Self-analysis to ensure that the initiative is of a type that the region's environment and characteristics can support
- Preparatory education and political consensus-building
- Dedicated people and infrastructure

- Adaptation to local culture and conditions
- A lead organisation (Regional Catalyst) with strong social capital among the target groups

Social Capital Is Crucial

SMEs are best persuaded to participate when their participation is requested by an organisation that they respect and trust, and perhaps know to be a source of funding. It is less a question of what is being asked and more a question of who is asking. It is only human to say 'yes' to a trusted friend asking you to do something you are unsure of, and 'no' to a stranger making the same request. If the Regional Catalyst or a Driver asks their customer to become involved and assures them that it will be of advantage, SMEs are far more likely to agree.

Social Capital can be manifested as network connections, and we found through network analysis that our Regional Catalysts tended to be well connected throughout their regions, both with SMEs and/or with government or other agencies that they could call on for support. An example of this is the Finnish Regional Catalyst for Tampere region, who were already in close contact, via a number of projects, including a national Open Source Centre, with TEKES, the national technology agency.

8 Specific Tools

There are a number of specific tools and techniques that were developed by DBE, OPAALS and PEARDROP[iii], a project specifically devoted to DBE regional deployment, which can be adapted to assist any region in preparing and managing a knowledge transfer. Examples are:

[1] Social network analysis and visualisation[iv]
This was extensively used in OPAALS and can provide a way for identifying who are the best-placed organisations to act as Regional Catalysts in terms of their connectivity to other regional actors. It was even produced as a software and used as a demonstrator for part of the research known as the 'EVESim', which can be combines with Google maps[v] to produce a geographical visualisation.

Balanced scorecard
A technique for regional self-analysis and inter-regional comparison based on a well-established management tool[vi] that measures several factors to produce a profile. This was used in DBE to analyse and benchmark regions.

Impact index[vii]
A definition of a Digital Ecosystem Impact Index (DEII) has been produced by OPAALS but not yet applied.

By researching and selecting appropriate tools, a region can self-analyse, prepare, manage, monitor, and evaluate.

9 Connection to Other Initiatives

A Digital Ecosystem is only a platform: an environment. To achieve something there need to be one or more applications mounted on that platform and used. The dream scenario is to use a complementary economic development initiative as the application to be mounted. One can then achieve more than the sum of the parts as the application is used to good effect and at the same time the benefits of the platform become better understood. To create a realistic exemplar of a Digital Ecosystem requires a critical mass of infrastructure and of participants. A regional development project can provide both, even if it is unrelated to ICT.

Examples are:

- A regional tourism initiative
- An inter-regional cooperation
- Inner Sourcing and Community Sourcing
- ICT infrastructure improvement or ICT adoption

10 Conclusion

Clearly regions have a very special role in knowledge transfer for economic development, whether in ICT or any other technology. They are able to overcome the significant barriers in very effective ways: ways that are not practical on a national or international scale. They have the social capital that is essential to bring target groups into participation, whether that is SMEs, micro-enterprises, or private individuals. Regions understand the local culture and current climate and can adapt and adopt, acting as technology translators. They are well motivated by the immediate and local economic improvements from their efforts, and may already have infrastructures and complementary projects that can be utilized and multiplied. Given a degree of intelligence in their approach, and a willingness to learn from the experiences of others, regions and their actors are uniquely placed to play a large and important role in harnessing the increasingly rapidly advances in technology for local economic and social benefits.

References

[1] Barber, B.R.: Jihad v McWorld. Random House Books (1996)
[2] UN FAO: World Summit on Food Security,
    http://www.fao.org/wsfs/world-summit/en/
[3] Peardrop project site, http://www.peardrop.eu/Pages/index.aspx
[4] Adelberger, Kurz, Eder, Heistracher: Deliverable: EvESim extensions for P2P simulation,
    http://files.opaals.eu/OPAALS/Year_3_Deliverables/
    WP03/D3.5.pdf
[5] Kurz, Heistracher, Eder: Deliverable: Visualisation Service for P2P Infrastructure and EvESim based on Google Maps,
    http://files.opaals.eu/OPAALS/Year_3_Deliverables/
    WP10/D10.7.pdf
[6] Balanced scorecard: Wikipedia, the free encyclopedia,
    http://en.wikipedia.org/wiki/Balanced_scorecard
[7] Rivera-Leon, L., Passani, A., Pennese, F.: Deliverable: Preliminary study on methodologies for DE socioeconomic impact analysis,
    http://files.opaals.eu/OPAALS/Year_3_Deliverables/
    WP11/D11.8.pdf

Fostering Social Technologies Sharing through Open Knowledge Space: A Brazilian Case Study of Knowledge Network

Lia Carrari R. Lopes, Saulo Barretto, Paulo Siqueira, Larissa Barros,
Michelle Lopes, and Isabel Miranda

Instituto de Pesquisas em Tecnologia e Inovação
Av. São Luis, 86 cj 192, São Paulo, Brazil
{lia.carrari,saulo,paulo}@ipti.org.br,
{larissa,michelle,isabel}@rts.org.br

Abstract. The paper presents a case study of ICT adoption for a Brazilian national program focused on enhancing the share of social technologies to improve socioeconomic development. The paper presents some experiences of IPTI in this field and describes the case study. The paper also presents the computational platform adopted for this experience. Finally it illustrates how the behaviour and configuration of the community can be studied with the use of social network analysis. The authors expect that this work will help other researchers in forthcoming projects and experiences of building knowledge communities and ICT adoption.

Keywords: Knowledge network, social technologies, web.

1 Introduction

Nowadays, one can say that the centrality of information and knowledge is no longer the means for the information society to generate knowledge or produce and distribute new goods and service. People now are required to understand this new paradigm as well as be active in the process. In that sense, they need to share knowledge, information, goods and services if they expect to be included in this society.

Knowledge is considered the driving factor for the economical, cultural and political growth in contemporary societies. In particular, the widespread diffusion of information and communication technologies provides societies and its different communities with tools to create, foster and disseminate knowledge. This process is independent from temporal and spatial boundaries or networked organisations and communities.

Real possibilities of collective knowledge construction are available from this new paradigm as the individual becomes the centre of the processes in which he or she is involved. The magnitude of collaborative networking has reached a new dimension. For contemporary societal trends, the notions of network and community are now vital terms. Information and communication technologies have an effect on coordination, communication and control in all societal networks and communities. In that sense,

F.A. Basile Colugnati et al. (Eds.): OPAALS 2010, LNICST 67, pp. 100–108, 2010.

those characteristics also grant significant potential as well as threats for collaboration and coordination.

Then, the use of information technology under a contemporary perspective implies that the focus shifts from having access to computers or Internet to making conditions for registering, editing and publishing multimedia contents as well as enhancing the sharing and networking. The main goal is to make people aware that they are able to publish and share content in an interactive way. Content may vary from texts, hyper-texts, images and music through videos. This project aims to help people understand the meaning of uploading, instead of downloading only, and to find new ways to face their socioeconomic problems.

Nevertheless, there is no defined model for building virtual knowledge communities as well as the maintenance of a high level of social capital inside those communities. Those topics constitute challenges for this area and have been the main fields of research at the *Instituto de Pesquisas em Tecnologia e Inovação* (Research in Technology and Innovation Institute - IPTI) in the past years.

The objective of this work is to share an experience about the adoption of Information and Communication Technologies (ICT) in order to support knowledge networks focused on disseminating technologies and experiences for socioeconomic development in Brazil. To achieve that, the knowledge network known as *Rede de Tecnologias Sociais* (Social Technologies Network – RTS) [1] was selected to build this use case. This use case has the potential to become a relevant study to show how such knowledge communities can be built, what are the key functionalities to increase its adoption by users, and what are the successful communication and articulation strategies that allow virtual knowledge communities to be effective and keep their values.

This paper aims to present the RTS use case describing the adoption of *Guigoh*, an Open Knowledge Space developed by IPTI to improve socio-economic development. The aspect of publishing and searching social technologies and their contextualization are emphasized. In the same fashion, the governance model adopted as an attempt to establish a high level of social capital for this virtual network is explained. The paper will explore the use of social networks' visualisation as a mechanism to illustrate how the network configures itself along the process of e-adoption.

The computational platform *Guigoh* and its functionalities will also be detailed, as well as the OPAALS project outcomes that are directly related to *Guigoh* and the RTS use case. The authors believe that this paper can be a helpful reference for forthcoming experiences of adopting virtual environments to build knowledge communities

Firstly, two of the most recent IPTI's experiences are presented, regarding projects on research and development in knowledge communities and in e-adoption as well as the RTS use case. Then, the customized virtual platform for knowledge community developed by IPTI for the RTS use case and the tools available are described. The technology adopted in the development of this computational platform is also detailed and will become available as an open source project soon. Finally, the results obtained up to now and the potential business benefits for such case of ICT adoption are shown.

2 Methodology

The experiment focused on knowledge sharing and was conducted by the adoption of a social network system. The methodology employed was inspired by some of IPTI's previous experiences on research and ICT adoption. The most significant experience for this project was the Digital Culture project and the OPAALS project.

IPTI has acquired ample experience of e-adoption between 2004 and 2007 on behalf of the Brazilian Ministry of Culture and the United Nation Development Program (UNDP). The project *Projeto Cultura Digital* (Digital Culture Project) had the intention of encouraging people to register and share their local cultural expressions to make their community stronger and sustainable.

The plan for this project was to distribute multimedia kits, containing all hardware and software necessary for people to register, edit and publish multimedia contents, for NGOs focused on cultural projects in all Brazilian regions. Besides the multimedia kit, free broadband Internet access was also provided. To disseminate vital concepts like "sharing", "intellectual generosity" and "networking" as new approaches to foster local development, conceptual workshops were ministered to spread this information.

Apart from the concept dissemination, technical workshops were planned to train people to register and edit multimedia contents using free software. Besides sharing their contents, users could also evaluate contents published by other users, discuss and create new contents collectively, new knowledge communities, and so on. The project involved more than 500 cultural Brazilian NGOs along three years.

The experience concluded that even though the awareness workshops made technology available, the desired effect was not produced in the majority of NGOs.

The success was related to the presence of an already consolidated social basis around the community serviced by the NGO. For those situations, the technology was adequate and the activities of the NGOs could be amplified by that.

For the other NGOs, what was observed was that technology was applied in the traditional way, based solely on the consumption of information. The Digital Culture project granted IPTI an invitation to become a member of a Network of Excellence financed by the European Commission in 2006. The project Open Philosophies for Associative Autopoietic Digital Ecosystem (OPAALS), aims to exploit virtual collaborative processes as a mechanism to promote socio-economic development. The purpose is to use the Digital Culture experience as a case study.

OPAALS network joins researchers from 23 institutions and mainly from the computer, social and natural sciences. The conceptualization and development of a digital space to promote sustainable knowledge communities and a computational architecture for digital ecosystem platforms is part of the objectives of this project. Those environments are called Open Knowledge Space (OKS) and provide online collaboration as well as communication integrated to other software components developed by OPAALS. The technological architecture provides a free and decentralized model that is not dependent on server farms and guarantees data consistency. Unfortunately, due the delay on OPAALS to deliver its OKS it was not possible to exploit the Digital Culture project as a case study as IPTI's participation in this project has finished in September 2007.

In 2007 IPTI decided to implement its own OKS fully based on Web and developed in Java. The OKS developed by IPTI was named *Guigoh* that is a Brazilian species of monkey. This system incorporate a number of functionalities that allows users to create their own social network, consisting in contacts and communities, to communicate to each other by textual chat and VoIP (Voice over IP), to edit documents collectively and to share their multimedia contents (audio, video, images and texts) by using open licenses (Creative Commons). Next, RTS organization and structure will be detailed.

3 RTS Structure

To understand the structure and organization of those communities, it is relevant to comprehend the constitution of RTS. Social Technologies are considered products, techniques and methodologies that are replicable and developed in interaction with local communities and that represent effective solutions for social transformation. For instance, a technique how to collect rain water in a dry region and keep it drinkable for long period is a Social Technology and certainly other dry regions have interest in applying such solution. With the aim of stimulating the sharing of Social Technologies between people and institutions in Brazil RTS was created in 2005 and now is being also adopted as a knowledge network by other Latin America countries.

RTS is sponsored by several Brazilian partners, mainly by the public sector as the Brazilian Ministry for Science and Technology. RTS unites nowadays more than 680 members, from which a few are networks themselves. They work with solutions in terms of social technologies considering twenty themes, such as water, energy, among others. The goals of RTS include stimulating the adoption of social technologies as public policies, enhancing the appropriation of social technologies by the communities and encouraging the development of new social technologies.

The adoption of *Guigoh* OKS by RTS intends to amplify the possibilities of sharing and searching of solutions by stimulating the members to publish their technologies in multimedia formats and by reinforcing the sense of knowledge community between them. Besides, RTS also aims to enrich the existent social technologies by motivating the sharing of experiences from the application and contextualization of technologies.

RTS knowledge community runs over *Guigoh* platform with few modifications. Beyond all tools from *Guigoh*, RTS is implemented with a customized interface and concepts. The main aspect of the customization was giving more relevance to the Social Technologies as well as to RTS themes, as the Social Technologies are directly referred to them. As mentioned before RTS works with solutions in twenty different themes and when users want to publish a new Social Technology they have to relate it to one of those themes. When publishing a Social Technology users have to fill a form with details about their solution and each Technology has a specific web page. Similarly, when users want to retrieve any Social Technology they can easier do it by firstly defining which theme the technology is related to. Next it will be detailed some of the customized interfaces of RTS.

First, the main webpage presented to users (Fig. 1) shows all Themes and indicates the quantity of Social Technologies that are available for each Theme. Above this the interface presents a toolbar and some main information (number of members, next commitments, among others). Below (Fig. 2) a set of users (with the option to show all users existent in the network and also a search feature) is showed and the most recent documents and conferences in the system. Finally, there is a Calendar where users can schedule meetings and other events.

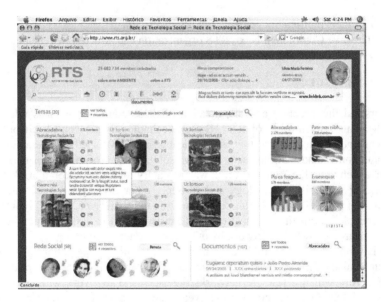

Fig. 1. Home page of RTS knowledge community

Fig. 2. Social Technologies area

When a specific Social Technology is clicked, its respective web page is reached (Fig. 3). In this page, all information about this Social Technology is presented, like its description, option for downloading, tags and about the author. The page includes a forum where users can make their comments about the technology and an option where users can recommend this technology to other users.

Fig. 3. Social Technology web page

The access to RTS knowledge community is open and free for any user which has only to register at its web site [1]. The RTS system is currently available in Portuguese and English but a Spanish version is also planned. Next, the specification of the tools developed on *Guigoh* platform is presented.

4   Guigoh OKS

The interface of Guigoh is designed to be easy to use and a complete environment for knowledge sharing and management. The currently version of Guigoh offers four main tools: Document Editing; Conference; Multimedia Publishing Environment; Social Network Analysis.

The Document Editing tool is used to build documents collectively. Users publish documents to the public and share it with others. All created documents are included in a group so that all users can find and access the content. In the main page of the Document Editing tool, users can see the documents that exist in the community as well as all documents this user is involved with. It is also possible to check the "Featured Docs" which consists in the most accessed documents. The most popular document can be viewed in the middle of the page. This tool was designed to promote the development of collective contents inside the OKS.

The Conference tool was developed because interaction is a key to the development of shared knowledge. This tool aims to manage conferences including participant administration, written and voice chat and whiteboard. To better understand each other and to improve interaction, different ways of communication are required. Voice chats help better and faster group coordination [2]. Besides managing their conferences, users can check their information and description. For past conferences, the system displays a log for the messages exchanged. If the whiteboard conference mode is selected, users can edit a whiteboard by drawing, writing and including images on a canvas. The canvas editing works on shifts, and every user that wants to edit must ask for the turn.

Another tool, available for the system managers are the sociograms (social graphs). Those are graphs representing relationships in a defined context. For RTS they are designed as Java Applets and divided in three different types: contacts sociogram, containing all users and their relationships with their contacts; document sharing sociogram, indicating document sharing and collaboration between users; Social Technologies sociogram, showing the collaboration by comments in each Social Technology. This tool is used by the system administration to observe and study the community evolution and content emergence.

Sociograms also make social network analysis available on the interface. As social networks are based in Graph Theory, the graphic representation is given by a graph, where actors are vertices and the relationship between then is an edge. Social network analysis is a method used to find patterns and key elements in the community. This method is employed to study many phenomena in real world, such as the behaviour of groups of people and communities and how different populations relate to each other [3-6].

In this sense, RTS Sociograms present system users as vertices (circumferences) and the relationships as edges (lines or arrows between vertices). The size of the vertex is defined by its degree (number of relationships for the vertex). Edges can also be displayed differently (dotted or dashed line and different colours), according to the relationship type. In the same interface on the bottom, users can access centralities and other information about the network. The centralities present the top 10 users with the highest degree, closeness or betweenness centrality.

The degree centrality is defined by the number of relationships each vertex owns, the more relationships, more central is the user [4]. On the closeness centrality, the most central user is the one that can interact with others with the lowest distance [4]. On the betweenness centrality, the most central vertex is the most stressed one, it means, the one containing the highest number of shortest paths passing through this actor [4]. The technology selected to develop Guigoh as well as the the OPAALS peer to peer architecture will be explored next.

5 Computer Architecture

Guigoh is built on top of a social network system developed by IPTI team, called *Primata*. This project contains the database model and logic to manage users, communities and relationships between users. *Guigoh* is also an umbrella over other projects from IPTI, to handle features such as authentication, document editing and publishing and user interface customization. All services are accessed through the Representational State Transfer (REST) architecture. The REST technique is used to transport data over default HTTP protocol methods, such as POST and GET.

This architectural style (REST) was chosen for its scalability, ease of use and portability (does not depend on a specific programming language or operational system). The requests are responded using JSON syntax. JSON syntax is plain text based and therefore lighter than XML – which is used by more traditional web services architectures.

The entire system, including the individual tools, was implemented using the Java language, version 1.6. The advantages that Java language provides include its multi- platform environment and also the variety of free tools and libraries used in the project. The application server JBoss was chosen because it is one of the most employed application server nowadays, it is free and maintained by an active open source community.

The database chosen was PostgreSQL, also an open source project with one of the most complete database management systems nowadays. The Hibernate Framework was employed as well for the efficient database handling, table mapping and querying manipulation. Finally *Java Server Pages* (JSP) and *Asynchronous Javascript And XML* (AJAX) technologies were responsible for the interface development. The option for AJAX is considered a friendlier interface that avoids reloading and handles request in a transparent way. JSP is the native JAVA technology for web systems.

Testing was also taken into consideration and that step employed the technologies JUnit, JMock and Selenium. Those tools are used for testing and bug fixing, as well as interface test integration. All project sources used Mercurial for version control that maintains a distributed repository synchronized with the server after solving the problems, minimizing server repository corruption. For database version control, the tool Liquibase were used, handling revisions through xml files.

The repository for the source code of the *Primata* project is already setup at http://kenai.com/projects/primata and the files will be uploaded and publicly available in the next few months. The same is also true for the "umbrella" project, *Guigoh*, the conference tool (http://kenai.com/projects/muriqui/) and the p2p framework that will be used underneath *Guigoh* soon.

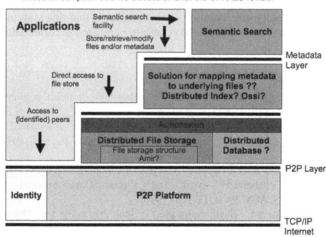

Applications have access to peers, files, metadata/semantic file overlay and semantic search

Fig. 4. Initial DE Architecture Diagram

As an example of an OKS *Guigoh* supports the principles and goals of Digital Ecosystems, except for the non single point of failure / control one – which is being worked on already through the p2p implementation efforts by IPTI and other partners from OPAALS DE Architecture. A DE architecture has been under development by the partners at OPAALS for a while now. Below is a diagram representing such architecture (Fig. 4).

This architecture is the one that will make the fulfilment all the requirements for a DE mentioned before possible, and also is where *Guigoh* will plug into.

6 Conclusions

This paper aims to present a computational system developed by IPTI to support knowledge communities as well as an introduction to a Brazilian experience of knowledge community (RTS) and its case study of e-adoption for improving socio- economic development. The RTS web environment for building a knowledge community on social technologies was developed and is freely available for users interested in publishing and/or searching for solutions to face their local problems.

The main business benefits of this project is the perspective of improving social and economical development by a process of a network stakeholders sharing their innovative solutions generated to face local problems. The RTS case itself is an experience of knowledge community which has helped several regions and communities to find solutions for their social and economical problems by face to face meetings (workshops, forums, conferences) as well as by newsletters and its web portal. In this project people that find and adopt a Social Technology are invited to share their experience of applying it, mainly describing the customizations needed in order to fit to the local conditions.

The knowledge community is now starting to be populated. As RTS is a very active knowledge network it is expected that the web environment can quickly became active. Finally, the intention of this project is that this system will be able to be adopted by open projects focused on building knowledge communities.

References

1. Rede de Tecnologia Social. IBICT - Instituto Brasileiro de Informação em Ciência e Tecnologia (2009), http://www.rts.org.br/
2. Halloran, J., Fitzpatrick, G., Rogers, Y., Marshall, P.: Does it matter if you don't know who's talking? Multiplayer gaming with voice over IP. In: Ext. Abstracts CHI '04, pp. 1215–1218. ACM Press, New York (2004)
3. Wellman, B.: For a Social Network Analysis of Computer Networks: A SociologicalPerspective on Collaborative Work and Virtual Comunity. In: Proceedings of SIGCPR/SIGMIS. ACM Press, Denver (1996)
4. Wellman, B., Garton, L., Haythornthwaite, C.: Studying Online Social Networks. Journal of Computer-Mediated Communication 3(1) (June 1997), http://jcmc.indiana.edu/vol3/issue1/garton.html
5. Molina, J.L.: El análisis de redes sociales: una introducción. Edicions Bellaterra, Barcelona (2001)
6. Wasserman, S., Faust, K.: Social network analysis: methods and applications. Cambridge University Press, Cambridge (1994)

Analysing Collaboration in OPAALS' Wiki: A Comparative Study among Collaboration Networks

Fernando A. Basile Colugnati and Lia Carrari R. Lopes

Instituto de Pesquisa em Tecnologia e Inovação,
Av. São Luis, 86 cj 192, São Paulo, SP, Brazil
{fernando,lia.carrari}@ipti.org.br

Abstract. This work aims to analyse the wiki tool from OPAALS as a collaborative environment. To achieve that, methods from social network analysis and statistics are employed. The analysis is compared with other collaboration networks. The results obtained here show the evolution of the tool and that the adoption was successful.

Keywords: Social Network Analysis, collaborative process, Virtual Communities, wiki.

1 Introduction

OPAALS Network of Excellence (NoE) "aims to work toward an integrated theory of digital ecosystems, to create new disciplines from the merging of old disciplines, and to do so by leveraging community values and principles." (Phase II Workplan, pp. 5). In this NoE, the comprehension of how networks behave and communities emerge can be considered a self reflection for the consortium governance. Another significant consideration is that this is an important research framework for the Social and Computer Sciences domain since SMEs scenarios in the adoption and use of Information and Communication Technologies (ICTs) for collaborative and democratic business is one of the main objectives of Digital Business Ecosystems (DBEs).

Information and communication technologies (ICTs) have an effect on coordination, communication and control in all societal networks and communities. OPAALS developed the concept of Open Knowledge Space (OKS) - as part of a broad definition - that is technically a set of computational tools for collaboration and knowledge construction by means of virtual communities. Created to be the first tool of the OKS, the wiki tool was until 2008 the most used environment for collaboration in OPAALS. The main knowledge repository for the consortium is formed by: Management issues, events calendar, dissemination material, Work Packages descriptions, among others.

F.A. Basile Colugnati et al. (Eds.): OPAALS 2010, LNICST 67, pp. 109–117, 2010.

So far there are only few studies analyzing in detail how virtual knowledge networks grow, how the process of communication and articulation between their nodes (local actors) happens, and the dynamics that are behind those aspects. More research is needed to generate knowledge about applicable strategies and mechanisms that are essential in order to make a newly established network successful and sustainable.

From the analytical point of view, the community emergence processes are nowadays understood as complex emergent systems [1-4], driving quantitative methods to look at them as physical and dynamic structures. The development of Social Network Analysis (SNA) methods, that in the beginning was static and purely deterministic, is now directed to dynamic stochastic processes where the whole of the actors are the central point, modelling the network emergence form a bottom-up perspective that will lead to a certain topology. The dynamic process led by actors by means of selection and social influence, and the topology of the network both provide interpretable parameters and metrics that quantify the intensity and velocity of the emergence and formation of communities or sub-communities.

This paper explores the evolution of collaborations in this repository using methods for Social Network Analysis over the four years of wiki existence. Also, the results are compared with other collaboration networks that used different ways for communication.

2   Methods

To verify the collaboration process over tasks defined in work packages, only the WP pages are considered, from the first publication in 2006 until June of 2009. Data used in this work were extracted with the toolkit Wille2 [2], developed by TUT.

In terms of SNA, two approaches are used. The classical approach using the methods based on degree, centrality and clustering [6] using these measures in time series and the final degree distributions was estimated to explore a possible scale-free behaviour [7]. Also, the dynamic of the network is modelled using methods presented by Snijders [8], and already presented for OPAALS consortium in [9] and [10].

3   Social Network Analysis for OPAALS' Wiki

Table 1 describes the OPAALS wiki network, which is not a significantly large network. It had a major growth between 2007 and 2008, with 37 collaborating researchers. Fig. 1 evidences that it does not follow a power-law, but show that there are two sets of actors in terms of degrees, above and below the horizontal line formed by the dots ("circled dots"). The set above the line are actors with few degrees, from 2 to 4. The "circled dots" comprises a few actors with degrees around from 10 to 15, and then actors with more than 20 ties.

Table 1. Number of actors and network degree for OPAALS Wiki

Year	Active Users	Network Degree
2006	15	46
2007	23	80
2008	34	222
2009	37	224

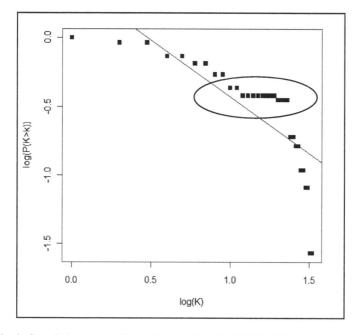

Fig. 1. Cumulative degree distribution for OPAALS Wiki collaboration network

The social graphs or sociograms confirm the described structure (Fig. 2). In 2009 in the accumulated network, a main core is formed by 12 actors (bottom part) with high degree values and many others spread out around this core. It is noticeable also the presence of many articulators, making bridges between smaller structures with the main one. Also, the interdisciplinary characteristic is a remarkable aspect; in Fig. 2 each colour represent a research domain and the blue one are consultants or administrative persons.

In terms of evolution, the increasing transitivity (Fig. 3d) claims attention, since this effect normally decreases in this type of network. The presence of many articulators can explain this high transitivity, increasing the probability of new ties mediated by them.

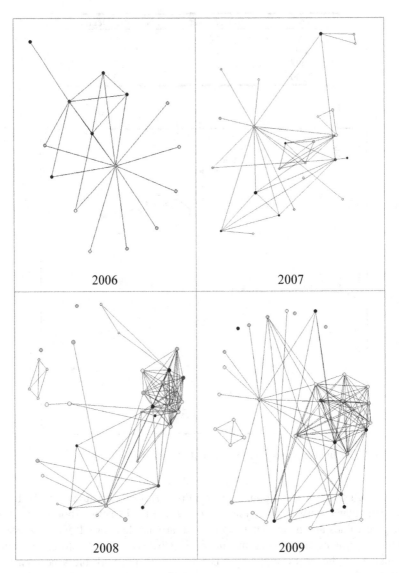

Fig. 2. Sociograms for the OPAALS wiki collaboration evolution. Yellow = Social Sciences, Green = Computer Science, Red = Natural Sciences, Blue = Consultants/Adm.

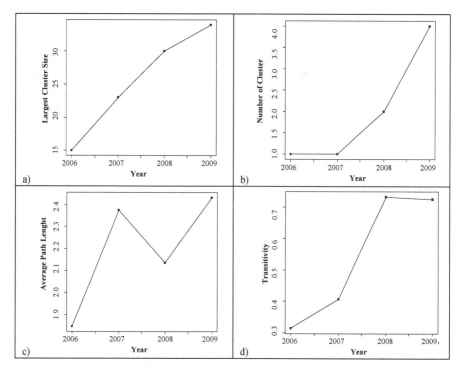

Fig. 3. a) Larger cluster size, b) Number of Clusters, c) Average path length and d) Clustering Coefficient evolutions for OPAALS wiki

4 Stochastic Dynamic Modeling

OPAALS wiki is modelled as a non-directed network (meaning that the relationships are reciprocal), adequate objective functions are introduced, as well as their effects interpretation. A covariate was introduced representing the domain of actors, as represented in the sociograms. Despite the fact that these domains have been coded as ordinal numbers, this was not taken in account in the analysis, being considered a categorical (nominal) variable. The rate parameters presented here have the same meaning and interpretation as in the directed case. Four objective functions were tested for this model, firstly without any covariate effect:

1. *Transitive ties*: is similar to the transitive triplets effect, but instead of considering for each other actor j how many two-paths i - h - j are there, it is only considered whether there is at least one indirect connection. Thus, one indirect tie suffices for the network embeddedness. It can be interpreted as a measure of clustering for the network, or even network closure.
2. *Betweenness*: represents brokerage: the tendency for actors to position themselves between not directly connected others, i.e., a preference of i for ties i - j to those j for which there are many h with h - i and h !-j ("!-"means not tied).

3. *Number of actors at distance 2*: expresses network closure inversely: stronger network closure (when the total number of ties is fixed) will lead to smaller geodesic distances equal to 2. When this effect has a negative parameter, actors will have a preference for having few others at a geodesic distance of 2 (given their degree, which is the number of others at distance 1); this is one of the ways for expressing network closure.
4. *Assortativity*: reflects tendencies to actors with high degrees to preferably be tied to other actors with high degree. If this parameter is positive and significant, preferential attachment is not likely.

One covariate effect is tested, the *Identity*, in this case the Domain identity. Positive values for this effect means that actors tend to choose similes in domain, a homophily representation. Due to the small number of actors in 2006, which means that too many structural zeros are present, the analysis will comprises 2007, 2008 and 2009 only. Table 2 present the parameters and different models adjusted. A parameter is considered statistically significant if the *t-statistic* is bigger than | 2 |.

Table 2. Dynamic Modeling for the Wiki collaboration

Effect	Parameter	Std. Error	t-statistic
Rate Parameters			
2007 - 2008	2.77	0.33	8.34
2008 - 2009	0.02	0.02	1
Objective functions (simple models)			
Transitive triads	0.63	0.54	1.16
Betweeness	-0.56	-	-
Assortativity	0.34	0.12	2.81
Multieffects Models			
Assortativity	0.34	0.12	2.81
Domain	-0.70	0.24	2.91

The estimated rates shows that in the first period, each actor had at least 3 chances to establish a tie, or at least 3 actors could collaborate with other, while in the second period this is almost 0, which is coherent with the data. Looking at the sociograms and the network evolution, few new actors entered and few new contributions were made from 2008 to 2009. So, the great part of the effects presented bellow happened between 2007 and 2008.

The only effects on objective functions statistically significant were the Nad2 and the *Assortativity*. Transitive ties effect was not significant and *Betweenness* did not achieve numerical convergence, so that standard errors could not be computed. This situation means a lack of stability for the function used, given the dataset, i.e., it is a numerical problem rather than a network topology problem. The negative value for Nad2 express the tendency to the network closure, such that few actors are separated at a distance two.

This is can be complementary to the positive *Assortativity*, where the selection of actors is lead by the degree centrality, such that, high degree actors are more likely to tie with other high degree actors. When adjusting for Domain, that is negative and significant, Nad2 loose statistical importance, but *Assortativity* remains the same. The negative sign means that the selection of ties happens among different domains preferably, and this selection does not depend on the *Assortativity* since this parameter is the same in the simple and multivariate model.

To interpret this composition of effects is not a simple task, and the definition of these effects above need to be verified. Nad2 expresses the effect of distance 2 actors, given the degree of that actor (number of actors at distance 1). Considering *Assortativity* a degree effect, where actors tend to choose others with same degree, the chance so that distance 2 links vanishes. The interesting conclusion is that it does not depend on the Domain homophily; actors collaborate with others with the same intensity of collaborative activity, and not by the research domain. By having the NoE as the interdisciplinary characteristic, this is an expected effect. Next, the collaboration process will be compared with other analysis.

5 Comparing to Other Collaboration Processess

In terms of information and communication technology, the OKS is a set of online collaboration tools such as chats, wikis, document editing tools, forums among others. The understanding of how the process of collaboration behaves in each of those virtual environments, in terms of the network formed by the researchers, can help one to follow the actual OKS dynamics. It also allows for intervention planning and strategies to keep this a sustainable community. Workshops, conferences, meetings, web-conferences are all ways to intervene in, foster or change direction in the use of the OKS.

In D 10.8 [11], twelve different collaboration networks were analysed, such as Forums, Wikis and co-authoring for scientific papers. All those networks could then be compared. The idea is to assess differences in terms of topology and clustering, having this last network property as an indicative for communities emergency.

It is clear from Table 3 that collaboration process differs basically in the environment in which it happens. Behaviour and properties are similar for both forums and co-authoring networks. All networks that follow approximately the power law distribution, present coefficients between 2 and 3, as found in many other studies about collaboration networks [12]. Transitivity is lower for Forums than for any other kind of text elaboration, as is suggested by the imposed and fixed treelike structure, forcing in a certain way the preferential attachment.

Then, the probability of new links among any 2 step actors is low, while for a structure free environment like a Wiki this tends to be high, as those actors have more chances to collaborate with any other actor. For the co-authoring network, the only one where ICT is not the environment, although email was and still is quite useful to write in a collaborative way, the behaviour is a little different. They are truly small- worlds but not entirely scale-free networks in the examples studied. This result does not match with other findings, like Barabasi *et al.* [13], who identified this behavior. Here, degree distributions tended to present heavier tales due to an effect sometimes called "finite population effect", in this case that dataset is not complete in terms of all collaborations or authors. The dataset studied just had ISI indexed papers, so journals not indexed by this database were not contemplated.

Table 3. Scale-free and small-wordness assessment parameters for networks in the last panel

Discipline	α	l	R (l)	CC	R(CC)
Mathematics	3.23*	3.17	0.48	0.55	**151.67**
Applied Mathematics	3.30	6.22	1.43	0.58	**237.95**
Interdisciplinary Applications	2.84	4.91	1.27	0.38	**115.23**
Computation Science	3.22*	2.95	0.54	0.70	**303.76**
Artificial Intelligence	2.76	2.66	0.59	0.63	**60.00**
Interdisciplinary Applications	2.70	1.38	0,29	0.85	**99.67**
Psychology	3.18*	3.36	0.73	0.76	**115.35**
Treatment and Prevention	3.10	2.49	0.66	0.78	**63.85**
Multidisciplinary	2.76	1.38	0.29	0.84	16.68
Supera	1.98	3.06	0.77	0.06	4.57
Converse	2.43	2.88	0.88	0.18	10.24
OPAALS Wiki	1.94	2.43	1.12	0.72	4.05

α: Adjusted exponent for the Power-law, **L**: Average path lenght, **R(L)**: ratio **L** / **L**$_{random}$, **CC**: Clustering coefficient, **R(CC)**: ratio **CC** / **CC**$_{random}$
* Do not follows power-law distribution

6 Conclusion

From this project it can be concluded that the wiki tool adoption was successful over time. Most active users found on the e-mail lists were using the wiki tool collaboratively by the end of the analysis. It was also observed that collaboration on the wiki tool does not follow a preferential attachment. The fact of the wiki tool usage was a migration from e-mail lists, were subgroups migrate together might suggest this result.

Another consideration for this includes the multidisciplinarity of the group as well as the level of homogeneity from most collaborators. The experiment results show different domains collaborating with each other in the wiki environment. Some actors that shower low collaboration and group integration could be temporary researchers that left the project or temporary consultants.

Finally, comparing the usage of the wiki tool with other collaboration networks and tools, one can conclude that media changes the network evolution. This can be suggested by the fact that different types of groups may obtain fluency to the media that is more adequate to their profile, and this is essential for the adoption success of the system [14].

References

1. Buckley, W.: Sociology and modern systems theory. Prentice Hall, New Jersey (1967)
2. Monteiro, L.H.A.: Sistemas Dinâmicos. Editora da Física, São Paulo (2006)
3. Mitchell, M.: Complex systems: network thinking. Artificial Intelligence 170(18), 1194–1212 (2006)

4. Salazar, J.: Complex systems theory, virtual worlds and MMORPGs: complexities embodied, `http://salazarjavier.mindspages.net/complex.pdf`
5. Tampere University of Technology. Wille (2009),
   `http://matriisi.ee.tut.fi/hypermedia/en/publications/`
   `software/wille/`
6. Wasserman, S., Faust, K.: Social network analysis: methods and applications. Cambridge University Press, Cambridge (1994)
7. Barabási, A.L., Albert, R., Jeong, H.: Scale-free characteristics of random networks: The topology of the world wide web. Physica A 281, 69–77 (2000)
8. Snijders, T.A.B.: The Statistical Evaluation of Social Network Dynamics. In: Sobel, M.E., Becker, M.P. (eds.) Sociological Methodology, pp. 361–395. Basil Blackwell, Boston (2001)
9. Colugnati, F.A.B.: Dynamic social network modeling and perspectives in OPAALS frameworks. In: 2nd OPAALS Conference, Tampere, Finland (October 2008)
10. Kurz, T., Heistracher, T.J., Colugnati, F., Razavi, A.R., Moschoyiannis, S.: Deliverable 3.7: Biological and evolutionary approaches for P2P systems and their impact on SME networks. In: OPAALS (August 2009),
    `http://files.opaals.org/OPAALS/Year_3_Deliverables/`
    `WP03/D3.7.pdf`
11. Colugnati, F., Lopes, L., Kurz, T., English, A.: Deliverable 10.8: Report on cross-domain networks. In: OPAALS (August 2009),
    `http://files.opaals.org/OPAALS/Year_3_Deliverables/`
    `WP10/D10.8.pdf`
12. Barabási, A.L.: Linked. Plume, Cambridge (2003)
13. Barabási, A.L., Jeong, H., Neda, Z., Ravasz, E., Schubert, A., Vicsek, T.: Evolution of the social network of scientific collaborations. Physica A 311, pp. 590–614 (2002)
14. Koh, J., Kim, Y., Butler, B., Bock, G.: Encouraging participation in virtual communities. Communications of the ACM 50(2), 68–73 (2007)

Towards the Complex Interface

Renata Piazzalunga and Fernando A. Basile Colugnati

Instituto de Pesquisas em Tecnologia e Inovação,
Av. São Luis, 86 cj 192
{renata,fernando}@ipti.org.br

Abstract. This paper introduces a new concept for interface based on user perception, according to its definition in phenomenology framework. Based on studies about fractal aesthetics and its effect in human perception, an analogy is made with the scale-free topology of collaborative networks. Properties of this scale invariance characteristic should then be used for a computational implementation of a dynamic and optimized interface, adequate to the user preferences, feelings and perception.

Keywords: Perception, interface, visualization, phenomenology, scale-free networks, fractal.

1 Introduction

The interface can be conceptualized as an entity that allows an interaction between living system and data sets. That is, designing an interface besides giving shape to information connected to a software program, explores the communicative processes between virtual entities revealed by this interface and actual entities revealed by the individual interacting with the program.

One of the lines of research in our group is concerned with investigating the relationship between space, from a point of view of an architect, and perception [1]. This is meaningful because it sheds some light on how we treat the interface.

The first relevant aspect about interface modeling is concerned with the graphic interface is nothing more than the space of action and interaction of the user. There is a resemblance between the creation process of concrete, architectonic space and digital spaces. In this way, we were able to conceive the interactive relationship between user and space in a more comprehensive way. We also noticed that the user of virtual spaces not only interacts with the environment through the interface but he/she also lives, experiences, perceives and "mentally inhabits" virtual space.

In addition to this, the development of an interface project for cyberspace must consider the peculiar dynamics of its configuration, as well as the properties to be implemented that are possible and accessible from a technological point of view. That is, from the spatial configuration point of view one of the peculiarities consists in the fact that this space is constantly submitted to new configurations and reconfigurations of its forms.

F.A. Basile Colugnati et al. (Eds.): OPAALS 2010, LNICST 67, pp. 118–130, 2010.

In the case of virtual environments, the interface serves as a gateway to access users' cognition of such systems, given that the mental processing involved in the interpretation of the informational contents depends on the stimuli that lead to perception motivated by this interface. So, we might state that interface design should be connected to "cognitive ergonomics", metaphorically speaking, that takes into account the interaction between the individual and the virtual environment, based on the exchange of stimuli between users and cognition, via the interface.

It is through the action of registering sensations that we can affirm we have a body that relates to the world. We can say we have a body by analyzing the acts we register (body sensations registered). Then it is possible to make references to physical objects and to space. This space is the basis to all concepts of space and we know our bodied presence because we register sensations. It follows then that there are three fields of perception: the interior, the external and the act of registering both the internal and the external. Intermingled with these fields we have acts from impulses/drives, instincts and reactions. That is to say that we have a body that feels, registers and reacts and is capable of decision, control, reflection/reason, evaluation, understanding and generation of concepts. This occurs at different levels, depending on presence and achievement, individually or in group, depending also on the level of consciousness we have of these acts (quality and amount). This fact points to the structure, to the constitution of the subjects. Consciousness stands as the dimension where human operations are confluent: body, perception, and mental acts are registered.

E.g.: to see (to see an object; here, you are in a room and you see a book) as an example of data assimilation, understanding and interaction in a physical environment.

The sense involved is vision. The act of seeing it is perception (you register what you see). If you remove the book from your vision field, you can remember (an act that makes present something that is not there). In examining the new reality you know the difference between perceiving and remembering.

You can imagine the book where it was. You can analyze it. You can pay attention to it. But you can divert from it by a sound of someone arriving.

What drove your attention to it: motivation?

Did you have the intention of touching that book? Do you evaluate the book? You know that you see the book, you register this.

Remembering, attention, perceiving, imagining, analyzing, diverting, motivation, touching, seeing, evaluating are all ACTS (embodied capacities, mental capacities) and the perceptual experience is also an ACT, the act of perceiving. This synergic system is the intentional arc. The intentional arc is understood as the personification of the interconnection between perception and action, in the phenomenological context. It is the intentional arc "that unites the senses, uniting the senses with intelligence, and sensibility with the motor functions" [2]. In other words, it is the intentional arc that certifies acts of consciousness, certifies experiences of the phenomenological being in the world. According to Merleau-Ponty: "the life of consciousness – cognizant life, the life of desire or the perceptive life – is supported by an 'intentional arc' that projects around us our past, our future, our human environment, our physical situation, our ideological situation, our moral situation, or else leads us into placing ourselves under all these aspects" [3].

From this viewpoint, we support the thesis that the designing of really qualified interfaces leading to significant experiences in the virtual space, should necessarily concern the correct understanding of the perceptive process, a trend widespread in contemporary science of relating cognitive science to phenomenology. We are therefore here discussing the understanding of perception through the perspective of phenomenology. Two other models should be remembered: the serial and the connectionist models. Although important, they do not interest us here as they do not consider subjective experiences on the part of the individual. They are conceived from a representational view of perception. They are based on the principle that perception consists of taking characteristics of a pre-determined world and constructing a copy, or internal image of this world in the perceptual apparatus. Reviving phenomenology seems an appropriate strategy in line with current expectations, given that the principle of its conceptual formulation: *the analysis of consciousness in its intentionality*, functions as a counterpoint to the representational formulation of perception.

The formulations made by researchers such as the philosopher Hubert Dreyfus and the biologist Francisco Varela are examples of this relationship between the cognitive sciences and phenomenology. The idea that forms the basis of the thinking of these researchers is that the spirit forms an organic unit with the body and with the environment. This assumption is very important and must be considered in the creation of virtual environments.

Hubert Dreyfus holds to the thesis that Husserl can be considered as the precursor of the classical theses on Artificial Intelligence. Varela develops a research program referencing classical phenomenology known as the "naturalization of phenomenology". At the beginning, the intention of the project appears to be to create a natural science, a "neuro-phenomenology", incorporating concepts of phenomenology and concepts of the cognitive sciences.

The fundamental assumption orienting phenomenological research subverts the natural attitude characterized by the belief in the absolute external nature of the world's things. For phenomenology, it is consciousness that constitutes the meaning of the world's objects. Consciousness is revealed through the phenomenon: the consciousness of the perceptive experience involved in the observation of the phenomenon and the opening of consciousness to the world, that is to say, the intentionality of consciousness. When for example, I observe a scene of a child riding a bicycle, from the start I am paying attention to the movement in this action: I observe that he turns the handlebars and pedals, etc. But, generally, I am not aware of the perception involved in watching the child's activities, in other words, my perceptive experience involved in the observation of the phenomenon: I am concentrated, involved, thoughtful, etc. My gaze is concentrated on the external action and, therefore, on the direct perception of the object, and not on the internal action, on the perception of the essence of the phenomenon. One could say that this change in our way of looking is, in a simplistic way, what the phenomenological method proposes.

As a philosophical school, phenomenology is not a localized movement nor can it be reduced to a system. "It is always thought of as, even among its philosophers, as a research in movement that can never be summarized as a finished list of precepts and rules" [3].

This being so, we can take the general fundamentals of the phenomenological doctrine introduced by Edmund Husserl (1859) as a reference [4]. In 1900, Husserl published *Logical Investigations I*, with which he became considered the founder of phenomenology as a philosophical method. The fundamental assumption of phenomenology, *the intentionality of consciousness*, was developed by Franz Brentano (1838) [5], of whom Husserl was a student. Husserl's originality consists of having adopted this assumption for the consolidation of a philosophical method. The general fundamentals of phenomenology were to be dealt with by starting with the presentation of the key concepts, around which Husserl based his method, related to the specific sense of how phenomenology conceives its philosophy: as *the analysis of consciousness in its intentionality*.

For phenomenology, consciousness is always the consciousness of something. It is the source or beginning of the other realities in the world. For Husserl, consciousness is a chain of lived experiences, with each experience having its own essence, which in turn defines the way in which the object is revealed to the consciousness. These essences are acts of the consciousness, such as perception, remembering, emotion, etc. Thus, "the analysis of consciousness is the analysis of the acts by which consciousness itself relates to its objects; or, the ways by which these objects are revealed to consciousness. The acts of consciousness, or the ways by which the objects are revealed to consciousness form *the intentionality of consciousness*" [6]. That is, the intentionality of consciousness is the way in which consciousness opens itself to the world of experiences in order to extract its essences.

Pure phenomenology is not a science of facts, but of essences (eidetic science). To get to the essences, once must avoid the affirmation or recognition of the reality and assume the attitude of a spectator, interested in only gathering the essence of the facts, through which consciousness reports to reality. Consciousness should, therefore, assume an attitude of a *disinterested spectator* in relation to the world. This distancing is possible through a methodological artifice created by: the *epoché*. The *epoché* consists of a phenomenological practice; an internal or mental gesture of reduction (eidetic), which leaves the existence of the world in suspension. Through *epoché* one can reach the realm of subjectivity, distancing oneself from the world and assuming the role of a disinterested spectator of the facts, thereby creating a methodological strategy to reach the essences. The attention of the disinterested spectator is turned not toward the world of his reality, but to the phenomena that reveal this world to the consciousness.

According to the phenomenological analysis, the object is not part of the lived experiences. The subject that gives intent to the object does not become an integral part of the object; neither does the object given intent become part of the subject. "The world that in the attitude of *epoché* becomes a transcendental phenomenon is understood from the start as the correlation of occurrences and intentions, of acts and subjective faculties, in which the sense of its unity is constantly changing and which progressively assumes other senses" [7]. Husserl attributes the crisis in the sciences to not taking into account the realm of subjectivity.

The trend widespread in contemporary science of relating phenomenology to cognitive science consists of consolidating a model that emphasizes a practical dimension of phenomenology, giving it potential from a pragmatic dimension.

Supporting an approach that reconciles at least three of these models in a new vision of the perceptive process, Hubert L. Dreyfus says in *The Current Relevance of Merleau-Ponty's Phenomenology of Embodiment*: "I will suggest that neural-network theory supports Merleau-Ponty's phenomenology, but that it still has a long way to go before it can instantiate an intentional arc" [8]. This suggestion by Dreyfus results from the analysis of how Merleau-Ponty thinks of his phenomenological subject. Merleau-Ponty's phenomenal body, our cognizant apparatus which makes us conscious of the world, is not only a psychic entity, and not only a physiological one, but rather it is a synergic system that takes on and connects the functions of a physical, psychic and cultural order, that define the general movement of a being in the world. "Consciousness projects itself in a physical world and has a body, just as it projects itself in a cultural world and has habits" [9].

The instantiation of the intentional arc of which Dreyfus speaks perhaps occurs in cyberspace, which would promote ideal conditions for its occurrence, as it is a domain in space where phenomenological action and perception gain the status of being one at the same time as they are constituents of this same space.

2 Complex Interface Design

Since the interface measures the informational content offered the user, putting into play his understanding, what he receives depends on the information displayed and on the user's perception of it. Therefore, within this perspective, the "ideal" interface would be that which is able to neutralize in the most possible efficient way the differences between the external world and the subject inner world, it would be that which manages to touch the user's perception most directly in such a way that the user feels his body to be an integral part of this simulated context. The experience brought about by means of the interface should touch the essence and being in the most organic, natural and intuitive way possible.

Interface representation in the context of OPAALS proposal is seen as an evolutionary and dynamic system. The conceptualization for understanding and creating a model related to this interface connects concepts as: complex and evolutionary systems analyzed from the chaos theory; cognition and perception analyzed by phenomenology and cognition science; art and scale-free network topology using fractal metaphor.

In OPAALS´ project we have been developing the conceptualization of an evolutionary and dynamic representation system as a synthesis of the relations developed in a knowledge environment, i.e., different knowledge linked with different domain areas and different levels of information; and all of this information also linked with people that provides or demanded it once in the history of use. We have been employing scale-free properties using stochastic dynamic models to forecast links of users and/or contents of interest for another user. The intention is that the system should identify user's behavior aspects and construct the interface as a graphical representation of these aspects, so that interface could be viewed as a phenomenological result of the system.

To develop this research about complex interface design applied for the OPAALS project implies understanding the most relevant aspects to be considered about the INTERFACE in the OPAALS´ scenario. The main aspect related to that should be the understanding of INTERFACE as an integrated environment of the three different domain areas that interact in the OPAALS´ project: human science, computer science and natural science. So, the challenge in this case was to create an evolutionary and dynamic representation system as a synthesis of the relations developed in a knowledge environment, i.e., different knowledge linked with different domain areas and different levels of information.

The problem was to find a way of creating a graphical representation of an evolutionary and dynamic interface based on the aspects described above. The conceptual connections were clear but the task to create the equivalence among graphical representation and theoretical understanding was very difficult.

In this sense two important connections were established. The first one was Jackson Pollock's work because of the allusion in his work of multiple dimensions and infinity space and also because of the scientific link between Jackson Pollock's painting and fractals [10], i.e., an evidence that in an abstract and apparently at random generated shape, in an artistic process, exists a component that allows the extraction of relationships inside the complex shape created.

The second one was the interconnection between fractal and scale-free networks. It was visualized noted that the concept and technique of scale-free networks could be used in the production of a self-generated interface based on the user's behavior. Using scale-free networks, introduced in 90´s by Barabasi and collaborators, has been applied in many different types of complex systems, from metabolic processes to the economic scenario as well as techno social networks. It could be a way of representing the conceptual idea of complex, evolutionary and dynamic interface that results and transforms itself as a consequence of each new input of people and/or content.

Further applying fractal metaphor for understanding complex interface, allows:

a) Developing a kind of module or mathematical/stochastic reference in a complex system environment (chaotic, evolutionary and dynamic: the 3 things together): a very important aspect because it is a strategy to improve the system's intelligibility, i.e., to make possible a intelligible representation of massive volumes of data;

b) Combination and the re-combination of the agents of the system infinitely, at random, without losing its conceptual and graphical coherence, and therefore this is a way of to keep the coherence of the system in respect to its original function, i.e., its constitutive unit. This coherency, technically speaking, should be guaranteed by the dynamic and stochastic network model developed for this purpose.

The interface design aspects described above take in account both the human visual information processing and the more abstract cognitive information management.

The important thing about complex interface design is to test the correspondence between the conceptualized model and graphical representation. It is important to amplify the realm of study related to the question of interface, as proposed at INDATA project, to obtain a more sophisticated and broad solution interface that we call here sensory interface.

3 Fractals and Scale-Free Networks

In this section, the aim is to connect the concept of fractals to the collaborative process described as a network system. Some concepts and historical facts are introduced.

Since their discovery by Mandelbrot [11], fractals have experienced considerable success in quantifying the complex structure exhibited by many natural patterns and have captured the imaginations of scientists and artists alike. The main property that made fractals so popular for nature and complex systems modeling purpose is the scale invariance. This property states that a same pattern (a geometric pattern, for example) is observed in different magnifications. The simplest example is the case of a line. If one picks a line of length L and split this line in N identical smaller segments, then each of them is scaled by a ratio $1/N$ with respect to the original line. The same can be made, for example, with a square with area L^2 splitting it in N identical squares that has a scale ratio of $1/N^{1/2}$ with respect to the original size. So, this is the scale ratio that governs the called scale-invariance, since we have identical line segments and identical squares that differs from the original only by a scale ratio.

But, as defined above, this ratio depends on the geometrical dimension of the pattern. An important parameter for quantifying a fractal pattern's visual complexity is the fractal dimension, D. This parameter describes how the patterns occurring at different magnifications combine to build the resulting fractal shape. For Euclidean shapes, dimension is described by familiar integer values - for a smooth line (containing no fractal structure) D has a value of one, whilst for a completely filled area (again containing no fractal structure) its value is two. For the repeating patterns of a fractal line, D lies between one and two and, as the complexity and richness of the repeating structure increases, its value moves closer to two [11]. For fractals described by a low D value, the patterns observed at different magnifications repeat in a way that builds a very smooth, sparse shape. However, for fractals with a D value closer to two, the repeating patterns build a shape full of intricate, detailed structure.

The number of patterns, N, one can find in a fractal structure of size L, is a mathematical function of D:

$$N(L,D) = L^{-D}(1) \tag{1}$$

In the nature, however, patterns are not always identical rather than similes in terms of structure and topology. So, D can be considered a statistic that describe this scale-invariance, or as some authors argue a self-similarity. These pattern repetitions in different scales can be modeled stochastically using the so called Power-law distribution, which describes the probability and consequently the expected number of patterns of size L.

Many other processes in nature are also modeled nowadays as network systems. Metabolic chains and food web, are just few examples of natural phenomena that are explained as complex network systems. The same applies for social processes too, like scientific collaboration, sexual contacts and social relations.

In terms of network modeling evolution, the first model based topology defined is the Random Network, by Erdös and Rènyi in 1960 [12]. Networks that follow this topology evolve as a set of actors that build social ties randomly. For this reason, at first the actors are connected to a small number of other actors. This can be explained because when an actor is chosen randomly and a tie is established with another actor, the likelihood is that this new tie connects the actor with a low number of others, even if the actor already has some connections.

However, when there is a certain quantity of ties so that the mean degree is one, the fraction of the graph occupied by the largest group of the network suffers a phase shift. It means that the largest group, which was not significantly large before, now occupies almost the whole graph area. This organisation is used to explain several phenomena in nature, where small sets join together to build groups of a higher dimension, gathering every actor until the ties are built [13], [14].

In 2000 Barabási [15] and his team defined a new model for network topology, called *Scale-Free* networks. The authors realised that the degree of distribution does not follow the Poisson or Gaussian distribution, commonly used to model random graphs, but a Power-law distribution instead. The Power-law distribution is differentiated by its format; the maximum point is not found in the average value, the curve starts at its maximum point and decreases towards infinity. There is an exponent that can be calculated in this distribution that shows how the distribution changes along with a variable.

A function that follows a power law distribution can be described by

$$p_k \sim C k^{\alpha}$$

(2)

k is the degree, p_k is the probability of actors containing degree of value k, as defined in (1) and α is the function exponent. Another way to represent the Power-law distribution uses the Cumulative Degree Distribution, defined in (2),

$$P_k \sim C k^{-(\alpha - 1)}$$

(3)

So, to assess scale-freeness in a network, usually the fit of the degree distribution to a Power-law is investigated. The simplest way to do this is taking the logarithm of P_k and k and analyse the scattering of the points. A straight line is expected if the distribution approximately obeys a power law, as shown in Figure 1a). The slope of this line is then used to estimate α, since $\log(P_k) = \log(C) - (\alpha - 1)\log(k)$.

This topology also presents a few nodes with a high quantity of ties, called *hubs*. Then, the majority of the actors have a lower number of social ties and a few (*hubs*) have a higher number of ties. Those networks are more tolerant to failure and node removal. As most nodes have a few connections, the probability of removing one of them is higher than a hub, and they do not impact significantly on the network structure. However, a coordinated node removal focusing the hubs can completely break the network structure.

The scale-free network evolution obeys a process called *Preferential Attachment*, a.k.a. *Rich Gets Richer* effect, such that new nodes have a higher probability of connecting to the older nodes with a high degree. Then, the older nodes have more chances to become hubs, but new nodes may also become hubs depending on its

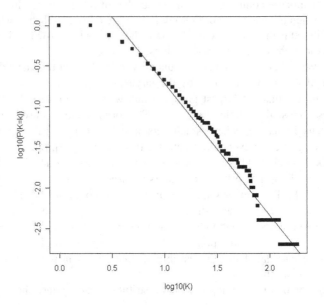

a)

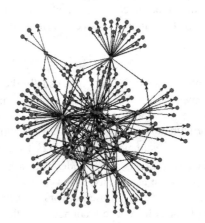

b)

Fig. 1. a) Cummulative distribution following a power-law havig $\alpha=1.98$, and b) the topology of the respective network

connectivity factor. This process usually leads to a hierarchical structure, where clusters are formed replicating the structure of predecessors, providing a topology like Figure 1b).

From the concepts introduced herein, it is straightforward the analogy of the Scale- free network and fractals. The number of degrees K is the analogue of L, and a parameter describes the degree distribution of Scale-free networks as the D describes the complexity of fractal patterns, so properties of two seemingly different models for different purposes are shared. It can even be said that Scale-free networks are stochastic fractals, if one consider the property of self-similarity described by a probability law, the Power-law distribution in this case.

4   The Role of Stochastic Fractals in Complex Interface

From the perception, discrimination and visual viewpoint, many studies have described the influence of fractal dimension in these psychophysical dimensions. [16]-[17] have shown high correlation between D and the pattern's perceived roughness and complexity, other studies found that discrimination of fractal curves by observers was maximal with dimensions closer to the terrain surfaces fractal models, suggesting that sensitivity of visual systems might be tuned to statistical distribution of environmental fractal frequency (ref). The aesthetic appeal is also closely related to the fractal dimensions. A seminal study conducted by Sprott [18], it was found that the figures considered the most aesthetically appealing had dimensions between 1.1 and 1.5, and the most aesthetically pleasing had $D = 1.3$. All these studies used experiments with volunteer observers and images, in general, generated by computers or models for real natural structures, like shore coasts and terrains. However, Spehara et al. (2003) [19] performed experiments using paintings by the American artist Jackson Pollock, presenting different parts of masterpieces. This study has showed that for the called Human Fractal the higher preferences for images presented $D=1.5$ in average.

In the development of a innovative user interface, those experiments supports the idea that a fractal structure might bring further motivation for users, but this is not the only feature that should be evaluated, neither the only quality that can bring advantages in the use of scale-invariance property.

In network research, it is well known that many collaborative processes follows power-law distribution for their degree distribution, and even for clustering coefficients, given the preferential attachment mechanism of evolution. Further, the a in these kinds of networks range from 1.2 to 1.8 [20], a range close to the fractal dimensions observed in the perception studies. If this intended interface should be the environment of a collaborative space, or an OKS, we should expect that collaboration among researchers, and even the semantic network among content tags, will also follow this model. If so, the aesthetic aspect is well done, at least technically. But the properties of a scale-free network can also provide some advantages:

- Order in chaos: the scale-free model is analytically defined and its evolution can be predicted, helping in the mechanism of visualization, since the topology is already known;

- For the same reason, scalability of contents to be shown in this interface can be adequate, allowing a navigation not only in a flat space, but allowing to explore the depth of the network, obeying the hierarchical scale originated by the network evolution;
- As a probability law, new links between researchers that might have a great potential for collaboration can be suggested, since they share any common interest in contents.

These are some prior hypothetical features that can be explored. The next topic presents a use case that will be the first experience in data collection for the development of this complex interface.

5 Use Case: Guigoh and the Social Technology Network

The first use case that will provide a framework to develop the interface model, is the Social Technology Network - RTS. Social Technologies are considered products, techniques and methodologies that are replicable and developed in interaction with local communities and that represent effective solutions for social transformation. With the aim of stimulating the sharing of Social Technologies between people and institutions in Brazil RTS was created in 2005 and now is being also adopted as a knowledge network by other Latin America countries.

RTS is sponsored by several Brazilian partners, mainly by the public sector as the Brazilian Ministry for Science and Technology. RTS unites nowadays more than 680 members, from which a few are networks themselves. They work with solutions in terms of social technologies considering twenty themes, such as water, energy, among others. The goals of RTS include stimulating the adoption of social technologies as public policies, enhancing the appropriation of social technologies by the communities and encouraging the development of new social technologies.

The adoption of Guigoh OKS by RTS intends to amplify the possibilities of sharing and searching of solutions by stimulating the members to publish their technologies in multimedia formats and by reinforcing the sense of knowledge community between them. Besides RTS also aims to enrich the existent social technologies by motivating the sharing of experiences from the application and contextualization of technologies.

RTS knowledge community runs over Guigoh platform with few modifications. Beyond all tools from Guigoh, RTS is implemented with a customized interface and concepts. The main aspect of the customization was giving more relevance to the Social Technologies as well as to RTS themes, as the Social Technologies are directly referred to them. As mentioned before RTS works with solutions in twenty different themes and when users want to publish a new Social Technology they have to relate it to one of those themes. When publishing a Social Technology users have to fill a form with details about their solution and each Technology has a specific web page. Similarly, when users want to retrieve any Social Technology they can easier do it by firstly defining which theme the technology is related to. Next it will be detailed some of the customized interfaces of RTS.

Using the data from the social networks, document collaboration, and content upload, we expect to have enough subside to develop and test the complex interface purpose.

6 Concluding Remarks

In the case of a complex interface design the scale invariance property will be applied in the development of a new paradigm of interface, using the fractal metaphor, once that the way which someone will navigate is a certain level of aggregation of information will not change if this one goes deeper in the content. Also, the Power- law probability distribution, followed by this type of network for connections, clustering among other network characteristics, can be used to forecast and suggest future connections for the user. Complex systems modelled by network models is a brand new area in science, so that the applications and development reached in this project will be a milestone in this research field, bringing a different way to associate art, human-computer interaction and knowledge management.

So, the interface design as understood at the OPAALS´ project take in account both the human visual information processing and the more abstract cognitive information management.

References

1. Piazzalunga, R.: A Virtualização da Arquitetura. Papirus, São Paulo (2005)
2. Merleau-Ponty, M.: Fenomenologia da Percepção. Martins Fontes, São Paulo (1999)
3. Begout, B.: Un air de famille. La Phénoménologie: une philosophie pour notre monde. Magazine Literature (403), 20–22 (2001)
4. Husserl, E.: Logical Investigations. Routledge, London (2001)
5. Brentano, F.: Psychology from an Empirical Standpoint. Routledge, London (1995)
6. Abbagnano, N.: História da Filosofia, vol. XIV, p. 93. Editorial Presença, Lisboa (1993)
7. Idem, ibidem, p. 93
8. Dreyfus, H.L.: The Current Relevance of Merleau-Ponty's Phenomenology of Embodiment (1996),
   http://hci.stanford.edu/cs378/reading/dreyfusembodiment.htm
9. Merleau-Ponty, M.: Fenomenologia da Percepção. Martins Fontes, São Paulo, p.192 (1999)
10. Taylor, R.: Personal reflections on Jackson Pollock´s fractal paintings. Hist. ciênc. saúde-Manguinhos 13(suppl.), 109–123 (Out 2006)
11. Mandelbrot, B.B.: The fractal geometry of nature. Freeman, New York (1977)
12. Erdös, P., Rényi, A.: The ő Evolution of Random Graphs. Magyar Tud. Akad. Mat. Kutató Int. Közl 5, 17–61 (1960)
13. Watts, D.J.: Six Degrees: The Science of a Conected Age. Norton & Company, Inc., New York (2002)
14. Barabási, A.-L., Albert, R.: Emergence of scaling in random networks. Science 286, 509–512 (1999)
15. Barabási, A.L.: Linked. Plume, Cambridge (2003)
16. Cutting, J.E., Garvin, J.J.: Fractal curves and complexity. Perception & Psychophysics 42, 365–370 (1987)

17. Knill, D.C., Field, D., Kersten, D.: Human discrimination of fractal images. Journal of the Optical Society of America 77, 1113–1123 (1990)
18. Sprott, J.C.: Automatic generation of strange attractors. Computer & Graphics 17, 325–332 (1993)
19. Spehara, B., Cliffordb, C.W.G., Newellc, B.R., Taylord, R.P.: Universal aesthetic of fractals. Computers & Graphics 27, 813–820 (2003)
20. Barabási, A.L., Jeong, H., Neda, Z., Ravasz, E., Schubert, A., Vicsek, T.: Evolution of the social network of scientific collaborations. Physica A 311, 590–614 (2002)

Knowledge Resources – A Knowledge Management Approach for Digital Ecosystems

Thomas Kurz, Raimund Eder, and Thomas Heistracher

Salzburg University of Applied Sciences, Austria, Europe
{thomas.kurz,raimund.eder,
thomas.heistracher}@fh-salzburg.ac.at

Abstract. The paper at hand presents an innovative approach for the conception and implementation of knowledge management in Digital Ecosystems. Based on a reflection of Digital Ecosystem research of the past years, an architecture is outlined which utilizes *Knowledge Resources* as the central and simplest entities of knowledge transfer. After the discussion of the related conception, the result of a first prototypical implementation is described that helps the transformation of implicit knowledge to explicit knowledge for wide use.

Keywords: Digital Ecosystems; Knowledge Management, Knowledge Resources.

1 Introduction

Starting with a very ambitious vision of Digital Ecosystems or even Digital Business Ecosystems, a couple of projects were initiated in the past and the respective results were evaluated and utilized by several regions around the world.

The status of the Internet as an information medium is subject to permanent change and evolution. Knowledge is no longer maintained by single experts but it is increasingly provided and supported by communities, which gives everyone the chance to publish knowledge on all kinds of different platforms.

This paper on the conception and prototypical implementation of Knowledge Resources is structured in two main parts. It starts with a reflection of Digital Ecosystems research and deals with the extension of service-focussed research to knowledge-focussed research guided by a general-purpose vision of a knowledge management approach. In addition, the idea of Knowledge Resources (KR) as the central entity in such a system is presented. The second part of the paper explains how such a knowledge management approach could be implemented based on an open source Web Content Management System.

2 Lessons Learned from the Digital Business Ecosystem Project

As the Digital Business Ecosystem (DBE) paradigm was a radically new vision a few years ago, there was no specific background on which the related research could

F.A. Basile Colugnati et al. (Eds.): OPAALS 2010, LNICST 67, pp. 131–145, 2010.
© Institute for Computer Sciences, Social Informatics and Telecommunications Engineering 2010

be built upon. Therefore, researchers very often faced the well- known chicken-or-egg causality dilemma. For example, if simulations, models and social analyses were required, a broad adopters community would be necessary but did not yet exist. In addition, real-world data would be needed for designing a proper model of digital ecosystems. On the other hand, a community was initiated in parallel from scratch and this community already was in need of some models and simulation results for setting up the required infrastructure accordingly.

For example it was recognized, that the work on a general-purpose simulation and the work on visualization capabilities of the infrastructure can dramatically improve the mutual understanding between different participating disciplines. Researchers can focus on their field of study and start with approximated data first and then – as the community grows – this data can be replaced by more realistic data from real life scenarios [1],[2].

One of the biggest challenges from the beginning on was the assertion of availability of a critical mass of services and a critical number of SMEs for running a *healthy* Digital Ecosystem [17]. This was accompanied by the discussion about definition of a *service* in a DBE. By widening the term *service* to *real- world services* as well as software services coming from every possible domain, the problem arose to have no benefit for a SME in the software sector finding a service for cleaning hotel rooms, for example. Therefore, the focus of the DBE project (http://www.digital-ecosystem.org) was mainly put onto the tourism sector and onto already built software components and modelling tools in the beginning. Nevertheless, the question of the critical mass for a Digital Ecosystem was left. First and foremost, research on scale-free-networks in general, P2P systems, and search algorithms for finding distributed services produced first results, specifically in the context of huge networks with many participants.

For the search of services, the use of genetic algorithms was suggested by DE researchers (see [3], [4]). However, genetic algorithms only make sense if there is a very large search space where users try to find the best service or service combination available. Optimizing algorithms and developing frameworks for large-scale and distributed networks bases on the assumption that such a network is available and the operational community is willing to contribute to such a network. Moreover, the participants need to publish their services in the Digital Ecosystem infrastructure, which is not a straight-forward procedure compared to other ways of deployment in the IT sector, having in mind the diversity of standards, approaches and technologies, unmentionable the individual solutions SMEs often use.

Publishing services in a Digital Ecosystems also raised the issue of identity and trust (see [5]). The present conception of an open knowledge space (OKS) has to deal with privacy and trust issues in parallel to the goal of an open, collaborative and distributed framework idea. Each user should have the full arbitrament about whom he wants to open access to knowledge. Simulating the benefits for SMEs to open information for others and demonstrating thereby the market benefits out of sharing information to them will be one success criteria for enforcing the idea of a really *open* knowledge space (see http://www.opaals-oks.eu).

3 Shift from IT Services to Knowledge Resources

It is expected from Networks of Excellence, such as the OPAALS project (see http://www.opaals.org) to show new paths for the future rather than to just apply existing technology. Having a totally new field of research like Digital Ecosystems at hand, the challenge herein is to create new knowledge and, in parallel, to set up the infrastructure so that people can get involved in the related community. Consequently, the idea behind an OKS has to be novel, simple to apply and very attractive to become a valid contribution, as there are hundreds of competitive *web 2.0* applications for knowledge management and social interaction.

Services in the DBE were modelled, for example, with the DBE Studio (see [6]) and consequently implemented as services on top of the ServENT infrastructure (see [7]). The ServENT is a P2P application container that abstracts the communication functionality providing a number of methods for lookup and service invocations. Such services usually are similar to standard web services and exhibit a DE-specific XML-based interface description as well as an optional business model for describing the service operations itself.

Following the development of DE research over the last years, it can be recognized that the initially business- and IT-centric *service* focus is now broadening toward Knowledge Services (KS) or Knowledge Resources (KR). We are not longer speaking about pure business or IT but rather about knowledge as such. Consequently, also the types of services need a modification to change from IT services to Knowledge Resources. But what is a Knowledge Resource? We define Knowledge Resources as follows:

A Knowledge Resource is any kind of knowledge a person possesses for private, shared or public use.

The most important point in this definition is the fact that we see knowledge always related to a person. Every Knowledge Resource has a holder or owner which holds the context of the information. The fact that communication with this holder or owner of the service is enabled by the system and he or she can extend the data with his or her know-how, the pure information at one peer qualifies for an upgrade toward a Knowledge Resource.

Hence in the simplest form, Knowledge Resources can be contact information like email addresses, bibliography entries for helpful references or web pages with useful information. More advanced services could be documents or reports, media files or pictures, and the upper end of complexity constitute software components, tools or code archives. Every item which can be stored as a file or archive on a local entity (storage media) can represent a Knowledge Resource.

In order to find services and to have a common format for services in the knowledge space, we suggest that each service is represented by an URI and a meta-description. The URI points to a local hard disc or a web space and the meta-description is provided as plain text. The variants of textual description are dealt with later on in this paper.

4 Where Knowledge Lives – A Vision of an Open Knowledge Space

In the following, the context and the basic requirements for a Knowledge Resource Framework are briefly outlined. In order to make the idea more demonstrative, the first-person narrative is used in the following. From my daily emails, on to my data on local discs and data I can access via the Internet, the amount of electronically accessible data gets bigger and bigger each day. Nevertheless, there is data I want to share with others and there is data I want to use for my private everyday life. For handling this data I have hundreds of possibilities for storing, structuring and retrieving. I personally often wanted to try out tagging but although there are many tools for tagging no one fits all my needs. The idea of Knowledge Resources should enable me to tag whatever I have on my machine or whatever site I have visited remotely. I just want an easy-to- handle web-application which lets me insert the tags or text I want to put as a meta-description and a URI for referring to the Knowledge Resources itself. For searching I want to have a text entry field (like in the Goggle search) for searching my private knowledge space *and* the open knowledge space. In order to fulfil these application requirements, the complexity of the data storage and the search has to be hidden behind a very simplistic user front end.

The minimum requirement for a Knowledge Resource is a tag and an URI. All meta-descriptions of Knowledge Resources are in text-form and are published to the user community. The user can decide how much of his meta-descriptions should be published As the link can point to a private space on a hard disc or a restricted area, access does not need to be limited by the OKS. Consequently, the public meta-information of all services is available and also the URI indicates sometimes which type of information can be found although the direct access is still limited. Nevertheless, there should be the possibility to send a message to the owner of a specific knowledge, to open access for the demanding party. The decision of opening the knowledge is again up to the owner of the knowledge.

In order to avoid ambiguity, the editing tool for the meta-information of the knowledge services needs references to a vocabulary or term definition repository. More detailed information on the editor can be found later on in this paper. Nevertheless, additional complexity like database connections and vocabulary should be hidden from the standard user.

The application areas for such a system are manifold, ranging from the pure private usage for organizing the local music folder up to a fully distributed search for commercial and free software components. We can also think of extending the search with automatic composition capabilities in future but at the moment the implementation described further below is merely for single service search.

What is the reason for setting up another search engine? The point is that the present framework is much more than an information search engine like Google and it is not necessarily governed by one company which reuses and sells this data.

The services can be both, private and public knowledge, and not just information which can be found in the Internet. Furthermore, the type of services outperforms in diversity the kind of information which can be found by a common search engine. Additionally, the introduction of visualisation features for the search results outlined later on will enhance the usability and navigation through search results considerably.

5 Preliminary Conception

Before we start to outline the web-based application for a Knowledge Resource framework, a few notes on the required infrastructure need to be provided. The basis for such an infrastructure system can only be a P2P network. A pure P2P network is governed by the community and it is the only convincing architecture model for setting up a highly collaborative framework for knowledge sharing. Other approaches could easily lead to centralization again. The simplest way of such a P2P node could be an SME with one computer including Internet access. It allows a secured part of its local information to be used by the system and thus shares data. The editor for inserting the URI / meta-description pairs is a web- application. As a company may decide to internally use the Knowledge Resource framework, it can put also an application server as a new node in the network and connect it to the internal network to that application- or web-server node in the OKS network. As long as there is the option of installing a lightweight P2P service for the connections, the decision which application server or which back- end technology is used should be up to the user. URI / meta-description pairs can be stored either file-based on the shared memory space or in a distributed database. Note that the decisions on the technological details are not relevant for this paper.

As already mentioned, the front end of the system should be an easy-to-use editor. The minimal idea is to have just three text edit fields: (1) the editor window where the meta-description of the service can be typed in, (2) a text- field for inserting the URI pointing to the service itself, whether on the local disc or on a shared place in the Internet, and (3) a search window for finding services similar to a common search engine.

The editor window allows entering tags in form of plain text or text-based notations like the Semantics of Business Vocabulary and Business Rules (SBVR) (see [8]). SBVR is an adopted standard of the Object Management Group (OMG) intended to be the basis for formal and detailed natural language declarative description of a complex entity and is chosen here as one possibility of a text-based but logical founded notation. Utility of SBVR was also investigated in connection with DE research, documented for example in [9]. If the user starts to type in a tag, a type assistant should superimpose in order to access already defined terms. That avoids ambiguity and the user can check if there is already a definition of a tag available. In order to store such definitions, we would suggest including a connection to a community repository for example. The repository could include

SBVR statements and definitions which are easy to read and therefore the user gets familiar with the syntax of SBVR even if he is just tagging the services. An alternative here would be a community ontology or set of ontologies. If there is no unambiguous definition available in the repository, the user should have the option to add a new definition. As such definitions can be formal (in SBVR syntax) and informal (plain text for example), it is up to the user to define terms in the way he or she wants.

Besides tagging, SBVR statements can also be inserted for describing services. We can also think of setting up configuration parameters for services as meta-descriptions in SBVR. The concrete software service identified an URI can then be configured automatically by parsing the SBVR meta-description.

The last option for inserting meta-descriptions for Knowledge Resource is natural language. For example, one could implement a small tool which automatically parses papers from the local disc and puts the abstracts as meta description in the editor including the URI path to the paper itself.

As an extra feature for natural language definitions, a transformation from natural language into SBVR could be offered (see [10], [11]). Using tools like the Stanford-parser, SBVR fact-types could be identified in the natural language text and listed for the user. A SBVR fact-type is a concept that conveys the meaning of a verb phrase that involves one or more noun concepts and whose instances are all actualities [8].

Through a simple manual check of the automatically transformed and respectively generated SBVR statements, the user can tell the system which transformation lead to a correct statement and which did not. Introducing a learning system here could feedback the manual corrections and therefore influence and enhance the next transformations.

The second prominent part in the standard user-interface is the search functionality. The first and simplest search could be implemented as a mere keyword search. Understanding whole SBVR expressions or *meanings* of full sentences in the search window would be a very interesting and challenging research question as such and could be an extension of the system for future releases.

Since the descriptions of Knowledge Resources are more meaningful than the contents of plain web pages, the results can be visualised as clouds of clustered results. Here is a short example for better illustration: The user wants to learn more on Genetic Algorithms (GAs) and types in *genetic algorithm* in the search window. The result is a visualization of clustered clouds of links or keywords: One cloud for contacts of people familiar with GAs, one other cloud for useful literature entries, another one for GA implementations and one for tools using GAs. The user has the possibility to click deeper into the implementation cloud, which zooms in and shows sub-categories like implementations in different programming languages for example. Using an intuitive and dynamic navigation, the user can browse through the Knowledge Resources found in the OKS. Then he can either access services directly via the URI or ask for further information by dropping a message to the holder / owner of the Knowledge Resource.

6   Relevance to the OPAALS Network of Excellence

Before we introduce a prototype of a Knowledge Resource based knowledge space, we want to sum up and stress the different ideas in the context of OPAALS. As the idea of Knowledge Resources arose from the discussion about the OPAALS Network of Excellence, it is straight-forward to see connection points to specific DE research domains. Furthermore, we always have in mind the current success of social networking tools of various types and the critical mass of services necessary for the OKS. Amongst others, we see here the following connection points to research questions in OPAALS:

- Tagging - SBVR - natural language
- Natural language evolution
- Evolutionary aspects - Simulations
- Social network analysis
- Advanced visualisation concepts
- Automated code generation
- Tools for automatic input, export or transformation
- Transactions and workflows

In the following we address these points separately.

First, the editor component should process *tags*, SBVR and *natural language* which leads to a natural convergence of the descriptive notations. Utilizing type assistants makes the user implicitly familiar with new notations and helps entering the first meta-descriptions. The community builds up a common vocabulary which becomes more and more advanced and structured in a better way.

The analysis of term definitions as well as the continuous *change in the language* used for describing services has a clear potential for research on the evolution of language. Besides the recognition of specifics for different domains, the open way of including also non-technical services and personal interests like music or photos could lead to a larger community and therefore more data for language analysis.

Data for the performed *simulations* originate from social analyses and questionnaires. Large-scale social network behaviour is most often estimated and extrapolated. With a larger user group, the networks and behaviour of the users can be modelled more precisely and therefore the outcomes of simulations will be more accurate.

Beside the usage in OPAALS and the focus on SMEs, other stakeholders could become interested in the *analysis of the DE social networks* in an Open Knowledge Space. Social science researchers can check their experience with existing social networks with a more structured and evolutionary approach in the OPAALS context. Moreover, the introduction of new volunteers in using the system and in contributing in one way or the other can ease the challenge of sustainability.

The *visualization and animation* of the search results opens a complete field of dynamic visualization options. Not only structure and static dependencies can be shown but the dynamic part of user interaction can make the work for the visualization research more interesting and challenging. It will become a new experience for the users to get structured clouds of community knowledge instead of a simple list of links.

Although, the move from IT-Services to Knowledge Resources means a change in the types of services under observation; Knowledge Resources still can be IT-Services or be represented by software components. The meta-description can either be a summary of the specification or a more detailed description of the business model in SBVR or even a full specification in SBVR. Here we can think of SBVR descriptions which are the basis for a *transformation tool*, transforming the SBVR model into a Grails web-application for example (see [9]). As people become interested in structured specification writing, more of these tools can be developed to add functionality to the Knowledge Resource Frame- work. Analogue to the Grails transformation, tools for *transactions and workflow generation* could be included as well.

To recapitulate, the ideas stated here can provide the basis of a prototypical implementation of a Knowledge Resource application, outlined in the following section. It partially implements the ideas stated so far in a Typo3 open source Web Content Management System (WCMS).

7 Prototypical Implementation

The average information content per page usually depends directly on the number of registered users and their activities. The more active users are, the more information is provided on a page. Therefore, the conception of a web-frontend is a crucial point of the implementation and the envisaged future success. For the first prototype, distributing the whole system is foreseen in the architecture, but it is not implemented in the initial stage. The following points are already carried out in the prototype and described in this section:

- Utilization of web 2.0 technologies and frameworks
- Frontend features like multilingualism and search engine optimization
- Insert, edit, delete and maintenance
- Reward system for placing more knowledge and additional functionalities in the system
- Easy navigation in the web-based system
- Geographical visualization of users and KR

In order to provide these functionalities in a self-explanatory way, Henrik Arndt published a comprehensive approach in [12]. In the centre of consideration is the user of a web page who is surrounded by the context and added-value of the page. In order to reach these surrounding requirements, the page has to fulfill the following aspects:

1. Usability: The two major points here are *traceability* and the ability of *self-description*. A user should have at any time the ability to know, where he/she is and how he/she can access whatever functionality.

2. Utility: Two important points here are relevance and being up-to-date. The page has to fulfil the personal requirements of the user and the user needs to find up-to-date knowledge.
3. Joy of Use: Static pages and complicated navigation reduces the attractiveness of web pages for end-users. The most popular Internet pages stand out because of innovation and the option of personalization, e.g. such has Facebook.

A page might have bug-free source and high-end database connections, but the user experience acts on another level. For the user, these three criteria are important. According to Andt, the decision whether the page will be visited again or not happens within the first 10 seconds. Therefore, even for the prototype, the focus is on easy usability, e.g. the positioning of items and the layout.

These requirements for the first prototype represent a considerable part of the conception of Knowledge Resources that is outlined in this paper. In the following subsections, we document how this implementation was done.

7.1 Technologies

For the implementation of the knowledge management prototype described in this section, we chose Typo3 (see http://typo3.org), an open source Web Content Management System (WCMS). It provides a full editor system, which allows easy maintenance of the page content. Although Typo3 is often cited as the most used WCMS, there are no official numbers available to support these statements. From the viewpoint of a developer, Typo3 is more of a framework than a WCMS. Contrary to other solutions, it needs to go through a few basic steps after installation in order to get a *hello world* page. On the contrary, it allows plugging in an enormous number of already-existing extensions and it supports the development of new extensions, like the knowledge management system for Knowledge Resources. Therefore we facilitated these extension capabilities for the prototype implementation. As there is extensive literature for Typo3, we refer here to the literature such as [13] or for the programming of the backend in PHP to [14] as well as to [15] for more detailed information.

For the implementation of AJAX functionalities, the Typo3 extension xajax was used. It inserts a PHP-AJAX framework. This extension is used to add AJAX functionalities also for other extensions of the prototype [16].

7.2 Frontend Features

For the optimization of the indexing for search engines, the existing Typo3 extension *realurl* is used. Typo3 assigns ID numbers for the single pages in the WCMS, which can be accessed with *realurl* also via text-based URLs. As a side effect, this enhances the ratings in Google searches for example.

The extensions *rgmediaimages* and *rgmediaimagesttnews* serve as additional add-ons for content elements. These allow the playback of videos in content elements. It is implemented by the facilitation of an open source media player.

Additionally, Youtube videos can be viewed directly inside the page. By the utilization of the JavaScript library *SWFObject* in the *rb_flashobject* extension, Flash movies can be included via direct back-end access.

Additionally, the following existing frontend extensions are used for the prototype documented here: *felogin* (blugins), *kontakt* (contact form), *tt_ address* (addresses), *tt_news* (news), *Static info Tables* (countries and currencies), *Static Methods for Extensions* (registration), and *templavoila* (templates).

For better user experience, four Typo3 extensions were adapted in order to fit with the needs of Knowledge Resources.

First, editors should not see passwords in plain text. This is implemented in *kb_md5fepw*. It ciphers the password before it gets inserted in a MD5 hash and is added to the back-end database. Here an adaptation to a W3C conforming code validation was necessary.

Second, the extension *perfectlightbox* was used for zooming of pictures. By darkening the screen and zooming the picture in the middle of the screen, the picture can be viewed much better because of the higher contrast. A small adaptation was needed to be compliant with the newest version of the other plugins.

Third, *sr_feuser_register* was utilised for the registration process of new users. It comes with support for all common registration steps including changing forms for user data. All registered users can be administered in the back-end. The personal factsheet and the assignment of longitude and latitude for geographical information was added to the standard functionalities of this extension.

Forth and finally, the assignment of tags as meta-descriptions for Knowledge Resources should be expressed as an animation of a tag cloud. Therefore, *t3m_cumulus_tagcloud* was extended so that users can not only add tags via the administrator in the back-end, but also in the frontend, when wishing to add a Knowledge Resource.

7.3 Data Insertion, Editing, Deletion and Maintenance

As soon as a user registers in the system, the system provides a personalized factsheet. This user profile information includes at least the registration information username, email address and country of origin. Additionally, a user photo can be added.

For inserting Knowledge Resources, the user gets an inserting form for the Knowledge Resource including title, textual description, a link to a file or web- page, optionally a file on the server and a set of tags. This information represents an extended version of the initial Knowledge Resource idea of textual description and URI. The reason for the additional fields for title, server file and tags is on one hand the familiarity of users with assigning titles and tags as well as the possibility of an easy bootstrap by providing server space for the early adopters. With an additional implementation effort, the title or tags can be incorporated in the textual description.

Both the user profile and the Knowledge Resources can be added in another template. The edit templates are implemented with AJAX in order to provide better usability.

Additionally to the existing Knowledge Resources, users can add so-called KR requests. This allows distributing needs for implementation, knowledge or services. Each KR request can be accessed by other users. When they open the details page of the KR, the system tracks the user ID and acts as observer of this KR.

Finally, also a search for KR is provided. At the moment this search is only implemented as a keyword search. A search can be started by registered and unregistered users. The search provides a free text search. Consequently, the text is searched in title, description and the assigned tags. All hits are listed then as search result. Similar to common search engines, during insertion of the search string, an AJAX-auto-completion helps the user with the search. Additionally to the free text search, a drop-down list with the available tags supports the user in parallel to a direct search functionality based on an auto-generated tag-cloud.

7.4 Reward System

In order to make it more attractive for the users to add more personal data, a reward system was implemented. As people might be deterred by needing to register at the first stage, a first level without registration was implemented too. In the following, the three access levels for the knowledge management prototype are listed:

Level 1 Frontend user without registration:

– Can use search functionality for KRs
– Can view all information, tags, SBVR or language, for every KR
– Cannot search for KR requests
– Cannot download KR attachments

Level 2 Frontend user registered and logged in with at least username, email address and location (city):

– Can search for KR requests
– Can download KR attachments
– Can add a personal profile and add his/her own KRs
– Cannot add files to a KR

Level 3 Frontend user with more than four registered KRs:

– Can add a maximum of 50 files to KRs with a maximum size of 2 MB, each

Level 3 is especially interesting for users without own web space who want to share knowledge with their colleagues or friends. With the restriction in file size and number of file uploads, a misuse is partially prevented for the prototype. Later releases will need more sophisticated mechanisms preventing possible fraud. Moreover, administration of the user data, tags and knowledge resources is possible via the Typo3 backend system.

7.5 Navigation

The navigation of the web-frontend of the prototype is a novelty for web pages. The basic idea is to imitate a keypad of a telephone. Consequently the navigation and access of certain pages can be done by dialing a number. The numbers can be dialed by mouse-click or by a shortcut on the keyboard.

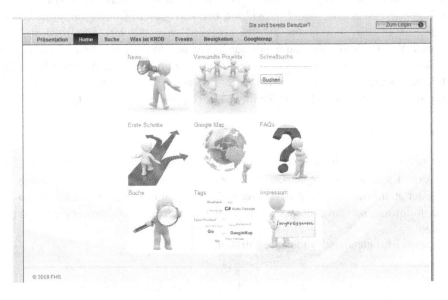

Fig. 1. Web-based frontend for the knowledge management prototype for Knowledge Resources

As can be seen in Fig. 1, the nine pads can be filled with all types of content. Starting with pure text or abstracts for the pages, up to pictures or flash animations can be inserted. The prototype can be navigated by the 3x3 matrix of the keypad OR alternatively also by a traditional WCMS menu on top of the page. The benefit of this navigation approach is the fast and direct access of pages for professional users in parallel with a step-by-step guidance for beginners. With three levels of a 3x3 navigation matrix, 9^3 (729) pages can be accessed with 3 or less clicks or key strokes.

7.6 Geographical Visualisation

The knowledge management prototype for Knowledge Resources includes two visualizations (see Fig. 2). First, a user is able to view all registered users locations on a map. Here a simple click on the position on the map triggers popping up of the basic user data and the registered KRs. This permits a simple search of users in the surrounding area. Second, a single Knowledge Resource can be visualized on a map, including the visualization of all related observers. With this simple feature, a company can for example visualize the potential customers of a resource, i.e. a market for a product or service.

Fig. 2. Visualisation of the Google maps based visualisation of registered users (left) and observers of an item (right)

The necessary latitudes and longitudes of all users are stored in the database of the Typo3 system. For showing connections between the observers and the item for example, polylines are used. After creation each item on the map is added by the system automatically via an overlay. If dense clusters of users are to be visualized within a region or the user zooms out on the map, the single items or users are visualized as clusters. While zooming further in again, the single elements of the cluster appear again as individual items on the map.

8   Conclusion and Outlook

In this paper a knowledge management approach in a Digital Ecosystem context was presented. Beside the broad conception of the system and the recommendation of Knowledge Services as the central entities in the system, a prototype was developed based on the open source Typo3 WCMS.

The implementation of the knowledge management prototype was focused on the transformation of implicit knowledge to explicit knowledge and on the ease of use for a broad user base. Users can easily publish and/or offer their knowledge and take advantage of a web-based system without more requirements than a simple browser.

As the focus at the moment is on a functioning system with easy accessibility for all kinds of potential users, some of the requirements like complete distribution of the system could not be implemented yet. In the following, a few potential future extensions toward the full implementation of a knowledge management system for digital ecosystems are outlined.

First, the distribution of the system is a core need toward a full implementation of the knowledge management approach outlined in the first sections of this paper. The next step for distributing the system would be a fully distributed file-system and database for the Typo3 (sub)system. We assume that at least for the next release some reliable rich web-based user interfaces would be necessary. Later on, the complexity of the user interface could be reduced and could be provided also for smaller clients and even mobile phones.

Moreover, currently a desktop application is developed for automatic upload of Knowledge Resources to the WCMS. It allows users to easily register their resources in the database of the WCMS. This desktop application features an automatic parsing of paper abstracts and an upload option for descriptions. Moreover an automated tagging mechanism will be introduced as well.

According to a request of some medium-sized enterprises, which want to use the prototype for their intranets, also a versioning add-on is currently developed. This add-on will connect each Knowledge Resource with a state-of-the-art versioning system.

Acknowledgments

This work was funded in part by the European Union's 6th Framework Program in the OPAALS Network of Excellence, contract number IST-034744.

References

1. Nachira, F., Nicolai, A., Dini, P., Le Louran, M., Rivera Leon, L.: European Commission - Information Society and Media, Digital Business Ecosystems (2007)
2. Kurz, T., Eder, R., Heistracher, T.J.: Multi-Agent Simulation Framework fr Interdisciplinary Research in Digital Ecosystems. In: Proceedings of International Conference on Digital Ecosystems and Technologies, IEEE DEST (2009)
3. Rowe, J.E.: Foundations of Learning Classifier Systems. Springer, Heidelberg (2005)
4. Briscoe, G.: Digital ecosystems: Evolving service-oriented architectures. In: IEEE First International Conference on Bio Inspired mOdels of NETwork, Information and Computing Systems (2006)
5. McLaughlin, M., Malone, P., Jennings, B.: A Model for identity in digital ecosystems. In: Proceedings of International Conference on Digital Ecosystems and Technologies, IEEE DEST (2009)
6. The DBE Consortium. DBE Studio, http://dbestudio.sourceforge.net/ (last accessed, 01.10.2009)
7. The DBE Consortium. DBE Servent, http://swallow.sourceforge.net/ (last accessed, 01.10.2009)
8. OMG. Semantics of Business Vocabulary and Business Rules Specification, Version 1.0, http://www.omg.org/spec/SBVR/L0/ (January 2008)
9. Eder, R., Filieri, A., Kurz, T., Heistracher, T.J., Pezzuto, M.: Model-transformation-based Software Generation utilizing Natural Language Notations. In: Proceedings of International Conference on Digital Ecosystems and Technologies, IEEE DEST (2008)
10. Eder, R., Kurz, T., Heistracher, T.J., Filieri, A., Russo, M., et al.: D2.2 - Automatic code structure and workflow generation from natural language models, OPAALS Project (April 2008)
11. Eder, R., Kurz, T., Heistracher, T., Filieri, A., Margarito, A.: D2.6 - Prototype of web-based User Interface with interfaces to existing SBVR tools, OPAALS Project (May 2009)
12. Arndt, H.: Integrierte Informationsarchitektur Die erfolgreiche Konzeption professioneller Websites. Springer, Heidelberg (2006)

13. Laborenz, K., Wendt, T., Ertel, A., Dussoye, P., Hinz, E.: Typo3 4.0 - Das Handbuch fr Entwickler Galileo Press (2006)
14. Kannengiesser, M.: Objektorientierte Programmierung mit PHP 5, 1st edn., Franzis Professional Series (2007)
15. Flanagan, D.: Mootools Essentials. Apress (2008)
16. White, J., Wilson, M.: xajax, http://xajaxproject.org (last accessed, 10.12.2009)
17. Kurz, T., Heistracher, T.J.: Simulation of a Self-Optimising Digital Ecosystemp. In: Proceedings of International Conference on Digital Ecosystems and Technologies, IEEE DEST (2007)

Launching Context-Aware Visualisations

Jaakko Salonen and Jukka Huhtamäki

Tampere University of Technology, Hypermedia Laboratory,
Korkeakoulunkatu 1, FI-33720 Tampere, Finland
{jaakko.salonen,jukka.huhtamaki}@tut.fi

Abstract. Web is more pervasive than ever: it is being used with an increasing range of browsers and devices for a plethora of tasks. As such, exciting possibilities for information visualisation have become available: data is potentially highly relevant, easy to access and readily available in structured formats. For the task of visualisation, the diversity of the Web is both a challenge and an opportunity: while rich data may be available in well-defined structures, they need to be extracted from an open ecosystem of different data formats and interfaces. In this article, we will present how context-aware visualisations can be launched from target Web applications. Methods for accessing and capturing data and web usage context are presented. We will also present proof-of-concept examples for integrating launchable context-aware visualisations to target applications. We claim that by maintaining loose coupling to target Web applications and combining several data and context capturing methods, visualisations can be effectively launched from arbitrary Web applications with rich usage context information and input data.

Keywords: Ubiquitous computing; information visualisation; context- awareness; mobile Web; augmented browsing.

1 Introduction

Web has become an intangible part of our everyday lives, and is now more pervasive than ever: online services and other Web resources are used for work and play, in many environments including desktop and laptop computers, mobile phones and embedded devices. Quite recently, Internet-connected smartphones and netbooks have gained popularity, enabling Web usage on mobile devices. In living room context, Web browsing is also slowly being enabled by new TV devices, but also by some latest gaming consoles.

As the contexts in which Web is being used are becoming increasingly varied, capturing and understanding Web usage context is becoming even more important. For example, as different devices have different capabilities, a Web application may be required to adapt to the device context in order to guarantee good quality of service. Also, location and social environment may have important effects on usage: for example browsing the Web in a crowded bus, with strangers potentially looking at your screen, may have a strong influence on Web usage.

F.A. Basile Colugnati et al. (Eds.): OPAALS 2010, LNICST 67, pp. 146–160, 2010.

Possibly due to the advent of Web 2.0 *mashup* scene, tools of *information visualisation* are gaining popularity in everyday Web applications. In general, a mashup is a Web application created "by combining content, presentation, or application functionality from disparate Web sources" [1]. In information visualisation, the underlying objective is to serve as an amplifier of the cognition of a user through expressive views giving insight on a certain phenomena represented by the data [2]. This objective is not far away from those of mashups: in fact, many mashups can combine and represent existing data in a novel fashion and thus may be regarded as information visualisations.

It would be beneficial to enable visualisations to have mashup-like features for context and data capturing. These features would not only make easier for visualisations to access data, but also to better support users in different domains (see e.g. [3]). At their best, web visualisations would not be available as detached tools but, instead, as integral part of web browsing experience in the fashion of augmented browsing. For instance, a web browser extension could be used to pass additional information from the browser to visualisations.

For smooth web browsing experience, it is important to design the integration of visualisations as unobtrusive: they should not interfere or annoy users. Instead, they should launch, when needed, to display relevant information or provide intelligent "cues" in relevant places. For this design task, capture and use of available web usage context information is potentially very useful. By adapting visualisations to captured context, users can be provided with relevant information and tools.

In this article, we will study and identify different means of launching visualisation tools within heterogenous Web applications. We will also discuss different possible approaches for automatically capturing information about Web usage context and methods for transmitting the information when launching visualisations in the Web.

The rest of the article is organised as follows: in section 2 we will define background of our work with regard to context information and context-awareness, related work and known challenges, in section 3 we present methods for capturing Web usage context, in section 4 we will go through an example of launching visualisations from web browser with different means and discuss their differences. We conclude our work in section 5.

2   Background

2.1   Context and Context-Awareness

Key motivation for capturing context is the ability to make applications *context-aware*. Dey and Abowd [4] define that "a system is context-aware, if it uses context to provide relevant information and/or services to the user, where relevancy depends on the user's task".

In their popular, widely accepted definition for context related to context- awareness research, Dey and Abowd state the following: "Context is any information that can be used to characterise the situation of an entity. An entity is a person, place, or object that is considered relevant to the interaction between a user and an application, including the user and applications themselves". As such, "if a piece of information can be used to characterise the situation of a participant in an interaction, then that information is context". [4].

When building context-aware applications, the important question is that what contextual information is likely to be useful. According to Dey and Abowd, there are certain types of context that are, in practise, more important than others: activity (A), identity (I), location (L) and time (T) [4]. These types are defined as the primary context types; they not only answer important questions of who, what, when, and where, but may also be used as indices to further, secondary contextual information.

While types time, identity and location are fairly self-explanatory, we see that the activity dimension of context needs to be elaborated. According to Dey and Abowd, activity answers the fundamental question what is happening in the situation [4]. We see that the problem of obtaining knowledge about activity context is a reflection of the general problem of context-awareness: situational and contextual information are often implicit, and thus not directly available for computers.

Related to automating the collections methods of metadata on media captured with mobile devices, Davis et al. [5] present an alternative categorisation for contextual metadata. While we recognise that the three categories - spatial, temporal, and social contextual metadata - are included in Dey and Abowd's classification, we see that the social dimension is worth mentioning explicitly.

2.2 Web Usage Context

When capturing Web usage context, we are interested in contextual information about usage of web documents and applications.

As a ubiquitous system supporting an ecosystem of devices, applications and formats, definition of web usage context is non-trivial. However, some assumptions can be made that valid often enough (for a human user):

- Web is accessed with a user agent device that displays content on a colour screen and has at least one input device
- A web browser with rudimentary support for browser extensions, HTML[1], CSS[2] and JavaScript is used
- In order to access web documents and applications, the browser connects to a (remote) web server
- Browser maintains state and history of the web browsing. This includes maintaining the state of currently loaded web documents and applications

All of the entities presented above – user agent device, web browser and target web documents and applications – are potential sources of usage context information. However, several challenges in accessing the context information exist.

First of all, context information is often fragmented across the different entities: relevant state information and data may be scattered across user agent

---

[1] HyperText Markup Language.
[2] Cascading Style Sheets.

device, web browser and (remote) web documents and applications. In order to obtain a clear picture of a usage context, information needs to be integrated from multiple sources.

Secondly, access to context information in the different entities is often limited to local APIs. For instance, relevant context information available on a user agent device may not be accessable from a web browser. Mechanisms for passing context data between web browser, local and remote systems are needed.

Finally, as a broad-spectrum of user agent devices and web browsers are in use, design of an all-inclusive web context capturing system is likely to be very demanding if not impossible. On the other hand, the use of single system or vendor- specific API cannot be considered sufficient. A considerably efficient approach would be to use (open) standard APIs whenever possible, and vendor-specific interfaces only when required.

The range of different user agent devices, operating systems and Web browsers make describing Web usage context especially challenging. Some typical choices, however, can be detected:

- Popular desktop operating systems for Web browsing include Windows XP, Windows Vista and Mac OS X, with Windows family systems taking easily over 90 percent market share [6]. On mobile OS markets, the most commonly used operating systems are iPhoneOS, SymbianOS (e.g. Nokia devices), and RIM (BlackBerry devices).
- It is estimated [7] that on desktop browser markets, Internet Explorer and Firefox take more than 80 percent share of the markets. Mobile markets are much more fragmented; Opera (incl. Opera Mini), iPhone, Nokia and iTouch browsers all four have a market share larger than 10 percent [6].

While specific configurations of Web browsers and devices are not easily defined, two different Web usage contexts clearly emerge: *desktop context* and *mobile context*. Statistics suggest that the configurations vary between these two settings: where desktop systems seem to be typically Windows systems with Internet Explorer or Firefox are mobile systems more typically smartphones using their vendor's own Web browser, or alternatively Opera. Only a fraction of Web usage can be detected from media devices like Playstation, suggesting Web usage from *living room context* is either very minimal or not visible in Web usage statistics.

2.3 Related Work

Numerous earlier and related approaches on mashup development and context information capturing influence and inform our work.

In project Mobilife, a context management framework was defined, according to which software called ContextWatcher was implemented as a mobile application for a limited set of Nokia 60 series phones (see [8]). ContextWatcher automatically collects, enriches and shares context information focused on visited places and people met. ContextWatcher is presented as a way to keep in touch with relatives and friends: use cases include managing social relationships, taking photos and receiving points of interests [9].

A fairly similar design in context management frameworks is introduced in Hydrogen. Similar to ContextWatcher, Hydrogen [10] specifies an architecture in which ContextServer is used as an intermediary component between applications that use context information and various adaptors providing context data.

Yahoo! Research's ZoneTag[3] presents a more practice-oriented solution for contextual data. ZoneTag is a rich mobile client that focuses on enabling context- aware photographs uploads from camera phones [11]. Additionally, the software supports media annotation via context-based tag suggestions. Also a mobile photo browser client, Zurfer[4], has been implemented and can be used to access photos based on geographic location as captured by ZoneTag.

Another example of a more specific context application is Nokia's Sports Tracker[5]. Sports Tracker is a GPS[6]-based activity tracker for compatible Nokia mobile devices and records information such as speed, distance, and time from a training routine. Context information also includes optional playlist from used mobile device.

A variety of tools and frameworks supporting the development of mashups with widget-like rich user interfaces and content processing pipelines exist. Examples include Yahoo Pipes[7], Intel Mash Maker[8] and Google Mashup Editor[9]. Also information visualisation tools including Adobe Flex[10] and Prefuse Flare[11] and web widget frameworks such as Dojo[12], Ext JS[13], Yahoo YUI library[14] and jQuery[15] contribute to the development of mashups.

2.4 Challenges

Several challenges in leveraging the potential of information visualisation in mashups exist. Means for accessing context information is often limited to simple attributes representing the capabilities of the device in use, user language and, recently, user geolocation. No harmonised solution to access context information is provided for either mashup or information visualisation developers.

Due to the limitations of web browser as a run-time platform, visualisations that use large data sets or are required to be highly interactive may not be feasible to be implemented as conventional mashups. Additionally, access to local and remote resources is very restricted in typical web browser security models. In effect, mashups or mashup systems based on web browsers may not be able to access and represent all information of interest.

---

[3] http://zonetag.research.yahoo.com/
[4] http://zurfer.research.yahoo.com/
[5] http://sportstracker.nokia.com/
[6] Global Positioning System.
[7] http://pipes.yahoo.com/
[8] http://mashmaker.intel.com/
[9] http://code.google.com/gme/
[10] http://www.adobe.com/products/flex/
[11] http://flare.prefuse.org/
[12] http://www.dojotoolkit.org/
[13] http://www.extjs.com/
[14] http://developer.yahoo.com/yui/
[15] http://jquery.com/

Other, more powerful visualisation tools such as Orange[16] and Vizster[17] or the ones implemented with Processing[18] often operate as separate applications: they take data from web applications as input and process it independently of the original application, often in a pipelined transformation process. In this line of thinking, web applications are seen as data sources that are only accessed in the initial step of the process. As many web applications expose similar application programming interfaces (APIs), it is in theory sufficient to create new data export component only once for each type of interface.

Based on our experiences, the creation of a general-purpose data export component has proven to be a difficult task. In a common pattern of practical visualisation development, a lot of information about application context, such as authorisation and authentication information, and location and format of the target data, has to be specified. An exporter component then needs to not only retrieve the desired data, but also perform tasks required to access the data, such as to perform user authentication. Also, data access process is often very application-specific: for instance, while two applications may share common data format, they may use incompatible authentication methods. In order to make data export components more general purpose, more parametrisation is to be added to make the use of these components more complex.

In a worst-case scenario, external export component may not be able to access data at all. For instance, it is common for web applications that require high level of security such as online banking system or institutional single-sign on services, to take additional measures to specifically stop non-human agents accessing any content at all. This kind of a "walled-garden" service provision policy found from many other domains as well may effectively block visualisation data export.

3   Context Capturing Methods

In this chapter, we will define and discuss various methods that are feasible for capturing web usage context.

3.1   HTTP Request Interpretation

HTTP Request contains multiple headers that are typically present and can be used for context interpretation. A commonly used, standardised request header is USER- AGENT, which contains information about the user agent originating the request [12]. Another widely supported standard, User Agent Profile (UAProf), specifies X-WAP- PROFILE header for transporting references to Composite Capability/Preferences Profiles (CC/PP) [13]. Also, proprietary header extensions may be used to obtain capability information. Opera Mini, a data transfer-efficient mobile browser system based on proxy architecture, for example, specifies several custom headers that are helpful in determining device capabilities [14].

---

[16] http://www.ailab.si/orange/
[17] http://jheer.org/vizster/
[18] http://processing.org/

A widely used method for extracting capability information is to map header data to device profile and user agent databases. Commonly used mappings and databases include resources provided by browscap.ini[19] and WURFL[20] .Capability information that is potentially easy to extract with the mapping approach includes the following: operating system or platform name, browser name and version, and browser capabilities including CSS support level, support for specific HTML elements such as frames and tables, cookies and scripting support.

By interpreting clients IP[3] address, HTTP request data can also be used for locating user. For example, hostip.info provides a public Web API (see [15]) that can be used to translate any given IP (Internet Protocol) address in to a geographic position. While not typically very accurate, this method provides us with a mean for making a rough guess of user's position and enabling at least some awareness of user's location context.

3.2 Generic DOM and JavaScript Access

Information on different capabilities may also be also obtained with JavaScript via Document Object Model (DOM), an API provided by Web browser and its various extensions. Information about browser capabilities is available via *window.navigator*. Screen capabilities (width, height, depth) can be accessed from *window.screen* (see e.g. [16]). Current browsing location is accessible via *window.location*. Also, an object containing information about browsing history is available, but does not actually expose any history data besides the number of elements in the history list (e.g. [17]). The objects are defined in DOM Level 0: a specification of JavaScript objects that existed in browsers before actual DOM specifications [18]. Most modern browsers still provide DOM 0 elements (e.g. [19]), thus making them a viable source of capability information.

Even some input metrics can be captured with scripting. Mouse coordinates relative to the screen can be obtained from property pair *event.screenX* and *event.screenY*, both available on all major Web browsers (see e.g. [20]). By creating a listener for mouse move events in JavaScript, input metrics from mouse can be captured within individual Web pages. In a similar fashion, by listening to *onkeydown*, *onkeyup* and *onkeypress* events, a limited set of events from keyboard can be captured.

APIs for accessing user's geographic position via browser APIs are also being actively developed. Google Gears[21], a browser extension for Firefox, Internet Explorer and Safari, implements a Geolocation API [21] that can be used to obtain positional data. Later on, this work has been progressed in the form of W3C's specification for Geolocation API that has been published as an Editor's Draft [12]. Also an experimental Opera build with Geolocation support exists [23]. In general, several browser vendors are supporting the API, making it a viable source for positional data.

---

[19] http://browsers.garykeith.com/downloads.asp
[20] http://wurfl.sourceforge.net/
[21] http://gears.google.com/

W3C's Geolocation API can be used to fetch user's position as a latitude-longitude value pair, but it also supports retrieving altitude, speed, heading and information about the accuracy of positioning (*accuracy, altitudeAccuracy*). A particularly convenient feature in Geolocation API, when properly implemented by a browser, is that it automatically tries to combine different sources of data to obtain accurate positioning. For example, in Firefox [24], information about nearby wireless access points and computer's IP address is gathered. This information is then sent to geolocation service provider, Google Location Services by default, to get an estimate of user's location. Data from GPS devices is similarly exploited when available. Note that without an established trust relationship between the user and a service requesting geolocation, user will be prompted for approval for sharing the location.

3.3 Context Scraping

Similar to *data scraping* or *screen scraping* where data is programmatically extracted from views that are primarily intended for humans to read, context information can be extracted from documents.

Web hypertext documents, however, already generally include information in formats particularly designed for machines to process. This information is often represented as *web feeds* in RSS (Really Simple Syndication and RDF[22] Site Summary) or Atom formats. Also, Web documents may include explicit information in Microformats[23] or other form of *semantic markup* that enables the automated processing of the content.

Further, the output of a particular content management system or other known platform follow a specific markup scheme. A popular wiki engine MediaWiki, for example, lays out the editing history of a given page as an HTML list structure. Detecting that a page is put out by an instance of MediaWiki can be identified on basis of document metadata, such as follows:

```
<meta name="generator"
content="MediaWiki 1.16alpha-wmf" />
```

Once MediaWiki is recognised, a revision history of a given MediaWiki page may be fetched, for example, in order to define a social context for a wiki page (if this is something that an application developer wants to do). In practise, this is simply done by adding a request parameter (*?action=history*) to the URI of the wiki page. Whereas interpretation surely is involved in making this kind of information explicit, a rough set of contextual information on social activity within a wiki can be automatically collected.

3.4 Application-Specific API Access

In many cases, accessing the content of a system via external components such as crawlers is either difficult of impossible. Examples include Facebook and other

---

[22] Resource Description Framework.
[23] http://microformats.org/

services that actively attempt to restrict crawler access for reasons of privacy and business. Injecting an intelligent probe may help in these cases. A probe implemented as a Greasemonkey[24] script, for example, is even able to access the Javascript functions of a specific service via its *unsafeWindow* object. Do note, however, that the use of *unsafeWindow* claimed to be data-insecure and, more importantly, does often meet the characteristics of a cross-site scripting (XSS) attack.

3.5 User Input

Direct or indirect input from user is also an option to collect context information. We see, however, that to be consistent with the unobtrusive approach of information visualisation integration, user input should always be considered as the last resort.

Some of the context information is always unavailable for automatic collection. Moreover, we agree with Dey and Mankoff [25] in that automatically collected context information is often ambiguous, thus a user should always have the option to review and refine the interpreted context information. We see that, when complemented with the option to refine the context information, visualising the automatically collected and interpreted context is a valid option for providing user with a *mediator* [25], a component that supports handling ambiguous information. With inspiration from the visualisation-driven, iterative process of *knowledge crystallisation* [26], in which the most relevant information in a given case is collected, often with the help of tools of information visualisation, we might want to refer to *context crystallisation* when collecting context information in co-operation with the user.

4 Launching Context-Aware Visualisations

In this chapter, we will go through an example on launching visualisations from web browsing context with different means.

4.1 Launching Visualisations as Stand-Alone Applications

Let us first consider how visualisations are launched – as how we see they typically are – as stand-alone applications.

As an example of a candidate visualisation to be launched, we present RSSVis. RSSVis is a visualisation application implemented with *Wille 2 Visualisation System*. The application consists of three widgets for representing RSS information: a timeline, a data table, and a map for RSS data that includes geolocation information in, e.g., GeoRSS[25] format.

Data source to RSSVis is given by adding parameter *rss* with full feed URL either as a URL-encoded parameter (GET) or encoded in HTTP request content (POST). The actual feed content is then retrieved by RSSVis. The aim of the visualisation is to serve as a general-purpose RSS visualisation tool: different feed formats and metadata fields are automatically detected and parsed. If an RSS

---

[24] http://www.greasespot.net/
[25] http://www.georss.org/

aggregator component was added either as independent component or integrated to RSSVis, data from several RSS feeds could be displayed in an aggregate data view.

In casual web browsing, RSS content is often stumbled upon, as an alternative representation of data primarily available in HTML format. When an autodiscovery mechanism such as RSS [27] or Atom Auto-Discovery [28] is used, modern web browsers are able to detect and inform users about available RSS content. By opening the actual RSS URL in browser, an HTML version of the feed is often displayed.

A simple option to enable the launch of RSSVis is via a regular HTML form that the visualisation user pastes the URL of the RSS feed to be visualised into. If user uses an RSS-enabled browser, it is sufficient for example to simply copy and paste the discovered URL to the HTML form.

4.2 Passing More Context with Bookmarklets

Launching a visualisation, such as RSSVis, as a standalone application, requires user to manually re-enter information about the RSS feed. A more handy option can be provided with a *bookmarklet*, a functional JavaScript-based bookmark capable of accessing the DOM of the document to be visualised. By using a bookmarklet, some of the context information available in browsing window may be passed automatically to the visualisation.

An example of a simple bookmarklet for launching a more context-aware RSSVis is presented in the following listing:

RSSVis Launcher Bookmarklet Code Example[26]

```
01  javascript:
02  appurl='http://localhost:8080/apps/rssvis/?uri=';
03  u=document.location.hostname+document.location.pathname;
04  t=document.title;
05  w=screen.width;
06  h=screen.height;
07  void(window.open(appurl+escape(u)+'&width='+
08  escape(w)+'&height='+escape(h),'_blank'));
```

Going through the listing, the following actions are performed by the bookmarklet:

1. URL of the target visualisation application is specified (line 2)
2. Document identifier metadata (URL and title) of the currently open document is retrieved via DOM (lines 3-4)
3. Some browser properties (screen width and height) are also retrieved (lines 5-6)
4. Finally, a new pop up window is opened with all retrieved variables encoded to target URL as URL arguments (lines 7-8)

Once tested thoroughly for browser-compatibility, bookmarklets operate in a wide range of desktop web browsers. Bookmarklets with limited functionality can be implemented also for Opera Mini and other mobile browsers.

---

[26] Note that line breaks have been added for clarity: in a functional bookmarklet, everything has to be encoded within a single line of code. Also, if the bookmarklet is encoded as a hyperlink (*a* element), some characters may need to be escaped and entity-encoded as well.

To make the launching process more straightforward, we can implement support for auto-discovery of RSS feeds into RSSVis. Auto-discovery is based on Web document metadata stating the existence of RSS feed(s) related to the document. Browsers and RSS aggregators, for example, implement auto-discovery to ease the subscription of Web feeds. With auto-discovery, a visualisation user can launch the visualisation directly from a Web document. The existence of RSS feeds can be detected either with JavaScript or by the actual RSSVis implementation.

4.3 Proactive Detection of Visualisable Contexts

For a true visualisation-augmented web browsing experience, bookmarklets may not be sufficient: a user needs to actively click a bookmarklet in order to activate a visualisation. In a more sophisticated solution, a visualisation could automatically capture the context and suggest possible visualisations for the currently active document or application.

In Mozilla Firefox, Google Chrome and other compatible browsers, Greasemonkey can be used for augmenting browsing with visualisations. Greasemonkey is a browser extension that allows implementing scripts that tailor Web documents on-the-fly as the documents are loaded into the browser. A URI space is defined for a given script to handle.

With Greasemonkey, we are able to implement a proactive mechanism for launching RSSVis where the browser automatically injects either an intelligent visualisation launcher or the actual visualisation into the document that falls into the defined URI space or, more generically, has an alternative representation in RSS.

In the following listing, an example of Greasemonkey's usage for visualisation augmented browsing is presented. The idea is to automatically discover data sources actively, while user is browsing, and inject a link to a visualisation whenever appropriate.

Greasemonkey code example

```
01   postdata = 'uri=' +
02   escape(unsafeWindow.document.location.href) +
03   '&representation='+ escape(new
04   XMLSerializer().serializeToString(unsafeWindow.document)
05   );
06
07   GM_xmlhttpRequest({
08     method: "POST",
09     url: "http://localhost/apps/rssvis/isvisualisable/",
10     data: postdata,
11     headers: {
12       "Content-Type": "application/x-www-form-urlencoded",
13       "Content-Length": postdata.length,
14     },
15     onload: function(response) {
16       if (response.status == 200) {
17         document = unsafeWindow.document;
18         link = document.createElement('a');
19         // Set up appearance and target of the link
20         // (Code removed for brevity)
21         unsafeWindow.document.body.appendChild(link);
22       }
23     }
24   });
```

Going through the listing, the following actions are performed:

1. Context data, including the whole content of the current page is captured and URL encoded (lines 1-4)
2. The captured context data is sent to an analyser application in order to determine if it can be visualised with RSSVis (lines 6-23)
3. In case visualisable data is available, a link to the RSSVis tool is injected directly to the page (lines 16-20)

A Greasemonkey script can be configured to execute on every page or to an enumerated set of URLs. Through configuration we can state, for example, that every user profile in a social bookmarking service delicious.com is equipped with a timeline representing the latest bookmarkings of the user. Further, when geolocation information is available in an RSS feed, the geolocation of a user can be resolved and used to set the original focus of the map visualising the items in an RSS feed.

4.4 Summary and Discussion

The presented visualisation launching strategies let users to maintain their current web workflows, while making potentially insightful visualisations available in right contexts. When an intermediary context collector and visualisation selector such as presented Visualise app is used, the approach is more proactive than actively browsing to an alternative location to operate the visualisations.

The most pervasive approach is to use a script injected with Greasemonkey or similar system. This approach, however, suffers from the lack of support in some browsers. Security considerations may also be biased: it is questionable if one wants an external application to actively monitor the web usage.

Bookmarklets, on the other hand, while less capable, are discrete and leave the decision for activation to the user. They can be also used as a fallback or as a selectable alternative to systems like Greasemonkey: if user-scripting is not supported by browser or disallowed by the user, bookmarklets can be used instead.

As a final fallback, an HTML form can be also used. In this case, we may lack much of the desired context data. This implies that potentially more user intervention is required for manual input of context information.

The coverage of data and context capturing can be also extended to environments that e.g. require authentication or do not directly support RSS. For example if the RSS feed requires user authentication, the visualisation application may initially try to retrieve data loaded to browser in client-side or as a fallback, ask user for credentials and perform data retrieval on server-side.

As the visualisation system supports local machine installation, local resources can be potentially accessed as well. Thus, an RSS representation of, for example, a local Zotero[27] article database could be implemented and visualised with RSSVis. Having the possibility to utilise context information in mind, new interesting applications are easy to invent. A researcher who is currently going through an academic article database, can be provided with a timeline representation of articles in the local database. The presented approach even makes it possible to mash up and display local

---

[27] http://www.zotero.org/

and remote data in browser: in this fashion, an integrated article timeline could contain both local and remote resources.

Access to documents and application APIs is complemented with DOM manipulation and access to the local files system is, when put together, a highly expressive combination that has to be dealt with care for reasons of data security. For now, we have focused to harness the full power of the technology stack; the production use of the defined approach, however, insists the availability of solutions increasing the user control. Most likely, the expressivity is compromised accordingly.

5 Conclusion

In this paper, we have described a model for Web usage context. Also, based on practical experimentation and literature review, we have identified and described generic methods for capturing Web usage context. Finally, we have presented a real- life example on launching information visualisation tools in heterogenous environment and on passing the context information to the application.

According to our approach, Web usage context capturing should be based primarily on exploiting information readily available with different automated means. If a specific piece of context information is not available but required, methods that may require user intervention can be used. Further, mediation of ambiguous context information can be supported with context information visualisation. As a last resort, user will be directly asked for direct input. User identification is an exception in this strategy: manual identification is always preferred.

An advantage of the given approach is that it very often minimises the need for both user intervention and creation of custom context capturing applications. As such, we could describe it as a lightweight strategy for Web usage context capturing. A limitation in the given approach is that, when followed, it does not itself guarantee a given coverage, accuracy or level of available contextual information. In the worst- case setting, this means that no contextual information is available for the application.

We see that context capturing is always restricted by several factors. Firstly, the more platforms we wish to support, the more work it always demands. Secondly, due to legislative and non-technical restrictions, capturing some context information may always require user intervention or is not possible at all. Finally, it seems that context information will be often left incomplete: a setting in which data is not available must be tolerated.

Future work includes extending the context-collection framework and enabling a more harmonised way to access context data for visualisation tool developers. On basis of this data, a more intelligent mediator application can be developed to support the visualisation-selection. Similarly to context data access, generic means to inform visualisation tools about the existing points of accessing input data should be developed. Our long-term goal is to enable a visualisation-launching ecosystem that different visualisations can be mapped into.

Acknowledgements. European Commission (IST network of excellence project OPAALS of the sixth framework program, contract number: FP6-034824) supports our work.

References

1. Yu, J., Benatallah, B., Casta, F., Daniel, F.: Understanding Mashup Development. IEEE Internet Computing 12(5), 44–52 (2008)
2. Ware, C.: Information Visualization, Perception for Design (Interactive Technologies), 2nd edn. Morgan Kaufmann, San Francisco (2004)
3. Galli, L., Guarneri, R., Huhtamäki, J.: VERTIGO: Find, Enjoy and Share Media Trails across Physical and Social Contexts, London, UK (2009)
4. Dey., A., Abowd, G.: Towards a Better Understanding of Context and Context-Awareness. Submitted to the 1st International Symposium on Handheld and Ubiquitous Computing (HUC '99) (June 1999)
5. Davis, M., King, S., Good, N., Sarvas, R.: From Context to Content: Leveraging Context to Infer Media Metadata. In: Proceedings of the 12th annual ACM international conference on Multimedia, pp. 188–195. ACM, New York (2004)
6. Statcounter.com: Top 8 Mobile Oss from July 1, 2008 to 28 July 28, 2009, `http://gs.statcounter.com/#mobile_os-ww-daily-20080701-20090728` (accessed January 8, 2010)
7. Wikipedia: Usage share of web browsers, `http://en.wikipedia.org/w/index.php?title=Usage_share_of_web_browsers&oldid=3364023` (accessed January 10, 2010)
8. Floreen, P., Przybilski, M., Nurmi, P., Koolwaaij, J., Tarlano, A., Wagner, M., Luther, M., Bataille, F., Boussard, M., Mrohs, B., Lau, S.: Towards a context management framework for Mobilife. In: Proceedings of the 14th IST Mobile & Wireless Summit (2005)
9. Kernchen, R., Bonnefoy, D., Battestini, A., Mrohs, B., Wagner, M., Klemettinen, M.: Context-Awareness in MobiLife. In: Proceedings of the 15th IST Mobile & Wireless Summit (2006)
10. Hofer, T., Schwinger, W., Pichler, M., Leonhartsberger, G., Altmann, J., Retschitzegger, W.: Context-Awareness on Mobile Devices - the Hydrogen Approach. In: Proceedings of the 36th Hawaii International Conference on System Sciences, HICSS'03 (2002)
11. Ahern, S., Davis, M., Eckles, D., King, S., Naaman, M., Nair, R., Spasojevic, M., Hui-I Yang, J.: ZoneTag: Designing Context-Aware Mobile Media Capture to Increase Participation. In: Mobile Systems, Applications, and Services, pp. 36-44 (2004)
12. Fielding, R., et al.: Hypertext Transfer Protocol - HTTP/1.1. Request for Comments (RFC) 2616. The Internet Society (1999)
13. Wireless Application Forum, Ltd.: Wireless Application Protocol WAP-248-UAPROF-20011020-a, `http://www.openmobilealliance.org/tech/affiliates/wap/wap-248-uaprof-20011020-a.pdf` (accessed January 10, 2010)
14. Von Streng Hsehre, K.: Opera Mini request headers (November 7, 2007), `http://dev.opera.com/articles/view/opera-mini-request-headers/`
15. Mozilla Developer Center: window.screen (August 25, 2009), `https://developer.mozilla.org/en/DOM/window.screen`
16. Community Geotarget IP Addresses Project: Using the Database - IP Address Lookup, `http://www.hostip.info/use.html`
17. Mozilla Developer Center: window.history (August 25, 2009), `https://developer.mozilla.org/en/DOM/window.history`
18. Koch, P.: JavaScript - Level 0 DOM, `http://www.quirksmode.org/js/dom0.html`
19. Aptana, Inc.: HTML DOM Level 0 Reference Index (January 8, 2010), `http://www.aptana.com/reference/html/api/HTMLDOM0.index.html`

20. Koch, P.: W3C DOM Compatibility – CSS Object Model View (March 29, 2009),
    http://www.quirksmode.org/dom/w3c_cssom.html
21. Google: Geolocation API – Gears API – Google Code (2010),
    http://code.google.com/apis/gears/api_geolocation.html
22. Popescu, A. (ed.): W3C: Geolocation API Specification. W3C Editor's Draft, June 30
    (2009), http://dev.w3.org/geo/api/spec-source.html
23. Mills, C.: Find me! Geolocation-enabled Opera build (March 26, 2009),
    http://labs.opera.com/news/2009/03/26/
24. Mozilla: Geolocation in Firefox (2010),
    http://www.mozilla.com/firefox/geolocation/
25. Dey, A.K., Mankoff, J.: Designing mediation for context-aware applications. ACM Trans-
    actions on Computer-Human Interaction 12, 53–80 (2005)
26. Card, S.K., Mackinlay, J., Shneiderman, B.: Readings in Information Visualization: Using
    Vision to Think. Morgan Kaufmann, San Francisco (1999)
27. Cadenhead, R., Holderness, J., Charles Morin, R.: RSS Auto-discovery. RSS Advisory
    Board (November 27, 2006), http://www.rssboard.org/rss-autodiscovery
28. The Internet Society: Atom Auto-discovery (November 11, 2005),
    http://philringnalda.com/rfc/
    draft-ietf-atompub-autodiscovery-01.html

A Practical Approach to Identity on Digital Ecosystems Using Claim Verification and Trust

Mark McLaughlin and Paul Malone

Telecommunications Software & Systems Group,
Waterford Institute of Technology, Waterford, Ireland
`mmclaughlin@tssg.org, pmalone@tssg.org`
`http://tssg.org/`

Abstract. Central to the ethos of digital ecosystems (DEs) is that DEs should be distributed and have no central points of failure or control. This essentially mandates a decentralised system, which poses significant challenges for identity. Identity in decentralised environments must be treated very differently to identity in traditional environments, where centralised naming, authentication and authorisation can be assumed, and where identifiers can be considered global and absolute. In the absence of such guarantees we have expanded on the OPAALS identity model to produce a general implementation for the OPAALS DE that uses a combination of identity claim verification protocols and trust to give assurances in place of centralised servers. We outline how the components of this implementation function and give an illustrated workflow of how identity issues are solved on the OPAALS DE in practice.

Keywords: Identity, trust, digital ecosystems, federated identity.

1 Introduction

Central to the ethos of digital ecosystems (DEs) is that DEs should be distributed and have no central points of failure or control. This means that failures are always local and that system-wide authorities are not necessary for the ecosystem to function. On the OPAALS project, it has been a policy to re-use existing technologies and standards, where appropriate, in building the DE infrastructure in order to focus resources on what is truly innovative. These requirements have essentially shaped the identity model and implementation that have emerged from the project. Thus we have reused existing authentication and authorisation mechanisms, as well as federated identity specifications and technologies, to build a solution for identity on DEs, which are not centrally governed or maintained.

Traditionally, an identity was merely a username and a password (or perhaps an X.509 certificate) supplied by an identity provider (IdP) for a system, and it applied universally across the system. As a result, there is a continued expectation that in order to access a system or service, a user authenticates somewhere and then is fully identified to all parties on the system and can consume any service. This is what users, administrators and service providers want and expect.

F.A. Basile Colugnati et al. (Eds.): OPAALS 2010, LNICST 67, pp. 161–177, 2010.

The problem is that on a DE, we do not assume a centralised authentication or authorisation infrastructure. If a user authenticates, the user perhaps has access to local services, but not automatically to services hosted on other systems. Other contractual or trust based mechanisms are required to bridge the inter-domain gap. Clearly, authentication on a DE is only the first step, other mechanisms must then be used to assure service providers (SPs) in one domain of the validity of an identity asserted by an IdP in another.

In this paper, we state our motivation for supporting decentralised identity provisioning; we outline what we mean by identity on a DE, what this identity can be used for. We also outline what steps an entity must perform in order to be identified. We describe how entities can have an identity on a DE that is derived from an identity that they previously used to authenticate with for another purpose. We describe how the OPAALS identity model implementation, based on the open source project, IdentityFlow, can be used to provide this functionality. Finally, we describe how IdentityFlow integrates with Flypeer, the P2P infrastructure on which the OPAALS DE is based.

2 Motivation

DEs are devolved network environments with no central points of control or failure. The power structure inherent in particular digital systems is often reflected in the identity provisioning scheme adopted. For example, if identity is provided centrally, the centralised IdP will exert a large amount of control over the entire system due to its control over sensitive user data and the processes of authentication and authorisation. If identity is provided by a number of parties exerting only localised control, no single IdP will be able to exert control over the entire system. Therefore, our primary motivation is to devolve identity provisioning to a large number of localised IdPs whilst still maintaining the possibility of an identity that is applicable systemwide.

Fig. 1 gives a typical identity domain with centralised identity provisioning, in this case, Google. All of the Google services illustrated use a common identity that is backed by the Google IdP (not shown). The more services within that identity domain, and the more users consuming those services, the more power that is centralised in Googles hands by virtue of their ownership of the IdP and control over all aspects of identity provisioning. Microsoft attempted to create a global identity domain, administered by them, called Microsoft Passport. Passport offered the possibility of single sign-on (SSO) to all participating services on the web, which would have centralised a huge amount of power and responsibility in Microsofts hands. However the service was cancelled due to concerns about user privacy and a change of strategy based around the user-centric ideals embodied in Kim Cameron's Laws of Identity[4].

Subsequent attempts to introduce SSO across heterogenous service platforms focussed on the federation of identity domains rather than the expansion of one dominant domain. Federated identity allows identities on one domain to be used

Google

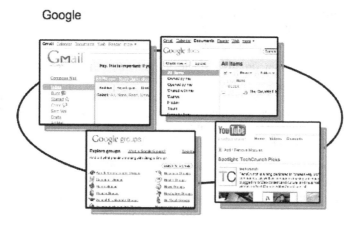

Fig. 1. An illustration of services in the Google identity domain

to access services on another. This is facilitated by inter-domain identity assertion protocols, such as SAML, and a business and/or legal agreement between
the identity domains, specifying how identity attributes are passed between the
domains. Fig. 2 gives a prominent example of a UK-based federation of academic
identity domains, the UK Access Management Federation for Education &
Research (UK AMF). Constituent identity domains host IdPs or SPs or both,
allowing, for example, university students to authenticate with their university
IdP and potentially use any service hosted by any university within the federation
(e.g. online academic databases). Thus, although identity provision is local to
the university, the federation agreement ensures that identities are applicable
throughout the federation.

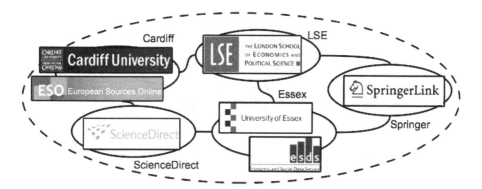

Fig. 2. Selected UK Access Management Federation for Education & Research domains

Although the federated domain scenario is a significant improvement on the large, monolithic identity domain scenario, there are still a number of issues that predicate against its immediate adoption in DEs:

1. All parties must sign up to a single agreement, which limits the scope and the scalability of the federation.
2. When federations become large, a single authoritative entity is required in order to moderate federation membership (e.g. Eduserv for the UK AMF).
3. Federation agreements are difficult to change because a consensus of all participants is required in order to enact proposed changes.

The focus of our work is to overcome these issues and arrive at a solution for decentralised identity provisioning in DEs, and to do this in such a way as to maximise the re-use of federation specifications and software. The particular approach adopted has been to replace the fixed federation agreements with contextual, transient agreements, based on measures of computational trust. These agreements strengthen and weaken in particular contexts based on the experiences of the IdPs and SPs in the DE, allowing unstable federations to be continuously created, destroyed and modified. Overlapping federations, or sets of federations with common participants, can potentially extend the applicability of identities to the entire DE (see Fig. 3), thus ensuring scalability and avoiding the need for authoritative entities.

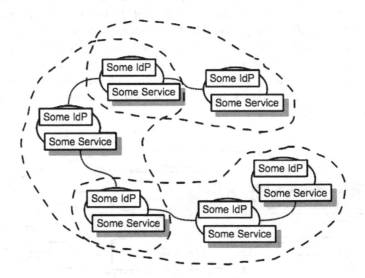

Fig. 3. Dynamic overlapping identity federations based on trust

The current work furthers this agenda by laying out in concrete terms how identity claims arising from a user-centric, local authentication can be asserted to other entities in a DE, and how these claims differ in scope from traditional, centralised or federated identity claims.

3 State of the Art

There is currently much work being conducted in the area of online identity, from a social as well as from a computer science viewpoint[7]. Many definitions have been put forward for digital identity[6],[4],[15],[24] and partial identities [8],[6] [5],[24]. When we talk of identities, we define them similarly to partial identities in [8],[6]: 'that which represents a person in a particular context in the online world', where not otherwise specified. There is also much work being conducted in the related areas of trust and reputation[10],[13]. In this work we use trust to refer to reliability trust in [10]. Federated identity[17] is concerned with federating previously separate identity domains, across large organisations and the enterprise, such that users in one domain can consume services in another. Federation specifications and protocols, such as SAML[1], facilitates federated identity management.

Digital Ecosystems can be described as "distributed adaptive open socio-technical system, with properties of self-organisation, scalability and sustainability, inspired by natural ecosystems" [3]. The term has been used in other contexts [2], however, we refer here to the body of research initiated under the heading of Digital Business Ecosystems (DBE), that is intended to promote the pervasiveness of ICT in SME and move organisations towards a "more, fluid amorphous and, often, transitory structures based on alliances, partnerships and collaboration" using biologically inspired metaphors [23].

Identity management in DEs draws inspiration and influence from the intersection of distributed/decentralised identity management[26],[17] and user-centric identity management[15],[18], and utilises the SAML federated identity specification. The implications of trust in these identity management configurations is explored in [11]. The initial work in DEs was performed by[16],[9], which form a precursor to the final OPAALS identity model, given in [21]. The identity model and implementation includes a generalised identity protocol framework (which includes SSO), and a number of bindings, including a JXTA[1] binding.

Trust has long been a topic of study in psychology, sociology, philosophy and economics; but in the nineties it has also found application in e-commerce, particularly in online markets such as eBay [25]. Trust can be described as "a directional relationship between two parties that can be called *trustor* and *trustee*," [10] where a trustor is said to trust, or not to trust, a trustee, in a particular context. Trust can be used as a form of 'soft security' [10] or, by reflecting the real world social relations, as an enabler of "trade, competition, collaboration and so on" [25]. There are numerous models for computing trust and reputation[2] [25, 10] on various systems and networks, including decentralised P2P networks [19].

---

[1] JXTA is a P2P platform: `https://jxta.dev.java.net/`
[2] "The overall quality or character [of some trustee] as seen or judged by people in general." [10].

4 Identity in Digital Ecosystems

An identity is something that is unique to an individual, or other entity, and is in some sense intrinsic to the entity in question. Traditionally, identities on computer systems have been represented by an identifier and a set of other attributes, describing that identity. The identifier is usually that attribute that is guaranteed to be unique systemwide. This guarantee of uniqueness is enforced by a centralised identity provider (IdP) on the system, which is responsible for maintaining a registry of all identities on the system.

In the real world, human identities are unique because no two human beings have identical biology and history. For the purposes of administration, the nation state will maintain identifiers and certain sets of attributes on behalf of each individual in the state, and it will guarantee the uniqueness of these by similarly using a centralised IdP or public service registrar(s). These attributes are linked to the real human being using some biometric data (e.g. a passport photograph) that are encoded along with them, as well as on identifying documents. However, in a DE, where there are no central points of failure or control, uniqueness of identifiers cannot be guaranteed, since there are no absolute identity authorities.

In the absence of a global identity authority, asserting the uniqueness of an identity, and maintaining a registry of unique identifiers, identity must operate on a different basis. Digital identities can no longer be unique, since there is no maintainer of a global identifier namespace. Our solution is to base identity on decentralised, inter-personal relationships, analogous to real life experience, where the following hold true,

1. Entities can be introduced to other entities.
2. Entities are recognisable to other entities.
3. Entities can make identity claims to other entities.
4. These claims can be verified by other entities.

In explaining 1., since no authority maintains the lifecycle of an identity, identities are effectively created, as far as other entities are concerned, by the process of 'introduction', analogous to the real world. Only when an entity has introduced itself, or represented itself to another entity with a given identity, can a history of interactions between the two begin. In explaining 2., entities must be 'recognisable' to other entities as being the entity that they have dealt with before, since they cannot rely on the system to 'announce' entity identities, which are accepted without question by the system as a whole. Recognisability can be affected by re-using the same public-private key pair for signing messages in a PKI[3], or by re-using the same shared secret passphrase.

1. and 2. allows entities to introduce themselves to others, to be subsequently recognised by those parties, and therefore any two entities can build a shared history between themselves. This shared history between entities is a basis for the decentralised uniqueness of identity and as a surrogate for the uniqueness offered by a canonical registry. It can also, itself, be later used for recognisability, since

---

[3] Public Key Infrastructure.

either entity can pose a question about their shared history that only the other entity can know the answer to. This shared history can also be used to evaluate trust between entities. Trust can then be used as a basis for accepting identity claims from asserting entities.

In explaining 3. and 4, we note that there are a number of identity claims that may be made about a subject, pertaining to issues such as 'age', 'address', 'fingerprint', 'portrait photograph', or 'occupation', such as "john murphy's fingerprint is represented by the following image" or "user longday11 is over the age of 18". These claims were formerly asserted by the 'system', however in decentralised environments, we must use a federated identity-like process of claim and claim verification, where any entity can make a claim about another entity and other entities can attempt to verify those claims. However, in the absence of a federation agreement, there is no straightforward mechanism for evaluating when the claims made by others are trustworthy. Our approach is to use computational trust and networks of transitive trust between entities in a DE, to evaluate the trustworthiness of these claims.

We discuss the process of claim and claim verification and their applicability to identifier claims in a DE in the proceeding two sections.

5 Identifier Claims from IdPs and the Role of Authentication

Here we discuss how identifiers, derived from authentication, can be used in a DE, even though there are no centralised authorities to back these identities or control the identity 'namespace', and therefore guarantee the uniqueness of these identifiers. The use of a pseudo-unique, 'naïvely distinguishing' identifier is still useful even though absolute uniqueness is impossible. For example, humans find it convenient to apply first name labels, such as 'john' and 'philip', even though these identifiers do not identify these individuals uniquely. The kind of identifiers we propose for DEs will be similarly provisional. The ability to verify that this 'john' is the 'john' that I know well means that I will not be deceived by a multitude of individuals calling themselves 'john', even though the name john is not unique.

These non-unique identifiers can be identifiers that are already used by an entity for another purpose, such as an email address (used by Yahoo! and Google), e.g. bob@bob.org, or an LDAP distinguished name, e.g. dn: dc=org,dc=bob,username=bob. If the IdPs for these identities are represented on the DE, then they can be used. However, these IdPs, which act as authoritative IdPs on their respective domains, are not authoritative on the DE, and therefore the identifiers that they are responsible for are only naïvely distinguishing on the DE. These identifiers are in effect self-asserted by entities, since they can choose an arbitrary IdP to 'assert' them on their behalf. (They could in fact assert their own identifiers if they had the functionality of an IdP.)

In order to make use of these identifiers, authentication mechanisms will usually be used, allowing entities to authenticate with their IdPs. Authentication may appear pointless when all identifiers are non-unique and self asserted, but it does serve the purpose of binding an entity to its IdP in an implicitly strong trust relationship, and provide a strong basis for trust between that entity and other entities using the same IdP. Hence authentication can be of benefit in building out a trust network. It also provides the benefit of outsourcing IdP functionality, such as generating and verifying identity assertions, to third parties equipped with this functionality, which also allows identities on legacy systems to be re- used, if the IdP in question has been modified to interoperate on the DE.

In order to refer to other entities, it is clear that some kind of identifier is required by referring entities, whether it is unique or not, however authentication is not a strict requirement for obtaining this identifier.

6 Verifying Identity Claims with Identity Operations

In federated identity, identity domains are federated such that identity claims made in one domain can be accepted in another. Single Sign-On (SSO) is the process by which a user can sign-on in one domain and consume services on the other domain using the same login or username. From a technical point of view, SSO is a protocol flow between the user, the service provider (SP), and the user's (domain) IdP, in which the IdP passes identity assertions to the SP, often via the user (see Fig. 4). These assertions are accepted by the SP because there is a federation agreement in place that allows IdP and SP to trust each other in the context of a certain set of identity claims. The process has been adapted to make identity claims possible in a DE.

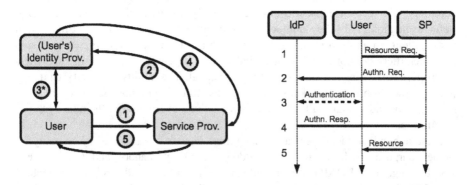

Fig. 4. High level view of federation SSO protocols (e.g. SAML)

IdentityFlow is a SAML-based, programmable framework for building arbitrary protocol flows for verifying identity claims. It is capable of operating in a DE composed of heterogenous platforms, by allowing separate bindings to be specified for portions of the overall protocol flow. The approach taken is to model DEs as a set of unstable federations between overlapping identity domains. However, rather than relying on federated agreements to provide trust between IdP and SP, we propose the use of a trust network for deriving explicit trust evaluations between two entities in the context of asserting identity claims. These trust evaluations would change over time based on the experiences of entities dealing with would-be trusted entities, as well as the referrals of other trusted entities. As a result, unstable federations are constantly made and unmade, and the claims supported by these federations are trusted or not trusted accordingly.

IdentityFlow allows us to modify the standard SAML SSO profile to include an extra actor, the SP's IdP. As a convention, we say that all entities have an IdP that is capable of generating and consuming identity assertions on its behalf, even if the entity can behave as its own IdP, just as the SP does in accepting assertions in the standard SAML SSO profile. This allows us to consider the trust between IdPs only, in the context of asserting identity claims, as the criteria for whether claims can be accepted. We term *IdentityFlow* protocols that use unstable federations based on trust to assert identity claims, *Identity Operations*. Identity operations were introduced in [21], and are based on SAML-like profiles and bindings.

A basic, but typical identity operation protocol flow is given in Fig. 5, which is equivalent to the typical federation SSO protocol flow illustrated in Fig. 4 except the SP is split into 'Entity B' and 'Entity B's IdP'.

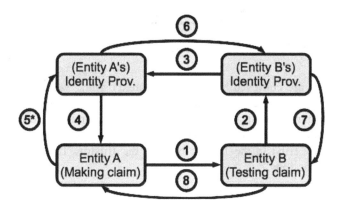

Fig. 5. Basic identity operation protocol flow

7 The Role of Trust in Identity Operations

During the execution of the protocol flow there will be appropriate points at which to measure the level of trust relationships to ensure they are sufficient for the operation to succeed, since messages from untrusted parties are useless. Fig. 6 gives an identity operation triggered by a service request. The User claims a given identifier, 'real world' identity or makes some other identity claim, and the SP verifies the claim before service access is granted. We see that the SP's IdP makes a call to the User's IdP (connection 3) and later receives a response (connection 6); it makes sense to test the level of trust that the SP's IdP has in the User's

IdP in the context of making identity claims, prior to making connection 3, since other connections are redundant if there is insufficient trust to accept the response sent in connection 6. Trust checks are particularly important where it is felt that trust relationships are non-absolute and/or evolving over time.

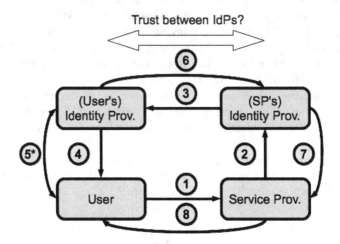

Fig. 6. Identity operation triggered by service access conducts trust checks

Trust ratings are conveyed in (trustee,context) pairs, where trustee is the trustee and context is the trust context. A component called the trust manager records 'performance' trust ratings based on direct experience and is capable of gathering 'referral' trust from third parties. Referrals are conveyed from an entity with performance trust in the trustee back to the trustor. Trust managers have the following functions,

1. Maintain a set of trust ratings with entities with whom the trustor has direct experience.
2. Discover trust transitive paths between trustor and trustee in the given context.
3. Aggregate these paths using appropriate strategies and algorithms to produce a trust rating.
4. Be capable of checking the integrity and authenticity of referrals from referees on the trust paths.
5. Have some mechanism for updating trust ratings based on experience.

2. is a challenge in decentralised environments, or otherwise, where a virtual trust network must be traversed in order to discover trust transitive paths. A number of options for discovering such paths in decentralised networks with various topologies are given in [22]. A scheme for aggregating trust paths and using belief calculus to produce a compound rating is given in [12]. [14] gives the

rationale and methodology for verifying the integrity and authenticity of referrals. Entities can update their trust ratings, and strategies for deriving trust ratings based on experiences from interacting with other entities. 'Experience reports' can be automatically or manually generated and submitted to the trust manager to derive an updated trust rating, according to some subjective scheme. The processes and algorithms for trust evaluation are described in [20]. The open source project *Trustflow*[4] is actively developing an implementation of a trust manager for use in P2P environments.

We leave a discussion of appropriate strategies for generating and incorporating experience reports on the OPAALS DE and the specification of the trust manager for a future work.

8   A Practical Approach to Identity on DEs

Thus far we have expressed a set of requirements and potential strategies arising from a critical appraisal of identity in DEs vis a vis identity in centralised and federated environments. In this section, we will outline a workable solution for asserting identity claims in the OPAALS DE environment.

The OPAALS DE environment is based on *Flypeer*[5], which in turn is based on JXTA. Two main steps are outlined for verifying identity claims on the OPAALS DE.

Step 1: Identifier Claims from Authenticating IdPs

JXTA provides a MembershipService framework for authenticating to Peer-Groups, which *IdentityFlow* implements as an extension to its core functionality. This allows *Flypeer* to authenticate users to the main Flypeer group. The IdP- MembershipService, which *IdentityFlow* implements, is available to each node on the JXTA network and operates as follows,

1. A set of login parameters, and the JXTA advertisement of a preferred IdP, is supplied via Flypeer as a JXTA Authenticator on the authenticating node.
2. The IdPMembershipService contacts the IdP node and supplies it with the correct authentication parameters.
3. The IdP returns an authentication result to the authenticating node, allowing the IdPMembershipService to complete it's PeerGroup 'join' attempt.
4. If authentication has been successful, the IdPMembershipService generates a credential from the result passed back from the IdP, which acts as proof that the entity's identity is indeed asserted by the IdP.
5. This credential can then be used as a means of filtering access to services in a JXTA-based environment.

---

[4] TrustFlow, http://sourceforge.net/projects/trustflow/
[5] http://kenai.com/projects/flypeer

The scheme for generating and using the credential assumes that all IdPs have a public-private key pair, and is as follows,

1. Following a successful authentication attempt, an IdP will sign the 'claimed identifier'[6] with its private key.
2. The identifier, the IdP's public key, the encryption method and the signed identifier are packaged with other metadata into a JXTA Credential object.
3. Other entities can verify that the IdP in question authenticated the authenticating entity by obtaining the Credential object and using the IdP's public key to decrypt the signed identifier and comparing it with the claimed identifier.

The Credential implementation used by *IdentityFlow* marshals and unmarshals to and from XML. An edited (for space) example of a marshaled Credential is given below.

```
<?xml version="1.0"?>
<!DOCTYPE jxta:Cred>
<jxta:Cred type="jxta:LocalCoordinatorCred" xml:space="..." xmlns:jxta="http://jxta.org">
<PeerGroupID>urn:jxta:uuid-FEF..102<PeerGroupID>
<PeerID>urn:jxta:uuid-596..403<PeerID>
<SignedPeerID>MCwpf4i...K3O2oEPw=</SignedPeerID>
<SignAlgorithm>DSA</SignAlgorithm>
<GroupPublicKey>MICC...MYboJk=</GroupPublicKey>
<Username>identityflow@Guigoh</Username>
</jxta:Cred>
```

Step 2: Verifying an Identifier Claim

Identity claims are verified using identity operations, as discussed in sections 6 and 7. The identity operation specifies[7]

1. A set of contingent ordered connections that are executed in sequence (from one actor to the next), potentially with conditional logic affecting the protocol flow.
2. A set of actor tasks for each possible state of each actor in the protocol flow, that are executed when an actor assumes a particular state.
3. A binding, or set of bindings, that specify connection transport functionality in particular environments (e.g. JXTA binding).
4. A set of trust checks that must be made at appropriate points during protocol execution to ensure that there is sufficient trust between actors for the operation to succeed.

Identity operations are typically triggered by resource access, where it is necessary for entity identity claims to be accepted by the SP before the accessing entity can

---

[6] This is currently the PeerID of the authenticating entity, but the entity's public key would be less spoofable.
[7] A detailed specification of identity operations is beyond the scope of this work.

be granted access. Claims concerning identifiers issued by IdPs are important examples of claims that must be verified. In the particular case of the OPAALS DE, operations that verify user identifiers are simplified by the fact that each user propagates a credential which effectively encodes an identifier claim made by the user's IdP on behalf of the user. Therefore, only connections 1, 8 (service request and response) and 2, 7 (identifier claim verification request and response) in Figs. 5 and 6 need to be executed.

For all other identity claims, pertaining to any property that an entity can claim to possess and that its IdP can verify, identity operations are necessary. Examples of other claims are "I am over the age of 18" or "I am a member of the OPAALS consortium" or "My real name is John Murphy". Whether or not these claims are accepted by other entities will depend on whether the accepting party's IdP trusts the claiming party's IdP in the context of asserting identity claims.

Since the OPAALS DE is a JXTA-based environment, a JXTA binding is used, which means that all operation connection messages (by current convention) are passed via JXTA pipes. Using JXTA transport functionality ensures entities operate in a pure P2P overlay, where entities can be contacted by name without the need for knowing a peer's underlying transport address (i.e. IP address); and also provides transparent firewall traversal.

Identity operations will tend to be designed to address a particular need, such as identifier verification, and will tend to be triggered by an actor seeking to verify a claim during the course of some activity. Therefore, it is anticipated the development of identity operations will be in response to the needs of SPs. However, it is likely that the template of the simple operation given in Fig. 5 will suffice for most purposes. An example of a more complicated identity claim would be a situation where an SP would not accept a particular claim unless two or more IdPs asserted the claim on behalf of the user. This would necessitate a more complicated protocol flow involving 5 rather than 4 actors. In general, if any of the items in the list above change, modification will be required, which should be simplified by the modular, extensible design chosen by *IdentityFlow*.

9   An Example of an Identity Workflow on the OPAALS DE

A user logs into his *Flypeer* node by specifying his work chat server node as his IdP and his usual username and password (e.g. using the authentication mechanims of the XMPP[8] chat protocol). The chat server is already on *Flypeer* and can act as an IdP on the DE because it has implemented *IdentityFlow* JXTARequestInterceptors, which intercept *IdentityFlow* messages. The user's local node supplies the authentication parameters to the IdP and the IdP node authenticates the user remotely by passing these parameters to the backend XMPP chat server. If successful, the IdP will then create an AuthenticationResult, which will be passed back to the user's node. The user's node will then create a Credential object from the AuthenticationResult. Because the IdP node has

---

[8] http://xmpp.org/protocols/

signed the claimed identifier it has returned to the user node, the user cannot choose another identifier for this Credential, because he will be unable to verify that his IdP signed it.

Once the user has authenticated, and has generated a Credential, he has general access to the DE, for discovering other users and services, and so on. Unlike a centralised system, the user does not have automatic access to resources determined by centrally maintained access control lists, permissions or an authorisation service. Instead, he has provisional access to resources determined by the parties to whom those resources are assigned and depending on whether sufficient trust between the user and those parties exist. Similarly, identities are not global and absolute, but merely local (on the user's trust network) and provisional (in that trust levels can change).

At some point the user attempts to join a collaborative document editing session, coordinated by an SP node. This SP node has an IdP that is responsible for verifying identity claims. When the user attempts to join the session, the SP triggers an identity operation to verify that "the user's claimed identifier (stored in the credential) is asserted by a trusted/competent IdP". The SP triggers the operation by requesting its IdP to verify this claim. The SP's IdP examines the user's Credential and retrieves the user's IdP's public key; the SP also retrieves the PeerID of the user's IdP that was sent as part of the user's request to join the session. The SP's IdP checks that the claimed identifier in the Credential was signed by the owner of the public key by decrypting it with the public key and matching it to the unsigned string. The SP's IdP then does a trust check on the user's IdP's in the context of asserting identity claims. The trust check indicates that it currently has sufficient trust in the user's IdP to proceed. If the SP's IdP does not already know the user's IdP's public key, it will contact the user's IdP to request it, and conduct a 'signed conversation' to prove that it is the true owner. Once the SP's IdP is satisfied, it will send a response to the SP informing it that the claim regarding the supplied identifier is verified.

If the SP were to later decide that only users over the age of 18 were allowed to join a particular document editing session, it could specify this to the users, and it could inform them that for the purpose of verifying age, it (or rather its IdP on its behalf) would only accept assertions from one particular IdP. This IdP might take measures to ensure that the users registered on its system are the age that they report themselves to be. In this case, users could register with this new IdP, and use this different IdP as their IdP when dealing with the SP's IdP in the context of asserting age claims. The SP's IdP will verify these claims by starting an identity operation, which entails performing a trust check on the user's IdP in the context of age claims, which ought to succeed since the SP's IdP recommended the user's IdP in the first place; and then initiating/continuing the protocol flow, as illustrated in Fig. 5.

10 Conclusion

In this paper, we state the critical differences between identity in a centralized environment and identity in a decentralised environment, such as a DE. These differences are so fundamental that they invalidate many accepted assumptions, such as the belief that an identifier, or name, can be unique on a decentralised system, or the belief that one can authenticate with a decentralised system and obtain a globally applicable identifier. (Unfortunately, the JXTA MembershipService framework itself tends to give this impression.) What can be achieved however, is a kind of 'local' uniqueness, based on the principle that entities that have been encountered before can be recognised again in the future, and that entities can form themselves into a trust network, which adapts to entity behaviour towards other entities, good and bad. This makes it possible to build trust into recognised entities, and to use the trust network to propagate referrals such that the benefits of experience can be shared.

We also give a sketch of how our solution to identity problems are being implemented on the OPAALS DE. We recommend a two step approach to identity problems in general: i) users authenticate against an IdP of their choice, ii) SPs trigger identity operations to verify identity claims that are necessary for allow users access to services. *IdentityFlow* has been integrated into *Flypeer* from version 0.6, which allows users to authenticate to IdPs using a number of authentication mechanisms, i.e. LDAP, XMPP and Guigoh[9]. *IdentityFlow* has provided a framework for building operations since the start, and has added HTTP GET/POST and JXTA bindings. A template operation has also been provided that is akin to that illustrated in Fig. 5.

We are currently in the process of integrating trust checks into operations, and in general integrating *IdentityFlow* and *TrustFlow*. We intend to give the full identity operation specification, the trust manager specification, and strategies for generating experience reports and updating trust evaluations in future works.

Acknowledgment

This work is funded by the EU FP6 Network of Excellence OPAALS, http://www.opaals.org

References

1. OpenSAML v2.0,
   https://spaces.internet2.edu/display/OpenSAML/Home
2. Briscoe, G.: Digital Ecosystems. Ph.D. thesis, Imperial College London (2009)
3. Briscoe, G., De Wilde, P.: Digital Ecosystems: Evolving service-oriented architectures. In: Conference on Bio Inspired Models of Network, Information and Computing Systems. IEEE Press, Los Alamitos (2006), http://arxiv.org/abs/0712.4102
4. Cameron, K.: The laws of identity, http://www.identityblog.com/?p=354
5. Damiani, E., di Vimercati, S.D.C., Samarati, P.: Managing multiple and depend¬able identities. IEEE Internet Computing 7(6), 29–37 (2003)
6. Glasser, U., Vajihollahi, M.: Identity management architecture. In: IEEE International Conference on Intelligence and Security Informatics ISI 2008, pp. 137–144 (2008)

---

[9] http://www.opaals.org.br

7. Halperin, R., Backhouse, J.: A roadmap for research on identity in the information society. Identity in the Information Society (2008)
8. Hansen, M., Berlich, P., Camenisch, J., Clau, S., Pfitzmann, A., Waidner, M.: Privacy-enhancing identity management. Information Security Technical Report 9(1), 35–44 (2004)
9. Ion, M., Danzi, A., Koshutanski, H., Telesca, L.: A peer-to-peer multidimensional trust model for digital ecosystems. In: 2nd IEEE International Conference on Digital Ecosystems and Technologies, DEST 2008, pp. 461–469 (2008)
10. Jøsang, A.: Trust and reputation systems. In: Foundations of Security Analysis and Design IV (2007)
11. Jøsang, A., Fabre, J., Hay, B., Dalziel, J., Pope, S.: Trust requirements in identity management. In: Proceedings of the 2005 Australasian workshop on Grid computing and e-research, vol. 44, pp. 99–108. Australian Computer Society, Inc., Newcastle (2005)
12. Jøsang, A., Hayward, R., Pope, S.: Trust network analysis with subjective logic. In: Proceedings of the 29th Australasian Computer Science Conference, vol. 48, pp. 85–94. Australian Computer Society, Inc., Hobart (2006)
13. Jøsang, A., Ismail, R., Boyd, C.: A survey of trust and reputation systems for online service provision. Decision Support Systems 43(2), 618–644 (2007); emerging Issues in Collaborative Commerce
14. Jøsang, A., Pope, S.: Semantic constraints for trust transitivity. In: Proceedings of the 2nd Asia-Pacific conference on Conceptual modelling, vol. 43, pp. 59–68. Australian Computer Society, Inc., Newcastle (2005)
15. Jøsang, A., Pope, S.: User centric identity management. In: Asia Pacific Information Technology Security Conference, AusCERT2005, Australia, pp. 77–89 (2005)
16. Koshutanski, H., Ion, M., Telesca, L.: Distributed identity management model for digital ecosystems. In: The International Conference on Emerging Security Information, Systems, and Technologies, SecureWare 2007, pp. 132–138 (2007)
17. Maler, E., Reed, D.: The venn of identity - options and issues in federated identity management. IEEE Security & Privacy 6(2), 16–23 (2008)
18. Maliki, T.E., Seigneur, J.: A survey of user-centric identity management technologies. In: The International Conference on Emerging Security Information, Systems, and Technologies, SecureWare 2007, pp. 12–17 (2007)
19. Marti, S., Garcia-Molina, H.: Taxonomy of trust: Categorizing P2P reputation systems. Computer Networks 50(4), 472–484 (2006) (management in Peer-to-Peer Systems), http://www.sciencedirect.com/science/article/B6VRG-4H0RYYJ-1/2/b6a612053c7546bd311548c2b642c541
20. McGibney, J., Botvich, D.: Distributed dynamic protection of services on ad hoc and p2p networks. In: Medhi, D., Nogueira, J.M.S., Pfeifer, T., Wu, S.F. (eds.) IPOM 2007. LNCS, vol. 4786, pp. 33–43. Springer, Heidelberg (2007)
21. McLaughlin, M., Malone, P., Jennings, B.: A model for identity in digital ecosystems. In: Proceedings of the 3rd International Conference on Digital Ecosystems and Technologies (DEST), IEEE, Waterford Institute of Technology, Waterford, Ireland (2009)
22. Mello, E.R.D., Moorsel, A.V., Silva, J.D.: Evaluation of P2P search algorithms for discovering trust paths. In: Wolter, K. (ed.) EPEW 2007. LNCS, vol. 4748, pp. 112–124. Springer, Heidelberg (2007)
23. Nachira, F.: Towards a network of digital business ecosystems fostering the local development. Tech. rep., Bruxelles (September 2002)

24. Pfitzmann, A., Hansen, M.: Anonymity, unlinkability, unobservability, pseudonymity, and identity management a consolidated proposal for terminology. version 0.26 (2005)
25. Sabater, J., Sierra, C.: Review on computational trust and reputation models. Artificial Intelligence Review 24(1), 33–60 (2005),
    http://dx.doi.org/10.1007/s10462-004-0041-5
26. Weitzner, D.: Whose name is it, anyway? decentralized identity systems on the web. IEEE Internet Computing 11(4), 72–76 (2007)

Implementing a Trust Overlay Framework for Digital Ecosystems

Paul Malone, Jimmy McGibney, Dmitri Botvich, and Mark McLaughlin

Waterford Institute of Technology, Waterford, Ireland
{ pmalone, jmcgibney, dbotvich, mmclaughlin }@tssg.org
http://www.tssg.org

Abstract. Digital Ecosystems, being decentralised in nature, are inherently untrustworthy environments. This is due to the fact that these environments lack a centralised gatekeeper and identity provider. In order for businesses to operate in these environments there is a need for security measures to support accountability and traceability. This paper describes a trust overlay network developed in the OPAALS project[1] to allow entities participating in digital ecosystems to share experience through the exchange of trust values and to leverage on this network to determine reputation based trustworthiness of unknown and initially untrusted entities. An overlay network is described together with sample algorithms and a discussion on implementation.

Keywords: Trust, digital ecosystem, security.

1 Introduction

In the real world, when people interact with each other and provide and consume services, well-evolved social and commercial standards and structures help to provide assurance of orderly behaviour. Our senses are well tuned to the subtleties of real personal contact. Furthermore, in dealing with service providers like banks or shops, physical structures and contact help to convince us that we are really dealing with the service provider (and not an impostor), that the service provider appears to be backed by some assets and will still be there into the future, that the transaction is genuine, that the communication is confidential, and so on. Previous experience is also a major factor for example, a bank is more likely to extend a loan to a customer with a reliable track record. Previous good experience with a service provider assures a consumer of the quality of future transactions. Laws of the land provide penalties that reduce the incentive to behave dishonestly.

Reference is also made to third parties where appropriate. Credit card and other financial transactions need to be correctly authenticated and authorised. Various respected bodies issue credentials to people to help with verification of identification, nationality, permission to drive a car, access to restricted locations, and so on. Service providers may also be certified as to professional

---

[1] OPAALS, FP6 Network of Excellence, http://oks.opaals.org/website/

F.A. Basile Colugnati et al. (Eds.): OPAALS 2010, LNICST 67, pp. 178–191, 2010.

competence, adherence to health standards, and so on. More fuzzy personal recommendations and reviews are also of value.

The digital world provides opportunities for similar (and in some ways more sophisticated) social and commercial interaction. A great benefit of the Internet is its openness and lack of centralised control. Anyone can provide a service with the minimum of fuss and invite others to make use of it. As well as in the strictly Internet world, mobile telecommunications networks are facilitating more open service provision and consumption. The increasing proliferation of wireless and ad-hoc networks using unlicensed radio spectrum is serving to further loosen control.

This digital world presents a wide variety of risks. Anyone can set up a web site that looks just like your banks site and use various tricks to get you there; online communications can be eavesdropped upon, or even modified; someone else may impersonate you; incorrect or misleading information may be provided. Legal protection is made difficult by the inter-jurisdictional nature of these networks. A major factor is that we can no longer rely on physical structures, social skills or intuition to provide assurance of security, and thus there is a much greater need for reference to third parties for identification of entities and verification of credentials.

Trust is a basic requirement for these interactions. A successful online communication or transaction requires that the parties involved sufficiently trust each other and the infrastructure that connects them. Consider for example the case where I use a web browser to navigate to an online retailers site to purchase a product or service. For this to work, I need to trust several entities: that my computer hardware and software is acting in my interest (trojaned web browser software could direct me to a bogus location); that the website is actually that of the vendor; that the vendor and any other parties collaborating with the vendor (e.g. for payments) will act responsibly with my personal information; that the vendor will deliver the product or service as expected. I may also need to trust, to some extent at least, other parts of the infrastructure. Where there is no strong means to authenticate the other party (e.g. an email correspondent or a non-secured website), I may need to trust that the domain name system (DNS) directs me to the correct IP address and that my Internet service provider and other infrastructure providers correctly route data packets.

The current approach to trust establishment and management on computer networks works in some situations, to some extent, but has significant weaknesses that limit its potential, especially in enabling rich peer-to-peer interactions and transactions.

Note that the value of building trust is not limited to commercial interaction. Even if we consider knowledge that is shared for free and without restriction, there are still threats, which can be reduced by having measures of trust, such as:

- The knowledge provided could be deliberately false. For example, free software could contain a Trojan horse or other malicious code.

- The knowledge provided could be erroneous or subject to misinterpretation, due to limitations of its creator or editor.
- The consumer of the knowledge could waste a lot of time on a facet of low value. It takes time to read a document. Software takes time to install. There may also be a learning curve, meaning that significant time and energy needs to be invested in its adoption, making it painful to rollback if it turns out not to be fit for its intended purpose.
- Some knowledge may be undesirable and unwanted, especially if the consumer does not specifically solicit or request it. Spam is an example of bad 'knowledge'.

2 The Meaning of Trust

The social concept of trust is very complex and sophisticated, perhaps deceptively so. Trust is closely related to many other social concepts such as belief, confidence, dependence, risk, motivation, intention, competence and reliability [1]. It is also interwoven with the areas of accountability[2] and identity[3].

Trust management has become a well-studied area in recent years, and several recent surveys summarise the state of the art[6][11][7][5]. Much of the work on trust in the computer science domain attempts to provide computational measures of trust so that it can be used in making decisions related to electronic transactions, in a sense mimicking peoples well-evolved forms of social interaction.

One of the difficulties of modelling trust computationally is that social trust is based on quite an intuitive and personalised subjective assessment. As trust is quite an overloaded term, most attempts to model trust start by defining what is meant by trust (at least for the purposes of that particular model). Thus there are currently several different definitions and interpretations of what trust means and how it can be used.

Trust refers to a unidirectional relationship between a trustor and a trustee. A trustor is an autonomous entity that is capable of making an assessment. A trustee can be anything. A frequently cited definition of trust is by Gambetta, a sociologist, as [4]:

"a particular level of the subjective probability with which an agent assesses that another agent or group of agents will perform a particular action, both before he can monitor such action and in a context in which it affects his own action"

This definition formulates trust as:

1. a probabilistic measure; i.e. implying that trust can be modelled as a single value between zero (complete distrust) and one (complete trust).
2. defined by the subject; each trustor may have a different view of the same trustee
3. relating to a particular action i.e. a particular service offered by the trustor; you would trust your bank more for financial advise and your doctor more for medical advice.

4. an a priori measure; trust relates to incomplete information and is thus an estimate.

5. relating to context; trust depends on the viewpoint of the trustor on how it might affect his or her action

Jøsang et al. in [6] adapt the work of McKnight and Chervany [9] to define decision trust as:

"the extent to which one party is willing to depend on something or somebody in a given situation with a feeling of relative security, even though negative consequences are possible"

In some cases, "to trust" is taken to mean making the decision itself. "I trust you" means that I believe that you will act in a particular way. Trustworthiness is a term closely related to trust, and sometimes used interchangeably with it. Solhaug et al. [12] have made a useful distinction between trust and trustworthiness, by accepting Gambettas definition of trust as a subjective probability and defining trustworthiness as an objective value:

"the objective probability by which the trustee performs a given action on which the welfare of the trustor depends"

Reputation is also related to trust, though differing views exist on precisely how they are linked. A common view is that reputation is one of the sources used by a trustor in making a trust assessment. Other sources of information for a trustor, besides reputation, typically include verifiable credentials of the trustee as well as any direct experience the trustor has had of the trustee. One simple definition of reputation is [10]:

"perception that an agent creates through past actions about its intentions and norms"

For reputation to be meaningful (distinct from the trustors own experience), it must be the collective perception of a community of observers that is somehow made available to the trustor.

3 Trust Overlay Model

For the purposes of this work, we adopt a two layer model for communications between peers(see Fig. 1). Peers can either interact for a transaction or to exchange trust information. For modelling purposes, each service usage interaction is a discrete event. A logically separate trust management layer handles threat notifications and other pertinent information. We mimic social trust by setting a fuzzy trust level. Each different service can then make an appropriate decision based on this trust level e.g. certain actions may be allowed and others not. In our system, we model trust as a vector. In the simplest case, at least if there is just one service, this can be viewed as a simple number in the range (0,1). Each peer may then maintain a local trust score relating to each other peer of which it is aware. If peer As trust in peer B is 1, then peer A completely trusts peer B. A score of 0 indicates no trust. Note that this is consistent with Gambettas widely accepted definition of trust as cited above. If trust is to be a probability measure, then the (0,1) range is natural.

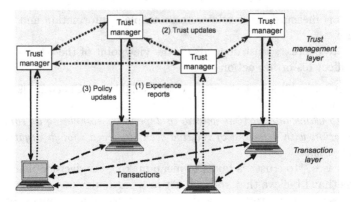

Fig. 1. Trust overlay helps to secure transactions in a digital ecosystem.

3.1 Algorithms

The way trust is updated based on both experience and recommendations has a profound impact on the usefulness of this kind of overlay system. Note that peers are autonomous in our model and each might implement a different algorithm for updating and using trust. It can also be expected that a peers behaviour in terms of handling trust may change if it is hijacked. Potential trust update algorithms include:

Moving average: Advanced moving averages are possible, where old data is remembered using data reduction and layered windowing techniques.

Exponential average: Exponential averaging is a natural way to update trust as recent experience is given greater weight than old values, and no memory is required in the system, making it more attractive than using a moving average.

No forgiveness: This in a draconian policy where a peer behaving badly has its trust set to zero forever. Even more extreme is where a peer that is reported by a third party as behaving badly has its trust set to zero forever. This could perhaps be used if a particularly sensitive service is misused.

Second chance (generally, nth chance): Draconian policies are generally not a good idea. IDS and other security systems are prone to false alarms. A variation on the no forgiveness approach is to allow some bounded number of misdemeanours.

Hard to gain trust; easy to lose it: To discourage collusion, there is a case for making trust hard to gain and easy to lose.

Use of corroboration: To prevent an attack by up to k colluding bad peers, we could require positive recommendations from at least $k + 1$ different peers.

Use of trust threshold for accepting recommendations: It is possible to model the ability to issue a recommendation as a kind of service usage on that receiving peer. Thus the receiving peer can apply a trust threshold to decide whether to accept that recommendation in the same way as any transaction attempt is adjudicated.

Next, we examine a simple exponential average algorithm in the context of both direct experience and referral.

Exponential Average Direct Experience Trust Algorithm. In the case of direct experience the exponential average algorithm, published in [8], will be written as:

$$T_{i,j(n)} = \alpha E + (1 - \alpha)T_{i,j(n-1)}$$

Where:

$T_{i,j(n)}$ is the nth value of trust placed by i in j
E is the latest direct experience report ($0 \leq E \leq 1$)
α is the rate of adoption of trust ($0 \leq \alpha \leq 1$)

The value α determines the rate of adoption. $\alpha = 0$ means that the trust value is not affected by experience. If $\alpha = 1$, the trust value is always defined as the latest experience and no memory is maintained. The algorithm behaviour is illustrated ihn Figs. 2 through 4. Each of these illustrations considers the case where experience is a measure of the dependability of a service and that dependability is measured is measured as a binary value (1 or 0).

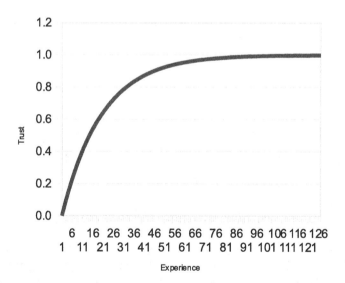

Fig. 2. Exponential Average Experiential Trust (100% Positive Experience)

When the experience E is always 1 (i.e. the service is always dependable) this trust evolves as in Fig. 2. Fig. 3 shows how this trust evolves where every

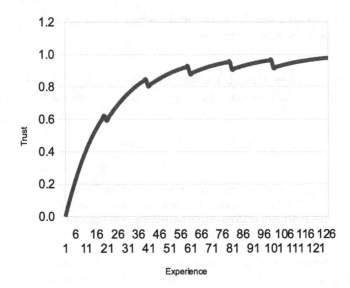

Fig. 3. Exponential Average Experiential Trust (95% Positive Experience)

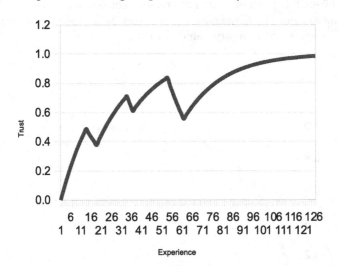

Fig. 4. Exponential Average Experiential Trust (with failure clusters)

20th experience is reported as a failure. There remains a trend of exponential recovery but the failures impact on the trust value while allowing for recovery in subsequent success reports. Similarly, Fig. 4 demonstrates how the algorithm performs when clusters of failures occur. The rate of adoption and reduction in trust in this algorithm is thevalue. In some cases it might be desirable that this value varies depending on the latest experience. For example in some cases it can be argued that while it is difficult to gain trust (slow adoption of trust) it is easy to lose it (fast reduction) based on bad experience. To model this we can vary the value as follows:

$$\text{if } (T_{i,j(n-1)} < E)$$
$$\text{then}$$
$$\alpha = 0.1$$
$$\text{else}$$
$$\alpha = 0.4$$

Exponential Average Referral Based Trust Algorithm. In the case of referral based trust the exponential average algorithm, published in [8], can be written as:

$$T_{i,j(n)} = \beta T_{i,k} T_{k,j} + (1 - \beta T_{i,k}) T_{i,j(n-1)}$$

Where:

$T_{i,j(n)}$ is the nth value of trust placed by i in j
$T_{i,k}$ is the trust i has in k's referral
$T_{k,j}$ is the trust k has in j
β is the influence that referrals have on local
trust $(0 \leq \beta \leq 1)$

The value of $T_{i,j}$ represents an indirect functional trust that i has in j. In this case i has received a recommendation of direct functional trust,$T_{k,j}$. $T_{i,k}$ is the trust that i has in k's referral. The larger the value of $T_{i,k}$, (i.e. the more i trusts k's referral), the greater the influence of ks trust in j on the newly updated value of $T_{i,j}$. Note that, if $T_{i,k} = 0$, this causes $T_{i,j}$ to be unchanged.

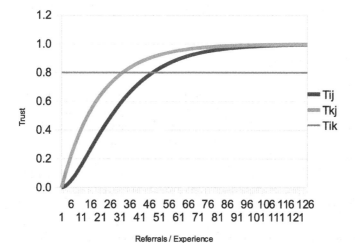

Fig. 5. Exponential Average Referral Trust (100% Positive Experience, $\beta = 0.1$)

Fig. 5 shows how this algorithm behaves for a fixed value for $T_{i,k}$ and $\beta = 0.1$. $T_{k,j}$ is a direct functional trust derived as described above and shown in Fig. 2. $T_{i,j}$ indicates how the indirect functional trust evolves. Fig. 6 shows how this is affected when the direct functional trust experiences one failure in every 20 experience reports. Fig. 7 shows what happens when the direct functional experience reports yield one failure in 20 experiences followed by a general failure after which only negative reports are received. Each of these examples consider the referral trust to be constant. The following figure (Fig. 8) show how the indirect functional trust evolves when the referral trust varies, in this case a linear increase.

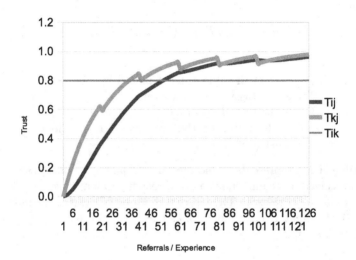

Fig. 6. Exponential Average Referral Trust (95% Positive Experience, $\beta = 0.1$)

4   Implementation

A project called *TrustFlow*[2] has been established on sourceforge to house the development of the OPAALS trust model. The project contains the source code as well as standalone examples showing how the framework is used. Sample trust algorithms are provided together with documentation on how to create and use further algorithms.

The OPAALS Trust Model for digital ecosystems is described in Section 3. The approach is to use a trust overlay network for providing a community based approach to trustworthiness based on the reputation of entities. A UML Class Diagram of this model is provided in Fig. 9. The Entity can represent a node, service, resource, a service provider or a service consumer. Each entity has a

---

[2] TrustFlow, http://trust ow.sourceforge.net

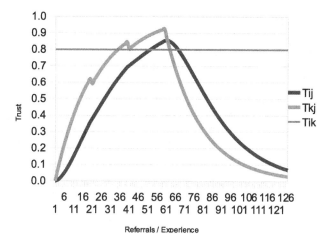

Fig. 7. Exponential Average Referral Trust (95% Positive Experience, general failure after 60 experiences, $\beta = 0.1$)

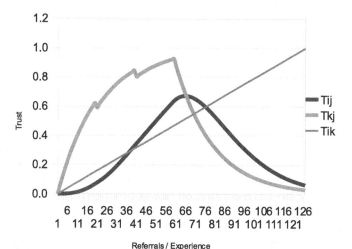

Fig. 8. Exponential Average Referral Trust (95% Positive Experience, general failure after 60 experiences, $\beta = 0.1$, linear increase in referral trust)

Trust Manager associated with it. The entities gain experience from interacting with other entities and publish reports of these experiences to the Trust Manager. Using a pre-defined context dependent algorithm the Trust Manager updates the entities' local trust and based on a policy of sharing trust information provides trust updates to other Trust Managers in the overlay network.

Entities are responsible for creating trust algorithms. Each Entity has an associated TrustManager. A TrustAlgorithm is associated on a per entity basis with an trustor and a context. Entities create these algorithms and publish them to the TrustManager. The TrustManager maintains a set of these algorithms and

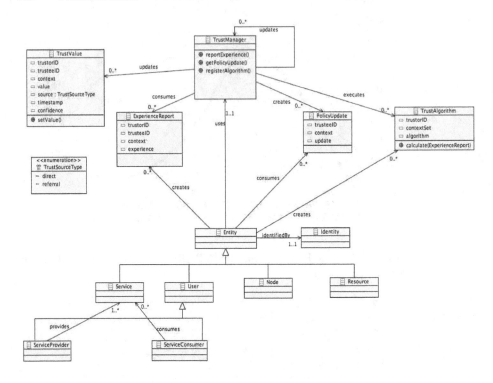

Fig. 9. UML Class Diagram of TrustManager and associated classes

when it receives an ExperienceReport, it uses the appropriate TrustAlgorithm by performing a lookup on trustor-context tuple.

The TrustManager also maintains a set of TrustValue objects which it updates after performing the algorithm on the ExperienceReport. The Entity can request PolicyUpdates from the TrustManager. These policy updates are derived from current TrustValues and are used by the entity in making choices about future interactions with other entities. The TrustValue class includes a 'source' element indicating whether the value was derived from direct experience or based on referrals. The class also has a timestamp attribute, as recently gathered trust might be of more value than older values. Also there is a confidence attribute, used to denote the confidence in this Trust Value (e.g. this value can be increased each time the Trust Value is updated, i.e. 100 updates to the Trust Value implies more confidence in the Trust Value than 10). The rest of this section describes the primary elements of this model and their role and implementation.

4.1 Trust Manager

The TrustManager acts as the core of the trust model. It has the following set of responsibilities:

1. Maintaining a set of algorithms for connected Trustor entities for specific contexts. Each context can have distinct algorithms for determining trust based on direct experience, referrals and reputation.
2. Receiving ExperienceReports from the Trustor entities and generating updated trust values according to the appropriate algorithms.
3. Providing updated TrustValues to other TrustManagers which can be used as inputs to referral or reputation algorithms. It is only appropriate that these published TrustValues have been derived from direct experience.
4. Providing PolicyUpdates to the Trustor entities based on current TrustValues derived from direct experience and referrals and reputation.

4.2 TrustValue

A TrustValue represents a single instance of a trust value a relying party has in a trusted party in a particular context. In effect each TrustValue instance is the output of a trust algorithm. The trust value itself is represented as float. Each TrustValue has a type associated with it. The type indicates the source of the data used to derive the value. Firstly, this can be direct experience, where the trust value was calculated using a direct experience report. Secondly, this can be referral, where the relying party has received a trust value from a third party and has used an algorithm to calculate a trust value based on referrals. Thirdly, the trust value can be derived from a reputation based algorithm, where a set of trust values is used as an input. Finally, the trust value can be calculated using a hybrid of direct experience, referral and reputation algorithms.

4.3 ExperienceReport

An ExperienceReport represents a unit of experience from an end user pertaining to an experience with a trustee in a context. The ExperienceReport forms the basis of all trust calculations as it is required to calculate any direct experience TrustValues. Referrals are only valid if their source is one of direct experience.

The creation of ExperienceReports is vital for the successful deployment of our trust network and two implementations are currently available. The first of these is called SimpleExperienceReport, which considers experience as a simple negative or positive value, i.e. the experience is either good or bad. The second implementation is called AccountedReport (see Figure 3.6) which makes use of data which will be created by our Accountability framework.

4.4 TrustAlgorithm

Trust algorithms have the role of taking some known information and deriving a new trust value according to some predefined logic. A discussion of suitable trust algorithms is provided earlier in this paper in section 3. An interface for TrustAlgorithm is defined with two methods. The first method, setTrustValues(String, float), allows the user to input hard values for some trusted parties.

For example, a service consumer might decide that it would like to overload the algorithm with some hardcoded values for some of the service providers it interacts with. The second method is setParameters(String, String) which is used to load parameters of the logic in the algorithm. An example of these parameters would be the rate of increase or decrease in trustworthiness for a specific context. Three further interfaces extend this base interface

4.5 Configuring the TrustManager

Each TrustManager manages a set of trust algorithms for various contexts. As each relying party has a TrustManager to manage its preferred algorithms, it is useful to devise a means of configuring the TrustManager to achieve this on a run time basis. To aid this a TrustAlgorithmConfiguration XML schema is designed. Trust Contexts are configured separately and each Context can define TrustAlgorthms for direct experience, referrals and reputation.

5 Conclusion

The work described here provides an implementation of a trust overlay network based on a model developed in the OPAALS EU FP6 Network of Excellence project. The implementation can be easily reused by third parties to introduce trust to existing software. Sample algorithms are provided in the implementation and further algorithms can be easily developed by developed and used by the framework through a plugin model.

Acknowledgment

This work is funded by the EU FP6 Network of Excellence OPAALS, http://www.opaals.org.

References

1. Aitken, D., Bligh, J., Callanan, O., Corcoran, D., Tobin, J.: Peer-to-peer technologies and protocols (2001), http://ntrg.cs.tcd.ie/undergrad/4ba2.02/p2p/
2. Dingledine, R., Freedman, M., Molnar, D.: Accountability measures for peer-to-peer systems. In: Peer-to-Peer: Harnessing the Power of Disruptive Technologies. O'Reilly, Sebastopol (2000)
3. Douceur, J.R.: The sybil attack. In: Druschel, P., Kaashoek, M.F., Rowstron, A. (eds.) IPTPS 2002. LNCS, vol. 2429, pp. 251–260. Springer, Heidelberg (2002)
4. Gambetta, D.: Can we trust trust (July 09, 1988),
   http://citeseer.ist.psu.edu/489470.html,
   http://www.sociology.ox.ac.uk/papers/gambetta213-237.pdf
5. Grandison, T.: Conceptions of trust: Definition, constructs and models. In: Song, R. (ed.) Trust in E-Services: Technologies, Practices and Challenges. IGI Global (2007)

6. Jøsang, A., Ismail, R., Boyd, C.A.: A survey of trust and reputation systems for online service provision. Decision Support Systems pp. 618–644 (March 01, 2007),
`http://eprints.qut.edu.au/archive/00007280/`,
`http://eprints.qut.edu.au/secure/00007280/01/JIB2007-DSS.pdf`
7. Li, H., Singhal, M.: Trust management in distributed systems. IEEE Computer 40(2), 45–
53 (2007), `http://doi.ieeecomputersociety.org/10.1109/MC.2007.76`
8. McGibney, J., Botvich, D.: Distributed dynamic protection of services on ad hoc and p2p
networks. In: Medhi, D., Nogueira, J.M.S., Pfeifer, T., Wu, S.F. (eds.) IPOM 2007. LNCS,
vol. 4786, pp. 95–106. Springer, Heidelberg (2007)
9. McKnight, D.H., Chervany, N.L.: The meanings of trust. MISRC Working Paper Series
96-04, Carlson School of Management, University of Minnesota (2006),
`http://misrc.umn.edu/wpaper/WorkingPapers/9604.pdf`
10. Mui, L., Mohtashemi, M., Halberstadt, A.: A computational model of trust and reputation
for E-businesses. In: HICSS. p. 188 (2002),
`http://dlib2.computer.org/conferen/hicss/1435/pdf/14350188.pdf?`
11. Ruohomaa, S., Kutvonen, L.: Trust management survey. In: Herrmann, P., Issarny, V.,
Shiu, S.C.K. (eds.) iTrust 2005. LNCS, vol. 3477, pp. 77–92. Springer, Heidelberg (2005),
`http://dx.doi.org/10.1007/11429760_6`
12. Solhaug, B., Elgesem, D., Stlen, K.: Why trust is not proportional to risk. In: ARES '07:
Proceedings of the Second International Conference on Availability, Reliability and Security, pp. 11–18. IEEE Computer Society, Washington (2007)

SPAM Detection Server Model Inspired by the *Dionaea Muscipula* Closure Mechanism: An Alternative Approach for Natural Computing Challenges

Rodrigo Arthur de Souza Pereira Lopes[1], Lia Carrari R. Lopes[2], and Pollyana Notargiacomo Mustaro[1]

[1] Universidade Presbiterian a Mackenzie, Programa de Pós-Grad. em Engenharia Elétrica
Rua Piauí, 130 – Prédio Modesto Carvalhosa, Anexo do 2 º Subsolo, São Paulo, Brazil
[2] Instituto de Pesquisa em Tecnologia e Inovação
Av. São Luis, 86 cj 192, São Paulo, Brazil
rodlopes@gmail.com, lia.carrari@ipti.org.br,
polly@mackenzie.br

Abstract. Natural computing has been an increasingly evolving field in the last few years. Focusing on the interesting behaviours offered by nature and biological processes, this work intends to apply the metaphor of the carnivorous plant *"Dionaea muscipula"* as a complementary defence system against a recurring problem regarding internet and e-mails: spam. The metaphor model presents relevant aspects for further implementation and debate.

Keywords: Natural computing; *Dionaea muscipula*; spam.

1 Introduction

Natural Computing has been an increasingly evolving field. By using nature as its inspiration, it has provided unexpectedly efficient solutions to problems that were before unsolvable. Issues concerning this area are inconstant discussion in the scientific community. In that sense, computer scientists are uniting technical and computational researches with biology specialists to achieve new sources for solutions and innovation. Originating ideas for methods and algorithms as well as applications in natural computing stimulates creativity and multidisciplinary research.

This work intends to take advantage of such characteristics to propose a parallel solution to the issue of internet spam, based on the closure mechanism of a carnivorous plant. Here, the theoretical research and models on the theme will be explored, inspiring future implementations and modifications on current systems. It may also be used on similar situations to avoid not only spam mails, but also repeated actions from bots and other sources with malicious intents towards a particular server. For this situation, one could see the OPAALS request for service building system as a potential target.

F.A. Basile Colugnati et al. (Eds.): OPAALS 2010, LNICST 67, pp. 192–198, 2010.

This work is divided in four parts. Firstly, a brief overview of the different branches and applications of Natural Computing. Secondly, an explanation about the closure mechanism of the *Dionaea Muscipula*. Next, a description of the proposed problem and its encircling subject – spam. Lastly, the application of the metaphor in the three different proposed levels, followed by the conclusion.

2 Natural Computing

Natural computing is a discipline that bases itself on the study, interpretation and reproduction of natural phenomena. This area of study produces methods, algorithms and approaches to solving problems that are computationally untreatable (problems that do not have an optimal solution under a polynomial time algorithm and belong to NP-complete or NP -hard classes [2], and proposes innovative, novel solutions to problems that already have efficient solutions [3].

It currently focuses on three different branches, each of them having their own approaches and methods for tackling different kinds of problems. Firstly, there is nature-inspired computing. This branch makes use of nature and natural phenomena such as the organization of an ant colony, the inner workings of the immunological system, or the human brain to propose algorithms capable of solving problems. As an example, today there is a rather extensive use of Neural Networks for pattern recognition and other similar problems that are not easily solved by conventional means.

Secondly, natural computing may be used to simulate the natural phenomena mentioned in the previous paragraph, like simulating the flow of a river, the collective behaviour of a school of fish, or even drawing the shape of a mountain. This area presents diverse techniques for modelling patterns and structures, such as Cellular Automata and Artificial Life.

Lastly, the third branch concerns computing with natural materials - more specifically, computing with materials other thansilicon. Moore's law predicts that the number of transistors in a chip doubles every two years [4] - this has held true until now, and is expected to hold true for at least an other two decades . Should that be the case, chips will very soon reach their peak miniaturization levels, and thus, peak processing power. From this situation the need to find new materials arises, and that is the task undertaken by this branch of natural computing.

One of the most popular methods originated by the natural computing is the Genetic Algorithm. The Genetic Algorithms are mainly used in search and optimization obtaining solutions though the evolution of populations of encoded feasible solutions (individuals) [5]. Then, the population is updated by mimicking the natural evolution mechanisms such as selection and reproduction processes. The algorithm also takes into consideration reproductive aspects, such as crossover and mutation.

Following that idea, this work proposes a nature inspired model that will focus on the first described branch, thus employing the metaphor to the proposed problem. To do so, one must understand the workings of the chosen metaphor, the closure mechanism of the *Dionaea Muscipula*.

3 Dionaea Muscipula

The carnivorous plant *Dionaea muscipula*, popularly known as Venus flytrap (Fig. 1), as been studied for over a century, mainly due to its rapid closure mechanism [6]. It is a species of carnivorous plant that captures its prey by rapidly trapping them between its leaves [7], in the order of one tenth of a second.

Fig. 1. Venus Flytrap (*Dionaea muscipula*) [8]

The plant consists of several leaves, each one divided into two parts that are held together by a midrib. These parts are called lobes, and have at least three sensitive hair triggers positioned around their surface (Fig. 2), which are covered with a red pigment which attracts insects [7].

Fig. 2. Sensitive hair triggers from the Venus Flytrap [8]

Whenever one of those hairs suffers stimulation – that is – once an insect or another creature touches it, an electrical signal is released, generating action potential that propagates across the lobes and stimulates its cells. What stimuli are generated is still under debate, but there are two generally accepted mechanisms, which may simultaneously play a part in the closure of the lobes.

The first one involves the moving of hydrogenions causing swelling by osmosis, and thus changing the shape of the lobe; the second one involves the formation of water by osmosis causing the collapse of the lobe cells [8]. But these are very simplistic descriptions and a detailed description of chemical reactions and mechanisms of this closure are not relevant to the development of this work.

Under normal temperatures, two stimuli are required to trigger the closure mechanism; however at high temperatures, close to 40° C, only one stimulus is sufficient. After trapping the prey (Fig. 3), further stimuli from the prey's thrashing about will tighten the grip of the mechanism, eventually closing the trap hermetically [6].

Fig. 3. Closed Venus Flytrap and its prey [8]

Through the abstraction of the mechanisms of the Venus Flytrap, this work constructs a metaphor, applying its methods to Internet spam e-mail identification. Considering that there are many different approaches for this problem, some ideas for spam definition and recognition will be explored next.

4 SPAM

Spam is an unsolicited message that is delivered to a user's mailbox. Generally, it contains some sort of advertising information about services or products, but it also may contain fraud attempts or harmful content, such as advertising false bank sites that contain trojans (malicious code that could steal personal information or offer access to one's computer), chain mails, or pirated software offerings [10].

Several ways of combating spam have been proposed over the years, such as filtering by content, blocking SMTP servers, greylisting and others, but this work will focus on the complementary solution (as it is not a stand -alone defence against spam, but one of many) proposed by Lieven, Scheuermann, Stini and Mauve [11].

The SMTP RFC [12] specifies that upon receiving a temporary error, a SMTP server should wait for a period of time before trying to re-send the message, and should not try to deliver other messages to that same domain during this wait time. A *spambot* (an automated spam-sending script or software) or spam agent (infected machine, script, etc.), however, focuses on throughput, meaning that it will repeatedly try to deliver a message in spite of receiving a temporary error [11].

According to [11], effective results could be obtained by observing such retry patterns. Basically, spammers are not RFC compliant and will try at every cost to deliver their messages. By observing the retry intervals, Lieven et al. was able to filter the spammers based on the frequency of their requests. One of the steps used in the proposed method was to whitelist (mark as trustworthy) all mail hosts that were the destination of mails sent by local hosts.

This potential flaw allows infected machines to utilize the whitelisted mail host to convey unwanted spam to the local users, and this is one of the places the metaphor comes in.

5 Metaphor Application

This work intends to apply the metaphor of the closure mechanism of the *Dionaea muscipula* to computational problems, mostly focusing on spam. This will be done in three different levels. Firstly, by "trapping" spammers after a predetermined number of stimuli has been reached; secondly, by effectively banning the spammers once a time interval is elapsed after the trap with repeated stimuli; and finally, by implementing an effective "heat –zone", when total numbers of stimuli are enough to compromise actual CPU load.

The potential flaw described in the previous chapter allows an infected machine to use the trusted mail hosts to deliver spam to local users. The change proposed in this work is to not whitelist those hosts immediately, but apply different types of observance on three different levels.

The *Dionaea muscipula* closes its lobes capturing its prey after two stimuli in a given interval of time. Therefore, it is proposed that not one central spam barrier (a proxy to stand before the actual SMTP server) be implemented, but several, each covering an array of different delivery addresses.

As a simplistic example of the metaphor, *spam barrier* I may cover addresses starting with A and B, while *spam barrier* II covers addresses starting with B and C, and so on. Other, better algorithms for covering e-mails, may and should be applied, executing actual load -balancing between each spam barrier but not compromising the final purpose of this spread. Each *spam barrier* keeps track of its blacklisted addresses (the ones which keep trying repeatedly to deliver the message), and this information should be synchronized at predetermined intervals of time. The separation of addresses by starting letters confers additional "proof" that the same host is sending mail to multiple addresses, further insuring that it is a suspicious address.

In the same way that the *Dionaea muscipula* closes its lobes after two stimuli in an interval of time, this system may stipulate a number of stimuli to block the e-mails from a given host for a period of time.

Each occurrence of spamming on a *spam barrier*'s record is then considered a stimulus; the synchronization process is responsible for taking into account each stimulus, and calculating the total number of stimuli received from each remote host to decide on an action course. This process is also assigned the task of keeping a reduced database of "totals", which indicate the total times a given host has been flagged as a spammer; these totals are to be considered part of the metaphor in what relates to additional stimuli, the second level of protection. Repeated stimuli after each blocking period will eventually hermetically close the trap – that is, impose a much larger period of blocking, or eventually even applying a permanent ban to such host based on its previous history.

After the blocking period, the host may be unblocked again (barring the previous issuing of a permanent ban), in the event that the occurrence was perceived as being related to an infected machine, and not a legitimate spamming source (an actual server dedicated to sending spam). This situation could be detected over a period of time, since infected machines tend to be cleansed of its viruses and Trojans periodically, so a certain host would send spam intermittently. That means it would be off the blocking period for larger timeframes than it would be blocked.

This step is related to the *Dionaea muscipula's* capacity to identify if the agent responsible for stimuli is a potential prey and not something else like a rain drop. Even though the stimulation happens and the mechanism is triggered a combination of factors (electrical and chemical reactions) is considered before "processing" the victim.

The third and final step of applying the metaphor is the temperature. The absolute amount of incoming delivery requests is considered equivalent to a certain temperature, where a small amount equates to a small temperature, and a high amount to a high temperature. Analogously to the *Dionaea muscipula*, in a high temperature environment, a reduced number of stimuli could be assigned to trigger the closure mechanism. This could help alleviate the CPU load compromised by a large amount of delivery requests coming from different directions, effectively filtering spam more immediately, and redirecting the CPU power to legitimate requests.

In the case of a more generic situation, such as a request spamming bot, the same system could be used. This could be illustrated by using the request for service system used in OPAALS, as mentioned before [1].

This system uses a requirement from the user interface to generate a query and combine atomic services in order to build a more complex variant, in which the user requirements should be met. For this task, valid combinations of resources are generated through the querying of bounded resources. Resources that are unbounded are then repeatedly queried, once for each of the previously generated valid combinations. These may generate the final combinations that compose a whole service [1].

This operation may be exploited by a user (in this case a bot) submitting requirements repeatedly and taking advantage of the processor-heavy queries to overload the server. The same mechanisms would apply in this situation.

6   Conclusion and Further Work

Albeit not technically efficient due to implementing several *spam barriers* instead of a centralized unit, this mechanism could prove useful in offering more accurate detection of spam agents. It also reduces CPU overload, consequently compromising the system less than other applications. The abstract implementation of the metaphor could be applied in three different levels, possibly indicating it as a valid alternative or complement to current spam solutions.

Implementation details and a deeper exploration of the chemical reactions as specific algorithms for detection and parameters can be studied in the future. A broader metaphor may also be suggested, considering the *Dionaea muscipula*'s biological environment applied to the system architecture.

References

1. Krause, P., Marinos, A., Moschoyiannis, S., Razavi, A., Zheng, Y., Kurtz, T., Streeter, M.P., Gabaldón, J.E.: Deliverable D3.3: Full Architecture Definition. In: OPAALS (July 2008),
   http://files.opaals.org/OPAALS/Year_2_Deliverables/WP03/D3.3.pdf
2. Ascia, G., Catania, V., Palesi, M., Parlato, A.: A Evolutionary Approach for Reducing the Energy in Address Buses. In: Proceedings of the 1st international Symposium on information and Communication Technologies. Ireland, ACM International Conference Proceeding Series, vol. 49, pp. 76–81 (September 2003)
3. de Castro, L.N.: Fundamentals of natural computing: basic concepts, algorithms, and applications. Chapman Hall, Boca Raton (2006)
4. Intel. Moore's Law: Made real by Intel innovation. Intel Corporation (February 2008),
   http://www.intel.com/technology/mooreslaw/
5. Holland, J.H.: Adaptation in Natural and Artificial Systems. Univ. of Michigan Press, Ann Arbor (1975)
6. Volkov, A.G., Adesina, T., Jovanov, E.: Closing of Venus Flytrap by Electrical Stimulation of Motor Cells. Plant Signal Behav. 2(3), 139–145 (May-June 2007),
   http://www.ncbi.nlm.nih.gov/pmc/articles/PMC2634039/
7. Forterre, Y., Skotheim, J.M., Dumais, J., Mahadevan, L.: How the Venus flytrap snaps. Letters to Nature 433, 421–425 (2005)
8. Botanical Society of America.: Venus Flytrap - Dionaea muscipula - Carnivorous Plants Online - Botanical Society of America (2009),
   http://www.botany.org/carnivorous_plants/venus_flytrap.php
9. Markin, V.S., Volkov, A.G., Jovanov, E.: Mechanism of trap closure by Dionaea muscipula Ellis. Plant Signal Behav. 3(10), 778–783 (October 2008),
   http://www.ncbi.nlm.nih.gov/pmc/articles/PMC2637513/
10. Indiana University: What is spam? Knowledge Base (2009),
    http://kb.iu.edu/data/afne.html
11. Lieven, P., Scheuermann, B., Stini, M., Mauve, M.: Filtering Spam Email Based on Retry Patterns. In: IEEE International Conference on Communications ICC '07, pp. 1515–1520 (June 2007)
12. Postel, J.B.: Simple Mail Transfer Protocol. RFC 2821,
    http://www.ietf.org/rfc/rfc2821.txt (April 2001)

Towards Autopoietic Computing

Gerard Briscoe and Paolo Dini

Department of Media and Communications
London School of Economics and Political Science
United Kingdom
{g.briscoe,p.dini}@lse.ac.uk

Abstract. A key challenge in modern computing is to develop systems that address complex, dynamic problems in a scalable and efficient way, because the increasing complexity of software makes designing and maintaining efficient and flexible systems increasingly difficult. Biological systems are thought to possess robust, scalable processing paradigms that can automatically manage complex, dynamic problem spaces, possessing several properties that may be useful in computer systems. The biological properties of self-organisation, self-replication, self-management, and scalability are addressed in an interesting way by autopoiesis, a descriptive theory of the cell founded on the concept of a system's circular organisation to define its boundary with its environment. In this paper, therefore, we review the main concepts of autopoiesis and then discuss how they could be related to fundamental concepts and theories of computation. The paper is conceptual in nature and the emphasis is on the review of other people's work in this area as part of a longer-term strategy to develop a formal theory of autopoietic computing.

Keywords: Autopoiesis, computing, computability, structural coupling.

1 Introduction

Natural systems provide unique examples of computation, in a form very different from contemporary computer architectures. Biology also demonstrates capabilities such as adaptation, self-repair and self-organisation that are becoming increasingly desirable for our technology [1]. Autopoietic systems (*auto* = self and *poiesis* = generating or producing) as a theoretical construct on the nature of living systems centre on two main notions: that of the circular organisation of metabolism and a redefinition of the systemic concepts of structure and organisation. This theoretical construct has found an important place in theoretical biology, but it could also be used as a foundation for a new type of computing. We provide a summary of *autopoietic theory*, before discussing the development of autopoietic computation [17].

2 Autopoiesis

Autopoiesis explores the consequences of the operation of a system when it possesses a circular organisation that separates it from its surroundings. The development of ideas in

F.A. Basile Colugnati et al. (Eds.): OPAALS 2010, LNICST 67, pp. 199–212, 2010.

this field inherits from a number of scientific sub-disciplines, first and foremost general system theory and 2nd-order cybernetics. These fields have been populated by researchers relying on a wide range of methodologies and epistemologies, often incompatible with one another. For example, some system theorists have followed a very mathematical approach [15], while others are more qualitative in their development [2, 35]. Maturana and Varela started from a biological perspective and followed a descriptive and qualitative approach that has had remarkably wide repercussions in many fields outside biology [7]. From the point of view of Digital Ecosystems autopoietic computing research, the different epistemologies underpinning systems theory and autopoiesis present a problem because it is difficult to develop a unified and self-consistent *quantitative* theory to be translated into computer science formalisms that can benefit from the many interesting *qualitative* insights and arguments to be found in the literature in this area.

2.1 Some Premises

The job of developing a theory of autopoietic computing, therefore, must begin by critically analysing some of the claims that have been made in the literature, assessing whether and how they might be integrated into the self-consistent body of theory that is gradually emerging around the concept of autopoietic computing. For example, the following three assertions warrant closer scrutiny:

- The core of biological phenomena arises from circular organisation, and not from information processing, reproduction, the generation of *the* correct response to an outside stimulus, or the optimisation of metabolic fluxes by minimising energy use [17].
- The essential turnover of components in autopoietic systems, as well as the destruction and creation of whole classes of molecules during ontogeny[1], means that these systems are best characterised by more than just Dynamical Systems Theory [19].
- Also, as their structure can change, without changing the organisation, autopoietic systems cannot be easily described by a *fixed*-state space [16].

These points will require more in-depth investigation than we have so far performed. In this paper we only provide a higher-level discussion of these and other points in order to begin framing the problem in a way that is compatible with the other threads of our autopoietic computing research agenda.

In our research we are also attempting to pose the problem in a sufficiently general way to account for the non-linear character of metabolic processes, for the fact that the cell is an open system, and for the mapping of these properties to automata [8]. For example, we are attempting to apply generalisations of Lie groups (e.g. Lie groupoids) to non-linear dynamical systems theory. Groupoids are like groups but with a partial function rather than a bijection from the group to itself, meaning that the

---

[1] Ontogeny (also ontogenesis or morphogenesis) (ontos present participle of *to be,* genesis *creation*) describes the origin and the development of an organism from the fertilised egg to its mature form [13].

group operation is not defined for all the elements. This results in the group properties being limited to a subset of the system, which we hope will capture the local character (in parameter space) of most biological symmetries[2]. Similarly, we regard finite-state machines as an idealised approximation that is meant to capture the computational aspects of the cell over a time scale that is small relative to the duration of the cell cycle. Under these conditions, the local dynamics (in time) can be captured by treating the cell as a closed system and neglecting the constant, but relatively small, flux of new material flowing through it as an open system. With the above provisos, we can begin to build a conceptual framework of the cell from the point of view of its organisation, which will translate into a form of computing, once a formalisation has been defined.

2.2 Circular Organisation and Identity

The notion of circular organisation, the central aspect of living systems, is the axiom in autopoiesis that:

> An autopoietic system (machine) is organised (defined as a unity) as a network of processes of production (transformation and destruction) of components which: (i) through their interactions and transformations continuously regenerate and realise the network of processes (relations) that produced them; and (ii) constitute it (the machine) as a concrete unity in space in which they (the components) exist by specifying the topological domain of its realisation as such a network. [24, 32, 22, 26].

This is a somewhat abstract and self-referential definition, which is, therefore, nicely represented by the artistic interpretation of autopoiesis in Figure 1. According to Letelier and co-workers [17], in an autopoietic system the result of any given process is the production of components that eventually would be transformed by other processes in the network into the components of the first process. This property, termed *operational closure*, is an organisational property that perfectly coexists with the fact that living systems are, from a physical point of view, energetically and materially open systems [17]. The molecules that enter the system influence the organisation of a system, which generates pathways whose operation produces molecular structures that determine the physical system and the organisation of the system [11]. So, an autopoietic system does not have direct inputs or outputs in the traditional sense, instead it constitutes a web of interdependent (and in fact for the most part non-linearly coupled) molecular processes that maintain autopoietic organisation. As the internal dynamics of an autopoietic system is self-determined, there is no need to refer any operational (or organisational) aspect to the outside. Therefore, the external environment with which it interacts does not inform, instruct or otherwise define directly the internal dynamics, instead it indirectly perturbs the dynamics of the system [18].

---

[2] For example, all living things can remain alive only over a rather narrow temperature range compared to general physical systems. Outside of this range the mechanisms of self-organisation and autopoiesis break down (i.e. the organism either freezes or disassociates).

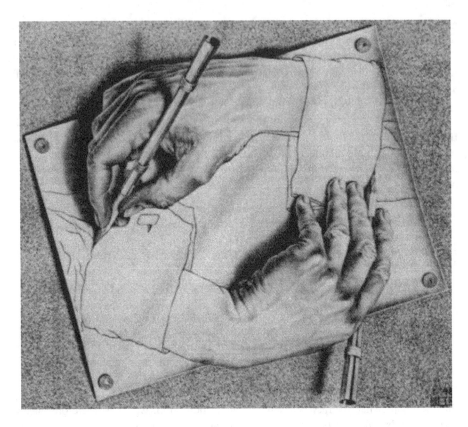

Fig. 1. Drawing Hands [10]: An artistic representation of the process of autopoiesis. The left hand draws the right hand, and the right hand draws the left hand. From two dimensions, and through an *exchange of complementary information*, emerges a third dimension. Also, while the two complementary hands can draw each other, one hand cannot draw itself.

The second clause (ii) in the axiomatic defintion of autopoiesis above implies, according to [18], that an autopoietic system has sufficiently complex dynamics to self-produce the boundaries that separate the system from the *non-system*. This apparently trivial clause has profound implications as it touches upon the need for autonomy, which implies that it is not possible to encode outside concepts or directly control its development. So, autopoietic systems would appear to be more than simple relational devices, conforming to an important topological property, in that their boundary (in the space where their components exist) is actively produced by the network of processes [18]. This property of autopoietic systems couples a purely relational property (operational closure) with a topological property, such that an autopoietic system becomes an autonomous unity, topologically and functionally segregated from its background [34]. In the realm of molecules, the coupling of these two conditions necessarily implies that the minimal metabolism compatible with autopoietic organisation must be rather more complex than the spatial coupling of a direct chemical reaction with its reverse reaction, and lends credence to the starting hypothesis that a profoundly complex mathematical structure must underpin the phenomenon. While difficulty exists in characterising precisely *complexity* in this context, it seems fairly evident that non-linearity is important. The fact that a full understanding of *non-linear behaviour* is in most cases context-dependent and very difficult to grasp motivates and justifies our continued efforts in the study of non-linear systems (e.g. [8, 14]).

In Letelier et al.'s understanding of autopoietic systems [18], it is important to distinguish between processes and components. Components interact through processes to generate other components. With this distinction, it is possible to define the organisation of a system as the pattern or configuration of processes between components that define the type of system. The structure, therefore, is the specific embodiment (implementation) of these processes into specific material (physical) entities. So, a particular structure reflects a possible instantiation or a particular organisation.

Where a mathematical or a purely qualitative approach at defining, modelling, or simply describing autopoietic systems falls short of the quantitative and predictive theory we ideally would wish to develop, we can resort to more synthetic attempts at formalisation. This methodology happens to be well-suited to the computer science perspective which tends to rely on a blend of epistemologies, as we discuss next.

3 Formalisations

Autopoiesis, as originally described [23, 24], lacks any mathematical framework. Many attempts have been made to provide one and simulate autopoiesis. The first tessellation[3] computer models [32, 36], redone in Swarm[4] [25], were a direct translation of a minimal autopoietic system into a small bi-dimensional lattice. Indicational Calculus [5] was used [33] to model autonomous systems, but progress with this approach has been limited. Other mathematical formulations have included the use of differential equations to model feedback [20]. Another attempt, a pure algebraic approach, used the theory of categories to understand systems operating with *operational closure* [18].

However, all these formalisations lack one aspect or another [27], not least because many do not manage the non-Turing computability aspect prevalent in autopoietic systems, as discussed further below. So, none of these quantitative, or semi-quantitative models have generated clear, satisfactory results.

3.1 Structural Coupling

Letelier et al. [18] suppose that autopoietic systems do not simply behave or react passively in an environment that is provided. Instead, a central aspect of autopoiesis generally has been a mechanism of structural coupling by which the living system and its medium determine, in a mutual way and resulting from a history-dependent

---

[3] A tessellation or tiling of the plane is a collection of plane figures that fills the plane with no overlaps and no gaps [4].

[4] Swarm is a platform for agent-based models (www.swarm.org)

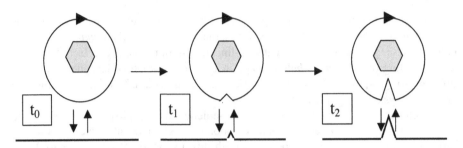

Fig. 2. Structural Coupling [17]: The autopoietic system, represented by a circle, defined by its structure and its organisation, initially confronts a medium without organised *objects*. As recurrent interactions between the medium and the system are stabilised, at t_1, an *object* begins to be configured. The *object* is made of two complementary parts. One part exists in the medium, and the other exists as a change in the autopoietic system structure.

process, some of their properties [18]. Figure 2 depicts the basis of this mechanism of structural coupling [17]. Summarised from [17], the autopoietic system, represented by a circle, and defined by its structure and its organisation (hatched area), initially confronts a medium without organised *objects* (at t_0). As recurrent interactions (represented by the arrows) between the medium and the system are stabilised, at t_1, an *object* (represented by the triangle) begins to be configured. The *object* is made of two complementary parts. One part exists in the medium, and the other exists as a change in the autopoietic system structure. Finally, at t_2 the *object* is totally configured. So, there is a change in structure, but not in organisation as the hatched area remains unchanged, which could be very important. From a computational perspective the important aspect is the existence of *objects* defined by spatio-temporal correlations, thus the change in the autopoietic system structure also contains these spatio-temporal correlations. Such spatio-temporal correlations take the form of complementary changes in interacting entities, which are called *congruences*[5]. So congruences are a consequence of structural coupling, meaning that the temporal changes in the structure of a system can potentially manage possible future changes in the environment, because future interactions with other spatio-temporal objects in the environment may be *congruent* with the then existing structure of the autopoietic system.

Letelier et al. claim [17] that in effect, as the organisation of a system is potentially maintained invariant, its structure can change in many dimensions that do not affect the organisation, for example its operational closure. However, such changes are not random, being neither an accommodation or adaptation to outside features (classical adaptationism), nor the result of the deployment of internal plans embodied in the structure of the autopoietic system (vitalism) [17]. Instead, the changes produced by structural coupling require the existence of recurrent interactions as well as a necessary level of plasticity (ability to change the structure) in the autopoietic system and its medium. During the ontogeny of the system (or the phylogeny[6] of the lineage) a

---

[5] A congruence, as opposed to equivalence or approximation, is a relation which implies a kind of equivalence, though not complete equivalence.

[6] Phylogenetics is the study of evolutionary relatedness among various groups of organisms (e.g., species, populations).

congruence between the system and its medium can be selected or stabilised, so that the medium gradually becomes the environment and, for external observers unaware of the buildup of the relationship, the organism appears to become adapted to some characteristics of the medium [17]. Therefore, Letelier et al. conclude that autopoietic systems do not only adapt to the defined ecological niche, the standard notion of evolution by adaptation, but also partly create their environment through the systematic production of congruences [17].

Letelier et al. go on to say that the structural change inside the autopoietic system is caused by the recurrent external trigger (perturbation – an indirect input) as well as its own internal, circular dynamics [17]. So, a given external perturbation will not induce an internal structural change that can be viewed as its representation or internal model. The relation between the internal structural change and the external trigger is one of correlation or congruence instead of identity or recognition. The external perturbation does not induce a one-to-one relational model in the autopoietic entity, instead a congruence is constructed to represent the complementary changes that occur [17].

In another paper, Letelier et al. argue that these concepts of autopoiesis could be the basis for a new type of autopoietic computation, which would probably not be program-based [19]. As autopoietic systems do not have *direct* inputs or outputs, only a circular dynamic which is perturbed (indirect inputs) and not directly defined by external agents, it would be difficult to encode outside concepts into autopoietic states and control a trajectory of states (like Turing machines). So, an external observer would have to define a computation for an autopoietic system as the particular ontogeny for that system. During the ontogeny of that system, a relation between it and its medium would be selected, and would eventually stabilise. This relation has meaning for the autopoietic system, which is structurally coupled to its medium, but not obviously for external observers [19]. Therefore, external observers (eventually users), if they should wish to use autopoietic systems to perform computations, would need a procedure before-hand to attach their desired meaning (computation) to particular moments and properties of an ontogeny for the system [17].

3.2 Computability

Autopoietic systems are intrinsically different from Turing machines, the structure of which is shown in Figure 3. They cannot be simulated by Turing machines as they are not Turing-computable, for the following reason. The self-referential nature of circularity that characterises autopoietic systems leads to the dynamic creation of an unpredictable number of states. According to [29, 30, 18], the dynamic creation of an unpredictable number of new states implies that no upper bound can be placed on the number of states required. As the Church definition of computability assumes that the basic operations of a system must be finite, e.g. recursive, the Church-Turing thesis[7] cannot be applied. Hence, autopoietic systems are non-Turing-computable This is difficult to prove using only the

---

[7] The Church-Turing thesis is a combined hypothesis about effectively calculable (computable) functions by recursion, by a mechanical device equivalent to a Turing machine.

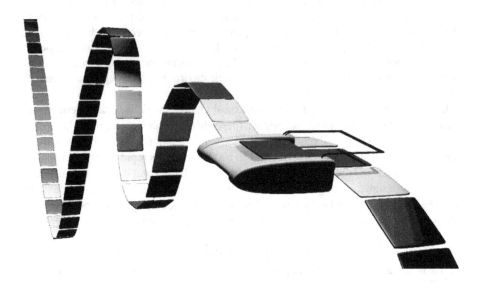

Fig. 3. Turing Machine [31]: A basic abstract symbol-manipulating device which, despite its simplicity, can be adapted to simulate the logic of any computer algorithm. The Turing Machine mathematically models a machine that mechanically operates on a tape for which symbols are written which it can read and write one at a time using a tape head. Operation is fully determined by a finite set of pre-determined elementary instructions contained within the Turing machine.

elements of autopoietic theory [23, 22], but it is claimed [18] to flow trivially from the inclusion of autopoietic systems in (M,R) systems.[8] The non-computability of autopoietic systems [16, 3] *suggests* (yet to be proven) that some intrinsic and fundamental part of their behaviour escapes our standard analysis based on phase states and/or evolution equations.

Letelier et at. explain that the non-computability of autopoietic systems by Turing machines has many important theoretical consequences [18, 17]. First, it limits the validity of mimesis (i.e. simulation) as a means to understand living systems, showing that the phenomenology that arises from the circularity of metabolism cannot be simulated with current computer architectures, those based on the Von-Neumann implementation of Turing machines [17]. Using different approaches this result has been hinted at on at least two occasions in the last decade [16, 3]. Using formal arguments, the impossibility of designing a living system without a real metabolism has been argued for [3].

A development for the concept of living systems, called *Component-systems*, has also been developed, for which it has been shown that equations of state, equations of motion, or evolution equations cannot be applied [16]. It is controversially argued [16] that component-systems are fundamentally uncomputable, because it is a pool of components

---

[8] A relational model, in which M stands for metabolism and R stands for Repair components or subsystem, such as for example active RNA molecules [29, 30].

that act on each other and combine with each other to produce new components. However Goertzel's self generating systems [12] are a model that arose as an explicitly computable analogue of component-systems. This non-computability of autopoietic systems would not seem to be supported by past simulation results involving tessellation automata [32]. However, new versions of this simulation show that the original report of computational autopoiesis was flawed, as it used a non-documented feature involving chain-based bond inhibition [25]. So, the closure exhibited by tessellation automata is not a consequence of the *network* of simulated processes, but rather an artefact of coding procedures (i.e. it did what is was programmed to do, produce the desired output, but not through the desired procedural methodology). So, the failure of closure in these computational models was not a conceptual failure of autopoiesis, but a reflection of the non-computability of autopoietic systems [18].

The non-computability of autopoietic systems could initially appear an overly strong claim or result [18, 17], but even in the restricted field of pure mathematics it has been possible to prove the existence of simple, but non-computable functions like the busy beaver problem[9] [28]. So, Turing non-computability is a property that does not require the complexities of circular organisation to be apparent, as it is already demonstrable in simpler systems or problems. The inapplicability of the Turing-Church thesis for autopoietic systems also opens some important new questions [17]. The first is to analyse whether an autopoietic system can implement a Turing machine. The second considers whether some Turing non-computable problems, like the busy beaver, can be computed by autopoietic systems [18]. The start to tackling these problems will require the use of *category theory* to represent autopoietic systems, so as to be able to understand and manipulate the operational closure of metabolism.

4   Autopoietic Computation

In this section we continue to follow mainly the concepts developed by Leterier and co-workers, who build on the concepts presented so far to advance a view of how autopoietic computation could be achieved. Although the discussion is qualitative and somewhat speculative, it is rather carefully put together and consistent with the concepts explained in the foregoing, thereby providing a useful conceptual framework for the continued development of autopoietic computing.

To consider the use of autopoietic systems as computational tools, Letelier et al. argue that it would be necessary to redefine the process of computation that we usually identify with the operation of Turing machines, which compute by performing symbol manipulation [17]. The symbol processing algorithm embodied in a Turing machine does not concern itself with the semantic content of those symbols, but only

---

[9] In computability theory, a busy beaver (from the colloquial expression for *industrious person*) is a Turing machine which, when started on an empty tape, runs as long as possible, but eventually halts. This machine attains the limits on the amount of time and space that a halting Turing machine of the same class can consume. The busy beaver function quantifies those limits and is an example of a non-computable function. In fact, it can be shown to grow faster than any computable function [28].

deals with the syntactical rules of symbol transformation. The semantics is left to the user who must map the string symbols to content. Autopoietic systems are significantly different to Turing devices, with their structure being variable, and hence lacking a phase-space in the traditional physics sense[10] [17], because of the intrinsic non-determinism of biological systems brought by changes in the phase space during ontogenesis (morphogenesis) [18].

So, we would require a new definition of *computing* that is not dependent on symbol processing, because the unfamiliar computing aspects of autopoietic systems arise primarily from the internal reference frame of the controller [19]. The control is through the logic (which could perhaps be modelled as symbols) of the maintenance of circular organisation [6] in the presence of structural coupling. The basic mechanism by which autopoietic systems would be able to compute is the history-dependent change in their structure, which is triggered by recurrent temporal correlations. This change is the consequence of the recurrent interactions between the autopoietic system and its medium. So, because of this relationship, every autopoietic system would transform the original medium of its deployment into an *environment* capable of computation.

Autopoietic processing would be rather different to reading a binary sequence from a uni-dimensional Turing tape. It would require a *mature* autopoietic system, where the necessary temporal correlations had been constructed, such that the structure of a system would manifest correlations to the environment. So, to compute with an autopoietic system, we would require [19] a history-dependent link between the autopoietic system and the medium (or space) in which we wish to perform computations. Therefore, an autopoietic system must be introduced in such a medium and a congruent lineage must be established via structural coupling, which would change the medium into an environment for this specific autopoietic system [19].

Such a combination of system and environment would operate such that as time passes the structural changes induced by structural coupling would become more and more ingrained in the structure of the autopoietic system, which would capture more and more temporal (and spatial) correlations from its environment [19]. This stage would be equivalent to the programming of a Turing machine, and once the autopoietic system is full of induced correlations it could be used for computation. So, ingrained in an autopoietic system, because of the structural coupling, would be the congruent dynamic of the environment. So, this procedure has the important advantage that programming is endogenously produced, by living inside the medium and forming a stable lineage. However, it may not be possible to specify exactly the type of computations we desire [17]. The intrinsic autonomy of autopoietic systems makes it impossible to force a system that has created its own relation with the environment to capture the temporal correlations that are important for us, as our relation (perspective) to the environment is different to that of the system [17].

A proposed solution [17] to this conundrum can be achieved by the simple expedient of brute force. So, instead of establishing a single lineage, we could simultaneously use many different initial autopoietic systems, ideally with rather different structures. Each

---

[10] Phase space is a concept from physics, where each *state* is given by the position and the velocity of the particle or system (one can talk about a system if there is more than one particle, so that the phase space becomes higher-dimensional).

lineage would transform the single medium (i.e. the space recognised by the observer/autopoietic-programmer) into its environment, and so a wider range of temporal correlations could be established and some of them would be useful in performing computations (user-desired processing) [17].

5 Discussion

We believe that the main contribution of the notion of autopoietic systems, in the endeavour of Autopoietic Computing, lies in constructing the notion of temporally correlated structural coupling in computing. Structural coupling is a mechanism by which a lineage of autopoietic systems can change their structure (i.e. components and processes), such that it can become progressively more congruent with the recurrent perturbations that arise in the medium. Taking inspiration from the architecture of biological systems like the cell, it is plausible to postulate, therefore, that new hardware architectures (i.e. other than von Neumann's) will be necessary to achieve this kind of computation.

On the other hand, the fundamental concept at the basis of such a postulate is the recognition that independent computing entities need to be set up in such a way that they can develop congruences through structural coupling. But such a require- ment is precisely the same as what is at the basis of interaction computing [8]. Interaction computing depends on achieving the emergence of ordered and meaning- ful behaviour from the interactions between autonomous entities, that we currently treat as finite-state automata. The spontaneous, happenstance, and bottom-up nature of biological interactions is emulated in interaction computing by allowing state transitions in any one automaton to trigger state transitions in the automata it is coupled to. Therefore, interaction computing achieves at a higher level of abstrac- tion the balance between autonomy and interdependence required by structural coupling that biological systems achieve *in hardware*. In other words, once the various processes have been *virtualised*, it does not matter whether they are being executed on the same processor, on a different processor, or on different machines across the Internet. What matters is to encode appropriate operational semantics in the dynamical behaviour of the various automata, so that the triggers caused by each on the others it is structurally coupled to can enable suitable congruences to emerge across the network of coupled automata.

In this manner we would achieve a kind of sub-symbolic communication between an autopoietic computing entity and its environment (composed of other autopoietic computing entities) that is inscribed in system *behaviour* at different levels of abstrac- tion. We believe this is how Maturana and Varela's term *languaging* can be interpreted: as a process of structural coupling that is common to low-level biological systems as well as high-level cognitive systems[11] and that is founded on a radically relativist construction of meaning that is compatible with a conception of computation based on the interaction between essentially reactive systems.

---

[11] See for instance [21]: p. 180 for low-level interactions, p. 186 for social insects, and p. 234 for languaging as a higher-level socio-cognitive phenomenon among humans; Mingers ([26]: 110) makes a similar point.

This perspective, which emphasises a conception of the construction of *order* that is fundamentally relativist and *binary* (i.e. based on interactions), does not preclude the same interactions from acting as a medium or *carrier* of higher-level information that could be symbolic and *absolute* (i.e. relative to a global rather than a local context). This claim will need to be proven (or disproven). Assuming it holds, it opens the possibility of developing an interaction protocol that acts as a multi-level language system that is self-consistent at different levels of abstraction. *Self-consistency* here could refer to some yet-to-be-precisely-defined correspondence between semantics and syntax. We already know that semantics depends on context, in other words construction of meaning is ultimately relative, whether the meaning construction process is locally or globally understood. Because, furthermore, we already know that low-level biological behaviour harbours non-trivial and computationally relevant algebraic structures such as simple non-abelian groups [9, 8], this perspective could imply a *structural* correspondence (in the algebraic sense) between the structure of reactive patterns of behaviour and higher-abstraction constructs. We cannot help noticing that such an abstraction hierarchy appears to reflect the correspondence between the symmetries of sets acted upon by permutation groups and the automorphisms of those same groups acting on themselves, exemplified most clearly and at the most elementary level by Cayley's famous theorem for finite groups.

6 Conclusion

In this paper we have attempted to provide a sufficiently complete discussion of autopoiesis from the computational point of view to be able to offer some proposals for what *autopoietic computing* could actually mean. We relied heavily on the interesting work in this area of Letelier and co-workers, and finally proposed a picture of autopoietic computing that builds on the emerging algebraic theory of interaction computing. No hard conclusions can be drawn yet, because many if not most of the claims made are still unproven. This is partly a consequence of the mixture of epistemologies that appear to be attracted by the field of autopoiesis and that make it such a difficult area to work in, and partly due to the fact that the formalisation of interaction computing is still in its infancy. We do believe, however, that the research questions we have raised here are in themselves an interesting contribution to the development of a comprehensive vision of autopoietic computing.

Acknowledgements

This work was supported by the EU-funded Open Philosophies for Associative Autopoietic Digital Ecosystems (OPAALS) Network of Excellence (NoE), Contract No. FP6/IST-034824.

References

1. Bentley, P.: Systemic computation: A model of interacting systems with natural characteristics. International Journal of Parallel, Emergent and Distributed Systems 22(2), 103–121 (2007)

2. von Bertalanffy, L.: General System Theory: Foundations, Developments, Applications. George Braziller, New York (1969)
3. Boden, M.: Is metabolism necessary? The British Journal for the Philosophy of Science 50(2), 231–248 (1999)
4. Bowyer, A.: Computing dirichlet tessellations*. The Computer Journal 24(2), 162166 (1981)
5. Brown, G.: Laws of form. Allen & Unwin, London (1969)
6. Casti, J.: The simply complex:" biologizing" control theory: How to make a control system come alive, the big problems of control theory. Complexity 7(4), 10–12 (2002)
7. Dini, P., Munro, A.J., Iqani, M., Zeller, F., Moschoyiannis, S., Gabaldon, J., Nykanen, O.: D1.2: Foundations of the Theory of Associative Autopoi-etic Digital Ecosystems: Part 1. In: OPAALS Deliverable, European Commission (2008), http://files.opaals.org/OPAALS/Year_2_Deliverables/WP01/
8. Dini, P., Schreckling, D.: A Research Framework for Interaction Computing. In: Proceedings of the 3rd OPAALS International Conference, Aracaju, Sergipe, Brazil, March 22-23 (2010)
9. Egri-Nagy, A., Dini, P., Nehaniv, C.L., Schilstra, M.J.: Transformation Semigroups as Constructive Dynamical Spaces. In: Proceedings of the 3rd OPAALS International Conference, Aracaju, Sergipe, Brazil, March 22-23 (2010)
10. Esher, M.C.: Drawing hands (1989)
11. Fleischaker, G.: Origins of life: an operational definition. Origins of Life and Evolution of Biospheres 20(2), 127–137 (1990)
12. Goertzel, B.: Self-reference and complexity. Component-systems and self-generating systems in biology and cognitive science. Evolution and Cognition 2, 257–283 (1993)
13. Gould, S.: Ontogeny and phylogeny. Belknap press (1977)
14. Horvath, G., Dini, P.: Lie Group Analysis of p53-mdm3 Pathway. In: Proceedings of the 3rd OPAALS International Conference, Aracaju, Sergipe, Brazil, March 22-23 (2010)
15. Kalman, R.E., Falb, P.L., Arbib, M.A.: Topics in Mathematical System Theory. McGraw-Hill, New York (1969)
16. Kampis, G., Kampis, G.: Self-modifying systems in biology and cognitive science. Pergamon Press, Oxford (1991)
17. Letelier, J., Marin, G., Mpodozis, J.: Computing With Autopoietic Systems. Soft Computing and Industry Applications. Springer, London (2002)
18. Letelier, J., Marin, G., Mpodozis, J.: Autopoietic and (m, r) systems. Journal of theoretical biology 222(2), 261–272 (2003)
19. Letelier, J., Marin, G., Mpodozis, J., Soto-Andrade, J.: Anticipatory Computing with Autopoietic and (M R) System. Soft Computing Systems: Design, Management and Applications, 205 (2002)
20. Limone, A.: L'autopoiese dans les organisations. Doctorat de Troisieme cycle. Universit (1977)
21. Maturana, H., Varela, F.: The Tree of Knowledge. The Biological Roots of Human Understanding. Shambhala, Boston and London (1998)
22. Maturana, H., Varela, F.: Autopoietic systems. Urbana, Biological Computer Laboratory University of Illinois (1975)
23. Maturana, H., Varela, F.: De mdquinas y seros vivos. Santiago, Chile: Editorial Universitaria. Translated as Autopoiesis and Cognition: The Realization of the Living. D. Reidel, Dordrecht (1972/1980)
24. Maturana, H., Varela, F.: Autopoiesis and cognition: The realization of the living. Springer, Heidelberg (1980)

25. McMullin, B., Varela, F.: Rediscovering computational autopoiesis. In: Fourth European Conference on Artificial Life, pp. 38–47 (1997)
26. Mingers, J.: Self-producing systems: Implications and applications of autopoiesis. Springer, Heidelberg (1995)
27. Nomura, T.: Category theoretical distinction between autopoiesis and (M,R) systems. In: Almeida e Costa, F., Rocha, L.M., Costa, E., Harvey, I., Coutinho, A. (eds.) ECAL 2007. LNCS (LNAI), vol. 4648, pp. 465–474. Springer, Heidelberg (2007)
28. Rado, T.: On non-computable functions. Bell System Technical Journal 41(3), 877–884 (1962)
29. Rosen, R.: Abstract biological systems as sequential machines: Iii. some algebraic aspects. Bulletin of Mathematical Biology 28(2), 141–148 (1966)
30. Rosen, R.: Life Itself. Columbia University Press, New York (1991)
31. Schadel: Maquina (2005)
32. Varela, F., Maturana, H., Uribe, R.: Autopoiesis: the organization of living systems, its characterization and a model. Currents in modern biology 5(4), 187 (1974)
33. Varela, F.: Principles of biological autonomy. North-Holland, New york (1979)
34. Weber, A.: The Surplus of Meaning. Biosemiotic aspects in Francisco J. Varelas philosophy of cognition. Cybernetics &# 38; Human Knowing 9(2), 11–29 (2002)
35. Weinberg, G.M.: An Introduction to General Systems Thinking, Silver Anniversary edn. Dorset House, New York (2001)
36. Zeleny, M.: Autopoiesis: A theory of living organization, New York (1980)

Flypeer: A JXTA Implementation of DE Transactions

Amir Reza Razavi, Paulo R.C. Siqueira, Fabio K. Serra, and Paul J. Krause

[1] Department of Computing, School of Electronics and Physical Sciences,
University of Surrey, Guildford, Surrey, GU2 7XH, UK
[2] Av. São Luiz, 86, cj 192, República, São Paulo - SP, Brasil
(a.razavi,p.krause)@surrey.ac.uk, (Paulo,Fabio)@ipti.org.br

Abstract. This paper introduces the Flypeer framework for transaction model in digital ecosystems. The framework tries to provide a fully distributed environment, which executes different type of order service compositions. The proposed framework is considered at the deployment level of SOA, rather than the realisation level, and is targeted to business transactions between collaborating SMEs as it respects the loose-coupling of the underlying services.

Introduction

The Digital Ecosystem is concerned with building an open source environment, through which businesses, in particular small to medium enterprises (SMEs), can interact within a reliable environment. The aim is to provide access to arbitrary services that help compose together to meet particular needs of the various partners. This collaborative software environment is being targeted primarily towards SMEs, who will be able to concatenate their offered services within service chains formulated on a digital ecosystem [1], [2].

Within a digital ecosystem a number of long-running multi-service transactions are expected to take place and therefore, of interest is the issue of providing support for long-term business transactions involving open communities of SMEs. A business transaction in this paradigm can be either a simple usage of a web service or a mixture of different levels of composition of several services from various service providers. We will argue that the current transaction and business coordination frameworks can lead to issues with tight coupling and violation of local autonomy for the participating SMEs.

Web services are the primary example of the Service Oriented Computing (SOC) paradigm [3]. The goal of SOC is to enable applications from different providers to be offered as services that can be used, composed, and coordinated in a loosely coupled manner. Web services in fact provide a realisation of SOC. Although recent years have seen significant growth in the use of Web Services, there are some very significant technological constraints that are stopping their full potential from being realised within a digital ecosystem.

The conventional definition of a transaction [4] is based on ACID properties. However, as it has been indicated [5], in advanced applications these properties can present unacceptable limitations and reduce performance [6].

F.A. Basile Colugnati et al. (Eds.): OPAALS 2010, LNICST 67, pp. 213–223, 2010.
© Institute for Computer Sciences, Social Informatics and Telecommunications Engineering 2010

The lack of consideration for the primary characteristics of SOC (such as loose-coupling) [7] or ignoring some important business requirements (such as partial results) [8], are important objections to the proposed transaction models, which also suffer from unnecessary complications of implementing a consistency model on top of the service-realisation boundary. In addition, the feasibility of a heavy coordinator framework causes some transaction models to use a centralised (or limited decentralised) coordination model [9].

The JXTA technology is a set of open protocols that enable any device on the network to communicate and collaborate in a P2P fashion, creating thereby a virtual network. It provides the peers, on this network, to share and discover resources, services and exchange messages with other peers, even if some of these peers are behind firewalls and network address translations (NATs). Thus, JXTA is an important key, in the Flypeer implementation, in the sense of give an effort only in the Transaction Model implementation.

In the first section, we have provided an overview for transactions in Digital Ecosystems; section 3 has explained the role of service oriented infrastructure of digital ecosystems for transaction model. We introduce the Flypeer framework in section 4 and section 5 provides the actual execution of transactions. Section 6 compares the model with similar models and the last section provides conclusion and clarifies future work and roadmap for the project.

Transactions in Digital Ecosystems

As we have already indicated, the very nature of business – as opposed to database – transactions, opens up a different angle from which to view transactions. For example, the specification of a transaction may involve a number of required services, from different providers, and allow it to be completed over a period of minutes, hours, or even days – hence, the term long-lived or long-running transaction. Indeed, a wide range of B2B transactions (business activities [10], [8]) have a long execution period. A business transaction between SMEs in a Digital Ecosystem can be either a simple usage of a web service or a mixture of different levels of composition of several services from various service providers.

It is important to stress that the long-term nature of execution frames the concept of a transaction in a digital ecosystem, since most usage scenarios involve long-running activities. In such cases, it is impractical, and in fact undesirable, to maintain full ACID properties throughout the lifetime of a long-running transaction. In particular, as we discussed in the Introduction, Atomicity and Isolation are questionable.

Within a digital ecosystem, a large number of distributed long-running transactions take place, each comprising an aggregation of activities which in turn involve the execution of the underlying service compositions. In such a highly dynamic environment there is an increased likelihood that some transaction (or one of its internal activities) will fail. This may be due to a variety of reasons (platform failure, service abort, temporary unavailability of a service, etc.) including the vulnerability of the network infrastructure itself (platform disconnection, traffic bottleneck, nodes joining or leaving the network, etc.).

The standard practice in the event of a failure is to trigger compensating actions that will effectively 'undo' the effects of the transaction – those effects visible before failure occurred. The objective is to bring the system to a state that is an acceptable approximation of the state it was before the transaction started. However, this is not a trivial task. Recovering the system in the event of a failure of a transaction (or an activity inside a transaction) needs to be done in a way that takes into account the dependencies both inside (to ensure all dependent activities are undone) but also outside the transaction (to ensure that any dependent activities of other transactions are also undone).

Further, and when considering SOA as the enabling technology for open e-business transactions, the recovery and compensation mechanism must respect the loosely- coupled nature of the connections, since interfering with the local state of the underlying service executions violates the primary requirement of SOA. In addition, access to the local state may not even be possible in a business environment with SMEs as service providers. This is an issue that has been largely ignored by current implementations of transaction models such as Web Services Transactions (WS-Tx) [11], [12] and Business Transaction Protocol (BTP) [13].

The abortion of a transaction, even if it is successfully recovered and compensated for, can be very costly in the business environment. Rolling back the whole system in the event of a failure may lead to chains of compensating activities, which are time-consuming and impact on network traffic as well as deteriorate the performance. For this reason it is important to build into the system capability or flexibility to deal with failure. In other words, it is key to design for failure by adding diversity into the system and allowing for alternative scenarios. The idea is to get some leverage in avoiding the abortion of a whole transaction (and all other dependent transactions) by means of allowing alternative paths of execution in cases where the path chosen originally encountered a failure.

It is also important to be able to preserve as much progress-to-date as possible. If an activity (sub-transaction) of a transaction fails, it is essential to undo (roll back) only those activities that have used results of this activity, i.e. are dependent on it. It is highly desirable to avoid rollback of other activities that have produced results (committed) and are not dependent on the failed activity. These are often referred to as omitted results and do no need to be undone as that would mean they will need to be re-done (re-started, re-calculated, re-computed) once the transaction is restarted (after abortion and recovery). Addressing omitted results can have significant benefits for SMEs in digital ecosystems in terms of saving valuable time and resources.

In earlier work, we have performed an extensive review of Transaction models, such as WS-Tx and BTP, that have been designed with web services in mind, and are currently widely used in practice [14]. Apart from certain issues regarding their coordination mechanism, which is geared towards centralised control, these models do not support partial results, do not provide capability for forward recovery, and there is no provision for covering omitted results.

Transactions and Service Composition

Based on the specification advocated in [15],[16],[17] service composition can be considered along the following dimensions: data, process, security, protocol. In this paper we are concerned with providing P2P network support for distributed transactions and hence we will be concerned with the aspects of data and process composition. In general, security and protocol compositions are usually addressed on top of the transactional layer.

In particular, process-oriented service composition is concerned with the following aspects:

Order: indicates whether the composition of services is serial or parallel.

Dependency: indicates whether there is any data or function dependency among the composed services.

Alternative service execution: indicates whether there is any alternative service in the service composition that can be invoked - alternative services can be tried in either a sequential or a parallel manner.

Following [15] these aspects can be seen within different types of service composition as follows.

Data-oriented service composition: The data generated at the service realisation level are released in terms of different data-objects. In this service composition, these data can be shared and manipulated between participants of a single transaction or, where partial results are concerned, be shared by participants of other transactions.

Sequential process-oriented service composition: This type of service composition invokes services sequentially. The execution of a component service is dependent on its previous service. These sequential dependencies can be based on commitment in which case we talk about Sequential with commitment dependency (SCD) where one cannot begin unless the previous service commits, or data in which case we talk about Sequential with data dependency (SDD) where one service relies on other service's outputs as its inputs.

Parallel process-oriented service composition: In this service composition, all the services can be executed in parallel. There may be data dependencies between them in which case we talk about Parallel with data dependency (PDD) or there may be differences in how and when the services can be committed (depending on some condition) in which case we talk about Parallel with commit dependency (PCD). When there are no dependencies between the parallel services we talk about Parallel without dependency (PND).

Alternative service composition: This type of service composition indicates that there are alternative services to be deployed and one of them is necessary. They are categorised to two different types: Sequential alternative composition (SAt) where there is an ordering for deployment of these services, and Parallel alternative composition (PAt), where there is no ordering (preference) between them and deployment of either service can satisfy the composition.

Generally, one or more service compositions can satisfy the user request. It can be seen that due to order and data in service composition, there can be increased complexity in composing services especially when transactions require a number of different services from different networked organisations. This means that there is a need for a context and data consistency model (at the deployment level) which can provide the correctness of the results.

Implementation Framework

The Flypeer project brings a few things to the table: web services technologies, long running transaction support and peer-to-peer. To avoid having to re-inventing a lot of wheels, JXTA is used to provide the basis of the peer-to-peer architecture. With this in mind, the main goal of Flypeer is the supporting Long Running Transactions.

To achieve that, Flypeer abstract the communication details of JXTA from the service developer. Flypeer exposes an interface, *FlypeerBusinessService*, which the developer implements. This interface defines the features available to service developers and, more importantly, allow the framework to correctly plug it into the execution of a composition – more on that in the next section.

Deploying services in a peer-to-peer node, in Flypeer, is just a matter of creating a jar file with the proper service implementation classes and a configuration file, which follows the mechanism defined in the *ServiceLoader* class available in Java, version 6 and above.

The project also tries to be as close as possible to WS-* standards. This means, for example, that we use WSDL (actually a simplified version of it) for service description, and should support SOAP calls from the "external" world in a not distant future version. More than that, with each new version of the framework we are trying to get closer and closer to commonly used standards – allowing for new developers to use the project with a lighter learning curve.

Modelling and Executing a Transaction

In order to model a long running transaction, a service composition has to be created with the required parameters of its services and then the transaction can be executed. A class diagram of this process can be found in the appendix A.

Regarding the service composition, Flypeer supports three types, SCD, PCD and Alternative Service compositions (recall section 3). All of these can be mixed to create from simple to complex compositions, using an xml file or programmatically, in what we call Transaction Context.

The *TransactionFlow* is the class which represents the Transaction Context and where is created the transaction flow. A *GenericFlow* implementation must be added into it to begin the composition. This can be a *SequentialFlow*, a *ParallelFlow* or an *AlternativeFlow* object. A *GenericFlow* object can receive Service or/and another *GenericFlow* objects.

The *Service* class represents the service which will be called, and its dependencies. The dependencies are mapped using the *addDependency* method, passing the input name only, if it has no dependency with another service, or passing also the dependent service object that will receive the output of this service as its input.

After composing the transaction, all of the required service parameters must be added into the composition. This is done using a Map where the key is the input name and the value is the parameter value which has to be a *Serializable* object. These parameters are validated automatically, before the transaction is started, according to the input mapping described in the service WSDLs.

After that, the transaction can finally be started using a *TransactionInitiator* object, which is obtained through the *CommandSimplePeer* object. Then, all service coordinators of this composition are notified and receive the Transaction Flow and the parameters definition. The transaction starts with the initiator's coordinator calling the first service's one. After that, each service coordinator calls the next one, until the last service coordinator on the transaction is reached.

In the end of the transaction, the initiator receives the result in a callback object defined prior to transaction start. This called implements the *FlypeerResponseListener* interface.

One of the missing pieces of JXTA is the transaction support. This provides us a nice window of opportunity to contribute a new feature back to the community – which is likely to happen when Flypeer is fully implemented.

A Simple Scenario

Fig 1 shows an example of a transaction in a tree structure with four services composed by three different composition types.

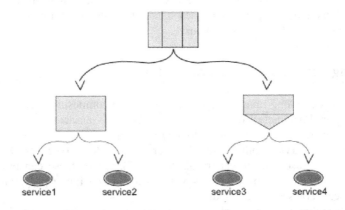

Fig. 1. Transaction in a tree structure

Fig 2 shows the transaction context (an xml file) for this example.

Executing a Transaction

The transaction in Fig 1 can be modelled programmatically using the Flypeer Framework as it is demonstrated in Fig 3 overleaf.

First of all, there is a parallel composition which executes at the same time a sequential and an alternative composition.

The sequential composition runs first the Service1 and when it's finished the Service2 is called.

At this same time, the alternative composition also is being executed. First, the Service 3 is executed and if it runs with no errors, this composition is finished, but if something wrong happens the Service4 is called as an alternative for the Service3.

One thing to note about Service2 is that, as mapped in the Transaction Flow (Fig 2), the Service3 receives the output of the Service2 as an input parameter. So, in order to Service3 to be able to start its execution, Service2 has to have finished and sent its output to Service3 already.

```
 1    <?xml version="1.0" encoding="UTF-8"?>
 2    <TransactionTree xmlns:xs="http://www.w3.org/2001/XMLSchema"
 3                     xmlns:xsi="http://www.w3.org/2001/XMLSchema-instance"
 4                     transactionId="0001">
 5        <SubTransaction subTransactionId="010">
 6            <Composition compositionType="Parallel">
 7                <SubTransaction subTransactionId="011">
 8                    <Composition compositionType="Sequential">
 9                        <SubTransaction subTransactionId="0111">
10                            <WebService serviceId="SERVICE1">
11                                <ServiceDescription>
12                                    Service 1
13                                </ServiceDescription>
14                            </WebService>
15                        </SubTransaction>
16                        <SubTransaction subTransactionId="0112">
17                            <WebService serviceId="SERVICE2">
18                                <ServiceDescription>
19                                    Service2
20                                </ServiceDescription>
21                            </WebService>
22                        </SubTransaction>
23                    </Composition>
24                </SubTransaction>
25                <SubTransaction subTransactionId="012">
26                    <Composition compositionType="Alternative">
27                        <SubTransaction subTransactionId="0121">
28                            <WebService serviceId="SERVICE3">
29                                <ServiceDescription>
30                                    Service3
31                                </ServiceDescription>
32                            </WebService>
33                        </SubTransaction>
34                        <SubTransaction subTransactionId="0122">
35                            <WebService serviceId="SERVICE4">
36                                <ServiceDescription>
37                                    Service4
38                                </ServiceDescription>
39                            </WebService>
40                        </SubTransaction>
41                    </Composition>
42                </SubTransaction>
43            </Composition>
44        </SubTransaction>
45        <Dependencies>
46            <InternalDependency paramName="param1" dependent="0111"/>
47            <InternalDependency paramName="param2" dependent="0112"/>
48            <InternalDependency paramName="param3" originator="0112" dependent="0121"
49            <InternalDependency paramName="param3" dependent="0122"/>
50        </Dependencies>
51    </TransactionTree>
```

Fig. 2. Transaction Context

When all compositions have finished the output is sent to the initiator as result of the transaction.

```
 1 Service service1 = new Service("SERVICE1");
 2 service1.addDependency("param1");
 3 Service service2 = new Service("SERVICE2");
 4 service2.addDependency("param2");
 5 Service service3 = new Service("SERVICE3");
 6 service3.addDependency("param3", service2);
 7 Service service4 = new Service("SERVICE4");
 8 service4.addDependency("param4");
 9
10 SequentialFlow sFlow = new SequentialFlow();
11 sFlow.addService(service1);
12 sFlow.addService(service2);
13
14 AlternativeFlow altFlow = new AlternativeFlow();
15 altFlow.addService(service3);
16 altFlow.addService(service4);
17
18 ParallelFlow pFlow = new ParallelFlow();
19 pFlow.addFlow(sFlow);
20 pFlow.addFlow(altFlow);
21
22 TransactionFlow flow = new TransactionFlow();
23 flow.setGenericFlow(pFlow);
24
25 Map<String, Serializable> params = new HashMap<String, Serializable>();
26 params.put("param1", "1");
27 params.put("param2", "1");
28 params.put("param3", "0");
29 params.put("param4", "1");
30
31 commandSimplePeer = new CommandSimplePeer();
32 commandSimplePeer.start(getParameter("username"), new MyNotifier());
33 TransactionInitiator transInit = new TransactionInitiator(peer);
34 transInit.startTransaction(params, flow, new MyResponseListener());
```

Fig. 3. Running the transaction

Comparison and Practicality

In Flypeer, the loosely coupled solution for a distributed transaction [18], [2] has been described in the context of digital ecosystems, and in particular we have been concerned with services, transactions, and network support within the digital ecosystem initiative. The structure of the interaction network within the architecture we have proposed emerges through the local interactions that take place in the context of long-running business transactions. In contrast with our model, WS protocols use a customised coordinator for avoiding inconsistency.

For providing a consistent environment, during concurrent actions (service deployment and compositions), WS-* (WS-AtomicTransactions and WS-BusinessActivity) and BTP, are using the two-phase commit (2PC) protocol, which requires synchronisation for the phases. This is applied through a centralised coordination framework, based on WS-Coordination [19]. Fig. 4 shows a simple example of the use of WS-Coordination for executing a transaction where the Initiator creates a coordination context and the Participants, based on their registered services, deploy their respective services. The synchronisation for concurrency control is done in a centralised manner. This causes a single point of failure as well as a single dependency on the provider(s) of the centralised coordinator framework.

In addition, a more careful study of this coordination framework, such as that reported in [9], shows it to suffer from some critical decisions about the internal build- up of the communicating parties - a view also supported in [20]. The Coordinator and Initiator roles are tightly-coupled and the Participant contains both

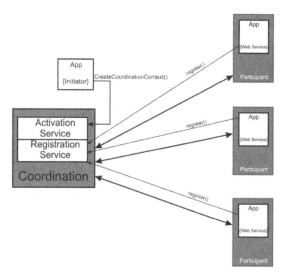

Fig. 4. WS-Coordination

business and transaction logic. These presumptions are against the primary requirements of SOA, particularly loose-coupling of services and local autonomy, and thus are not suitable for a digital business ecosystem, especially when SMEs are involved. This is because smaller organisations tend to be more sensitive in revealing their local design and implementation precisely because this is often where their model lies [2], [16].

Conclusions and Further Work

In this paper we have presented a framework for executing distributed long-term transactions in Digital Business EcoSystems. Various forms of service composition have been considered in order to provide a closer representation of business transactions within a service-oriented architecture. Our model is considered at the deployment level of SOA while respecting the loose-coupling of the underlying services. Further, it addresses omitted results, through forward recovery, in a way that does not break local autonomy.

Another strength of our approach is that it provides a light distributed coordinator which does not require to violate the autonomy of the local platforms or making any presumptions at the realisation level. This is particularly important in a business environment, especially for SME's. Further, these behaviour patterns can be applied dynamically on transactions in our approach, as was partly demonstrated in our example.

The work on progress of Flypeer will support the previous locking mechanism of the digital ecosystems transaction model (purposed in [8]). This is something we are keen to explore further as it lays the groundwork for a coordinated and collaborative service invocation specification to support long-term and multi-service transactions in Digital Ecosystems. The other part of Flypeer roadmap will provide a 'Trust framework' for businesses in digital ecosystems.

Acknowledgement

This work was supported by the EU-FP6 funded project OPAALS Contract No. 034824.

References

[1] Razavi, A.R., Krause, P., Moschoyiannis, S.: Deliverable D24.5: DBE Distributed Transaction Model (2006)
[2] Razavi, A.R.: Digital Ecosystems, A Distributed Service Oriented Approach for Business Transactions. PhD Thesis, University of Surrey (2009)
[3] Papazoglou, M.P.: Service-Oriented Computing: Concepts, Characteristics and Directions. In: Proceedings of the Fourth International Conference on Web Information Systems Engineering, December 2003. IEEE Computer Society Press, Washington (2003)
[4] Date, C.: An Introduction to Data Base Systems. Pearson Education, London (2003)
[5] Razavi, A.R., Moschoyiannis, S., Krause, P.: Preliminary Architecture for Autopoietic P2P Network focusing on Hierarchical Super-Peers, Birth and Growth Models (2007)
[6] Elmagarmid, A.K.: Database transaction models for advanced applications. Morgan Kaufmann, San Francisco (1992)
[7] Razavi, A.R., Moschoyiannis, S., Krause, P.: A Coordination Model for Distributed Transactions in Digital Business EcoSystems. In: Digital Ecosystems and Technologies (DEST 2007). IEEE Computer Society Press, Los Alamitos (2007)
[8] Razavi, A.R., Moschoyiannis, S., Krause, P.: An open digital environment to support business ecosystems. Peer-to-Peer Networking and Applications Springer Journal (2009)
[9] Furnis, P., Green, A.: Choreology Ltd. Contribution to the OASIS WS-Tx Technical Committee relating to WS-Coordination, WS-AtomicTransaction, and WS-BusinessActivity (November 2005),
   http://www.oasis-open.org/committees/download.php/15808
[10] Cabrera, L., Copeland, G., Johnson, J., Langworthy, D.: Coordinating Web Services Activities with WS-Coordination, WS-Atomic Transaction, and WSBusinessActivity (January 2004)
[11] Cabrera, L., Copeland, G., Feingold, M., Freund, R., Freund, T., Joyce, S., Klein, J., Lang-Worthy, D., Little, M., Leymann, F.: Web Services Business Activity Framework (WS-BusinessActivity). In: 2005 IBM Developer Works (2005)
[12] Cabrera, L., Copeland, G., Feingold, M., Freund, R., Freund, T., Johnson, J., Joyce, S., Kaler, C., Klein, J., Langworthy, D.: Web Services Atomic Transaction (WS-AtomicTransaction). IBM, US (2005)
[13] Ceponkus, A., Dalal, S., Fletcher, T., Furniss, P., Green, A., Pope, B., Inferior, A.: Business transaction protocol V1.0. OASIS Committee Specification (June 3, 2002)
[14] Razavi, A.R., Krause, P.: Integrated autopoietic DE architecture - D3.10 (2009)
[15] Yang, J., Papazoglou, M.P., van den Heuvel, W.J.: Tackling the Challenges of Service Composition in e-Marketplaces. In: Proceedings of the 12th International Workshop on Research Issues on Data Engineering: Engineering E-Commerce/E-Business Systems (RIDE-2EC 2002), San Jose, CA, USA (2002)
[16] Singh, M.P., Huhns, M.N.: Service-Oriented Computing: Semantics, Processes, Agents. Wiley, Chichester (2005)
[17] Papazoglou, M.P., Traverso, P., Dustdar, S., Leymann, F., Kramer, B.J.: Service-oriented computing: A research roadmap. In: Service Oriented Computing, SOC (2006)

[18] Razavi, A., Moschoyiannis, S., Krause, P.: An open digital environment to support business ecosystems. In: Peer-to-Peer Networking and Applications. Springer, New York

[19] Cabrera, L., Copeland, G., Feingold, M., Freund, R., Freund, T., Johnson, J., Joyce, S., Kaler, C., Klein, J., Langworthy, D., Little, M., Nadalin, A., Newcomer, E., Orchard, D., Robinson, I., Shewchuk, J., Storey, T.: Web Services Coordination (WS-Coordination) (August 2005)

[20] Vogt, F.H., Zambrovski, S., Gruschko, B., Furniss, P., Green, A.: Implementing Web Service Protocols in SOA: WS-Coordination and WSBusinessActivity. In: CECW, vol. 5, pp. 21–28 (2005)

A Research Framework for Interaction Computing

Paolo Dini[1] and Daniel Schreckling[2]

[1] Department of Media and Communications
London School of Economics and Political Science
London, United Kingdom
`p.dini@lse.ac.uk`
[2] Institute of IT-Security and Security Law
University of Passau, Passau, Germany
`ds@sec.uni-passau.de`

Abstract. This paper lays out an interdisciplinary research framework that integrates perspectives from physics, biology, mathematics, and computer science to develop a vision of interaction computing. The paper recounts the main insights and lessons learned in the past six years across multiple projects, gives a current definition of the problem, and outlines a research programme for how to approach it that will guide our research over the coming years. The flavour of the research is strongly algebraic, and the bridge to specification of behaviour of automata through new formal languages is discussed in terms of category theory. The style of presentation is intuitive and conceptual as the paper is meant to provide a foundation widely accessible to an interdisciplinary audience for five threads of research in experimental cell biology, algebraic automata theory, dynamical systems theory, autopoietic architectures, and specification languages, the first four of which are represented by more focussed technical papers at this same conference.

Keywords: Bio-Computing, Interaction Computing.

1 Introduction

This research is motivated by the fundamental question whether a biological ecosystem, or a subset thereof, could be used as a model from which to derive self-organising, self-healing, and self-protection properties of software. This research question is premised on the assumption that such biological properties can increase the effectiveness of information and communication technologies (ICTs) in various application domains, from ubiquitous computing, to autonomic communications, to socio-economic processes aimed at regional development, simply on the basis of their greater and spontaneous adaptability to user needs. Thus, this research addresses some of the non-functional requirements or software qualities of the underlying technology, which we refer to as software ecosystems [17].

This paper presents a research framework that aims to achieve a usable model of bio-computing, based on several years of research across several projects [16] [17] [18]. The application areas of interest ultimately are:

F.A. Basile Colugnati et al. (Eds.): OPAALS 2010, LNICST 67, pp. 224–244, 2010.
© Institute for Computer Sciences, Social Informatics and Telecommunications Engineering 2010

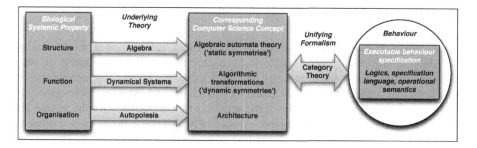

Fig. 1. High-level view of the theoretical research framework

- Service composition in the context of dynamic business workflow instantiation
- Biologically-inspired RESTful interaction framework
- Symbiotic security

Figure 1 gives a high-level view of the theoretical research framework that will be discussed and justified in more detail in the rest of this paper. The most important aspect of the theory that is emerging is that it needs to address three fundamental aspects of biology: structure, function, and organisation. Our preliminary results and insights point to algebra, dynamical systems, and autopoiesis, respectively, as the theories that can explain and/or model these aspects of biology and that need to be unified by a common mathematical framework that can effect a mapping to computer science. The target of these mappings appears to be a unification of the algebraic and algorithmic structure of automata, and novel ideas in software architectures and biological design patterns inspired by autopoiesis. Category theory is then able to relate any of the structures thus defined that have algebraic character to automata behaviour (which is also some kind of algebra) and from behaviour into a language which may be used to express (specify) some particular structures. Part of this language may be some kind of logic. Instantiation of this framework in modern distributed and web-oriented computing environments may be expressible compatibly with the Representational State Transfer (REST) architectural style [29]. It is important to emphasise that the term "structure" is quite overloaded in our work. It can refer to biological (physical) structure or to algebraic structure. Hopefully the different meanings will be clear from the context.

This paper outlines a research framework that is explored in greater depth in the following four companion papers at this same conference:

- A Research Framework for Interaction Computing (this paper)
- Numerical and Experimental Analysis of the p53-mdm2 Regulatory Pathway [62]
- Lie Group Analysis of a p53-mdm2 ODE Model [35]
- Transformation Semigroups as Constructive Dynamical Spaces [23]
- Towards Autopoietic Computing [9]

We now retrace the arguments and rationale that we have developed over the past six years in this area of research.

2 Historical Recap

The complexity and interconnections of the research activities that are gradually unfolding in the two projects make it necessary to provide a summary of past activities and to retrace the arguments that have led to the present research rationale. Hopefully this context will make it easier to understand and assess the relevance and validity of the current activities and of the activities that are planned for the remainder of the OPAALS project, and beyond. Accordingly, Figure 2 provides a graphical overview of our research in bio-computing over the past several years. The figure shows the main points that each report addresses (in some cases this is the title of the deliverable) along with the corresponding deliverable number, where by "main" we mean the topics that, in hindsight, were found to be most relevant in later deliverables, as a plausible theoretical and mathematical framework began to emerge.

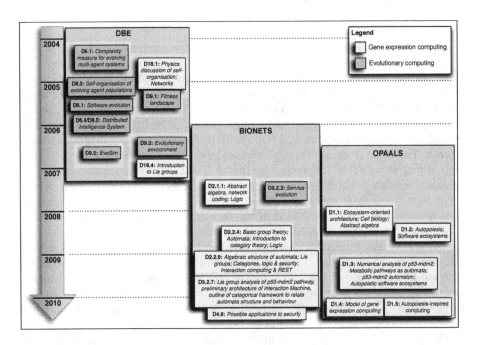

Fig. 2. History of relevant bio-computing reports across several projects

During the preparation of the DBE project, we proposed that the concept of ecosystem could be used not only as *metaphor*, but also as *model* for biologically-inspired computing. Ecosystems are characterised by self-organising and evolutionary processes. Whereas, strictly speaking, evolution is a form of self-organisation, by the latter term we refer to the order construction processes associated with cell metabolism and morphogenesis. In developing our theory of bio-computing, thus, we prioritise ontogeny over phylogeny.

2.1 Evolution and Self-organisation

The current research thread in gene expression or interaction computing began with a discussion of self-organisation through the minimisation of free energy, in DBE D18.1 [16]. Although the concept of free energy is very useful for understanding and modelling self-organisation in physical systems, unlike physical systems software systems are abstract. Thus, the successes of statistical physics are not readily transferrable to software due to the absence of an interaction potential energy and of the concept of temperature in the latter. Of course, the wealth of probabilistic methods based on uniform and nonuniform probability distributions do a good job at achieving an analogous effect; but such effect is contrived in the sense that it is imposed on the digital information which, if left to its own devices, would forever lie still in the 'current state'. However, the users provide a continuous input of information, which we can regard as analogous to the Sun's energy as the fundamental driver of the biosphere. Thus, even if we do not have a proper 'temperature', we do have a continuous flow of information through the system and a continuous poking and prodding by the users that can be seen as analogous to a certain level of thermodynamic 'mixing'. If we abstract a complex distributed computation and communication system as a set of coupled finite-state machines, user inputs become 'waves' of signals that propagate through the system, carried by the interactions between the state machines. The puzzle of self-organisation, thus, could be cast as the problem of deriving appropriate constraints in the execution paths of the state machines that can lead to the construction of ordered structure and behaviour by harnessing the 'energy' (information) flowing through the (open) system.

Clearly the problem posed in this manner is not trivial. In the DBE project we therefore developed an Evolutionary Environment ("EvE") in parallel with more mathematical research [32, 39, 55, 5, 6]. Although we were able to achieve some level of optimisation of the distribution of services in the ecosystem through a neural networks-based Distributed Intelligence System [7, 8], the evolution of the services to satisfy a particular user request was not achieved. It appeared that using services as the atomic units of evolution was not sufficiently granular to respond adequately to different contexts. On the other hand, breaking services down to apply genetic algorithms to the code itself is still too difficult for engineering applications.

The problem seemed to be a lack of understanding of the structural and dynamical features of ecosystems that need to be satisfied in order to support an effective evolutionary framework. Put simply, because evolution is a weak and slow process that, in order to avoid instabilities (death of the phenotype), can only make extremely small modifications to a given genotype, the ecosystem itself must already be highly performant, in the sense that its 'components' must already be quite compatible with one another and must already be close to satisfying a given fitness requirement. This implies the need for a holistic approach, whereby the ecosystem is in some sense 'bootstrapped' all at once through a massively parallel process in which hundreds if not thousands of requirements are satisfied simultaneously and compatibly with one another.

Our objective, therefore, is to find a balance between evolutionary computing and what we are calling gene expression computing. We seek an integration of the two approaches that is analogous to what DNA has been able to achieve: the same molecule is a carrier of hereditary traits across generations whilst also guiding the morphogenesis and metabolism of the individual organism. Based on our experience in these projects, we feel that the problem of gene expression computing must be solved first, before we can hope to achieve effective evolutionary behaviour. Figure 3 shows how the abstract concept of Interaction Computing can be instantiated into different contexts.[1] Gene expression computing refers to the nuts and bolts of cellular pathways and how they are able to construct order and exhibit stable and robust behaviour; so it is a model oriented towards a *local* perspective. Autopoiesis-Inspired Computing, on the other hand, looks at *global* properties of the cell and of autopoietic systems, and tries to map these properties to computer architectures that replicate autopoietic behaviour or its subsets (such as operational closure). Autopoiesis-Inspired Computing is discussed in another paper in this same conference [9]. Finally, Symbiotic Computing is more specifically focussed on the ecosystemic properties of interdependence and synergy, and it is being pursued in the BIONETS project in particular as regards software security.

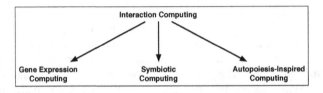

Fig. 3. Different possible models of computation derived from Interaction Computing

This prioritisation of ontogeny over phylogeny implied that an in-depth invest-tigation of the physics and mathematics of (non memory-based) self-organisation was necessary in order to understand what features could be transferred to software. Because, in addition to the minimisation of free energy, both cell biology and ecosystems are characterised by non-linear processes, we realised that we faced a 'double jeopardy': not only does it seem challenging to translate non-linear behaviour into automata or algorithmic constraints, as above, but the non-linear behaviour itself is in most cases the signature of systems that are not even integrable. In spite of the daunting stack of challenges that was taking form, we kept focusing on the fact that biological systems at all scales *are* able to cope with these challenges: they do an extremely good job at producing ordered structures and behaviour, in spite of their complexity and of the non-integrability of most mathematical models of biological phenomena (which could be related to their non-computable aspects). This was encouraging (if a biological system can manage this, there must be a way to formalise it), even if it suggested to us that new ways to think of complex physical and biological phenomena were likely to be needed.

2.2 Symmetry

Based on our previous experience in applied mathematics and physics of the use-fulness of the concept of symmetry, our starting point was to assume that the same

---

[1] No references are given for these terms because we invented them – and are in the process of developing formalisations for them.

concept was likely to play an important role also here. Our intermediate results so far have confirmed this hunch. Symmetry is a very general concept in mathematics that formalises the notion of invariance or regularity. In mathematics, a symmetry is a *transformation* that leaves some property of a mathematical object invariant. Now, it is a truth universally acknowledged (and easily proven) that the invertible transformations of a mathematical object that leave some property of its structure invariant form a group.[2] Therefore, the mathematical study of symmetries and regularities must necessarily rely on algebra.

The above statement should be taken as a *necessary* rather than as a *sufficient* condition. In other words, a technical system that interfaces at some level with human users and that is meant to support socio-economic processes must be open to new information and must allow for the emergence of new structures and patterns. Even if such a requirement were not enforced or relevant (i.e. if all we were trying to do was to develop an artificial life environment), the wish eventually to replicate and support evolutionary behaviour implies that the emergence of new forms must be supported. Our current understanding of algebra is not necessarily sufficient to develop the best mathematical framework for the formalisation of emergent behaviour and open-ended evolution. By the same token, however, the system must also be stable and reliable, since it is meant also to uphold robust (self-healing!) engineering applications and non- functional requirements. It must behave similarly in similar contexts; hence, it must embody a fair amount of regularity and predictable behaviour. This is what mathematics, and algebra in particular, formalises. Again, we wish to emulate the delicate balance between order/reliability and unpredictability/openness that biology has been able to fine-tune and leverage to produce stable but ever- changing life-forms of unbelievable complexity.

2.3 Lie Groups

In DBE D18.4 [15] we therefore began a discussion of the method of Lie groups for the solution of differential equations, since it is the most general method that applies equally well to linear and non-linear systems. At that time we were aware that a method developed for continuous systems would be difficult to apply to discrete automata, but we were also aware of the fact that generalisations of Lie groups have been applied to discrete dynamical systems.[3]

The relevance of an algebraic perspective was strengthened by observing how finite ring and field theory has been used in network coding. An examination of network coding was motivated initially by the BIONETS project, where we thought that the ability to reconstruct missing information from a bitstream might have been extended towards self-healing properties of software, or perhaps the reconstruction of the whole phenotype from a partial specification. However, it soon became apparent that the value of the exercise was more as an example of abstract algebra that was relatively accessible to computer scientists than as a technique that could be directly relevant to evolutionary or gene expression computing. Because this algebraic theory deals with discrete finite sets, it not only demonstrated another area of applications

---

[2] Paraphrased from Stewart ([61]:xxvii).
[3] See Maeda [42] and Peter Hydon's work at
http://personal.maths.surrey.ac.uk/st/P.Hydon/sym.htm

where algebra is relevant but, by providing a basis for the more difficult group theory, it also brought us one step closer to the mathematical formalisation of symmetries in the context of computer science. This abstract groundwork was reported in both projects [20, 17].

At about the same time we ran across the work of the Cuban HIV researchers Sanchez, Morgado, and Grau [57, 58, 56], an interdisciplinary research team composed of a biochemist, a mathematician, and a computer scientist. AIDS research is concerned with, among other things, mutations in the DNA of the HIV virus. Mutations that impede the ability of this virus to function are good news for us. The operational effectiveness of a particular strand of DNA is dependent on the geometry of the proteins (enzymes) that are synthesised from it through gene expression, because this geometry has to match the complementary geometry of its substrate for the enzyme to be effective. The 3-D shape of an enzyme depends on the folding of the strand of aminoacids built by the ribosomes from the corresponding tract of DNA, by applying the genetic code.[4] Protein folding depends to a large extent on polar bonds which, in turn, depend on the hydrophobicity of the aminoacids along the chain. The hydrophobicity of an aminoacid depends on the second base of the corresponding 3-base codon. We know empirically that mutations are most likely to occur in the middle or second base of a codon. Now the surprising fact is that, if a codon undergoes a mutation (most likely to happen in its second base) to a new codon, the hydrophobicity of the new aminoacid will be very similar to the original aminoacid's. Furthermore, it turns out that if the 20 aminoacids are arranged in order of increasing hydrophobicity the corresponding codons form a partial order, in fact a 64-node Boolean lattice.

Thus, a particular assignment of the bases to the field extension GF (2^2) (represented by the 4 nucleotite bases) leads to a Boolean lattice (as a third direct product of the 2×2 base lattice due to the fact that each codon is formed by three bases) whose minimum and maximum elements are the codons that correspond to the least and most hydrophobic aminoacids, and this assignment leads to a self-consistent partial order for the rest of the codons that matches corresponding levels of hydrophobicity. The relevance of this finding is that this particular algebraic structure corresponds to what amounts to hydrophobicity as a *continuous function* of codon mutation. In other words, the operational semantics of the DNA code are fairly robust with respect to mutations. This is not good news for AIDS research, because it confirms the observation that mutations of the HIV virus are likely to remain as deadly as the originals. However, the same effect underpins the stability of any other organism with respect to pertur- bations brought by genetic mutations, i.e. it takes a relatively improbable large mutation to upset the functioning of a particular phenotype. In other words, the robustness of the most fundamental 'architectural' feature of biology, the DNA code, is formalisable through an equally fundamental algebraic structure. Boolean algebras are not uncommon, however. So the fact that a particular data set forms a partial order or even a Boolean lattice (slightly more restrictive) is not necessarily of great significance.

---

[4] The genetic code is a many-to-1 map from the 64 codons to the 20 aminoacids. Each codon is composed of 3 bases, each of which can assume one of the 4 values A, G, T, E. Hence, 4 bases occupying 3 possible slots: $4^3 = 64$.

In their more recent work Sanchez, Morgado, and Grau [56] report that the codons actually carry additional structure, in particular they form a Lie algebra. A Lie algebra is a vector space whose elements satisfy an additional binary operation, the Lie bracket. Because the set of codons can be seen not only as a Boolean algebra but also as the Galois field extension $GF(2^6)$, it already was isomorphic to a (discrete and 3-dimensional) vector space over the finite field $GF(2^2)$, so this means that the codons also satisfy the Lie bracket, as an additional constraint. The physical significance of this fact is not clear; however, we know that a Lie algebra can also be seen as the tangent space to a Lie group at its identity, and a Lie group is the only algebraic structure that can sometimes help us in solving non-linear dynamical systems – for example the non-linear dynamical systems that formalise cell metabolic and regulatory pathways. Therefore, once again not only does the algebraic approach seem justified, but the need to develop a unified theory between (discrete) finite group theory and (continuous) Lie group theory around dynamical systems arising from cellular processes appears increasingly likely.

The investigation of DNA as a Lie algebra will be performed in future projects because first we need to assess the feasibility of the Lie group perspective in the solution of cell metabolic and regulatory pathways. Thus, our shorter-term objective is to extend the work begun in DBE D18.4 and perform a Lie group analysis of the p53-mdm2 regulatory pathway (see [35] in this same conference, which is based on [18] and [40]).

2.4 Functional Completeness

There is one more topic that provides an important background to our research: functional completeness [34]. The interesting aspect of this point of view is that it resonates with the physics and engineering research literature around a concept that seems at first unrelated to our discussion: choice of variables.

It is well-known in the modelling of physical phenomena that a judicious choice of coordinate system and/or of the representation of the dependent and independent variables can simplify the mathematics greatly, at the same time providing useful insights into the nature of the problem under study.[5] The choice of coordinate system is perhaps easier to see, for example when choosing cylindrical coordinates to describe fluid flow through a circular pipe. Many physical problems, however, can also be characterised by groupings of variables that also simplify the mathematics considerably. This was first noticed in the 19th Century by experimental researchers in a variety of applied and scientific disciplines, who noticed that particular dimensionless groupings of variables could sometimes lead to the collapse of data clouds and families of data sets onto single curves. The practical usefulness of this fact was soon to be investigated more rigorously, leading eventually to Lie's group-theoretical methods for differential equations.

The general epistemological principle we can derive from this is that in many complex problems increasing complexity of the variables used to describe them often appears to simplify the mathematical model, in some cases leading to an analytical solution. This same principle could be relevant to the problem of bio-computing

---

[5] E.g. see the famous Buckingham Pi Theorem [10] and generalisations thereof.

when, as Horvath has done, we generalise the fundamental structures of computer science to more complex structures.

In particular, digital computers today are able to perform any computation because they are functionally complete. This means that there is an algebraic structure, in this case a Boolean algebra, such that any n-ary function can be represented by a corresponding propositional logic expression (or 'polynomial') that is implementable as logic gates. It has been known for many years that one can use more general algebraic structures to achieve equally functionally complete computational models. Horvath investigated whether a semigroup can have the functionally complete property expressible as more general 'polynomials' than propositional logic. He proved that the answer is Yes, as long as the semigroup is a finite simple non-abelian group (SNAGs).[6] In group theory, SNAGs play a role similar to prime numbers in number theory, thus the possible ramifications of this fact are quite intriguing. Because, even though they are somewhat special, there are infinitely many such groups, this means that we could build a 'more complex' computer science using more complicated fundamental structures.

What does this ultimately mean and what would this buy us? In terms of Turing computability, these different ways of thinking of computing would not change anything. We would compute problems of the same complexity class. However, we argue that it is worth investigating what kind of computations we might be able to perform, and how, but using SNAGs rather than Boolean algebra as the fundamental starting point for computing. Another analogy that may help clarify this point is to compare the use of Assembly language versus objects. One can program anything in Assembler, and in fact any program is eventually compiled down to binary code, but it's a lot easier to program classes and let the compiler do the hard work.

With this historical background in mind we now turn to the problems we are currently facing in our research.

3   Current Research Questions

3.1   Abstraction Level

Cell metabolism relies on ultimately undirected bottom-up and random/stochastic processes that can only 'execute' through the spontaneous interaction of the various components. The interactions are driven by a combination of electrostatic forces (usually conceptualised as minimising the potential energy of interaction) and most probable outcomes (maximisation of entropy), which can be modelled together as the minimisation of free energy. In spite of this fundamental randomness, however, a healthy cell behaves in an organised and finely balanced way that is more evocative of a deterministic, even if very complex, machine than of random chaos. The cell in fact has a definite physical structure and executes well-defined 'algorithms' in the form of cellular processes (several hundred per cell type) such as metabolic or regulatory biochemical pathways. This suggests a description and modelling of cell behaviour at a level of abstraction that is higher than the molecular, and through mechanisms or constraints that are complementary to stochastic processes.

---

[6] Every group is also a semigroup, but not conversely of course.

In particular, our perspective views the stochastic nature of cell biochemistry mainly as a mechanism of dimensional reduction[7] that does not necessarily need to be emulated in any detail. For example, a gene expresses hundreds of mRNA molecules which, in turn, engage hundreds of ribosomes for no other reason than to maximise the probability that a particular, *single* genetic instruction will be carried out, such as the synthesis of a particular enzyme. As a consequence of this dimensional reduction (hundreds to 1), a higher level of abstraction than that at which stochastic molecular processes operate is justified in the modelling approach – in particular, a formalisation that retains, and builds on, the discrete properties of cell biology.[8] However, even the resulting lower-dimensional system can't plausibly be imagined to perform the complexity of a cell's functions driven simply by a uniform distribution of interaction probability between its (now fewer) components. Additional structure and constraints must be at play.

3.2 Dynamic Stability

The presence of additional constraints is evident from the internal physical structure or topology of the cell. For example, the citric acid cycle that metabolises energy from sugar takes place within the mitochondrion, isolated from the rest of the cell. But cellular macrostructures such as the mitochondrial membrane are too coarse to explain the bewildering complexity of parallel processing that takes place even within the mitochondrion itself. There must be constraints operating at a finer granularity that support specific reaction pathways over others and that prevent the cytoplasm from becoming a well-mixed solution of compounds of uniform concentration reacting indiscriminately with one another. In other words, even if the precise form of these additional constraints that keep cellular processes running smoothly is far from evident, their existence is implicit in the complex and *dynamically stable* operation of the metabolic and regulatory pathways.

Dynamic stability is only an intuitive concept at this point, which can be thought of as the signature of certain types of non-linear behaviour and for which a precise mathematical definition does not exist yet, although research in related fields is growing ([66, 43]). However, we can say that dynamic stability is a generalisation of the well-trod engineering principle of stable design, which tends to keep human machinery within its linear regime in fear of catastrophic failure if instabilities or resonances are allowed to grow. But linear systems are information-poor and cannot sustain rich and complex behaviour. Biology has been able to harness the expressive power of non-linear behaviour whilst maintaining adequate stability, thereby capturing the 'sweet spot' between order and chaos. From the point of view of information theory, linear systems tend to have a discrete power spectrum, whereas

---

[7] In dynamical systems theory, dimensional reduction refers to a reduction in the number of degrees of freedom of a system. Since biochemical systems are composed of thousands to millions of elements, the time evolution of each of which is governed (for the sake of argument within a Newtonian framework) by at least three separate ordinary differential equations (ODEs), successful abstraction and dimensional reduction can lead to significant theoretical insight and savings in CPU requirements.

[8] Notice that the statistical nature of the metabolic step carries a built-in robustness, i.e. if something is wrong with one of the proteins being generated, the metabolic cycle as a whole can proceed unhindered.

chaotic systems have a flat or continuous 'white noise' spectrum. An example of a human creation that strikes a balance between these two extremes and that is at a similar level of abstraction as software is music, which was discovered to be uniformly 1/f-noise, 30 years ago [63]. This provides motivation for why we think that mapping the greater expressive power of non-linear behaviour into computer science concepts will lead to a correspondingly greater power to 'compute' unprogrammed behaviour in real time.

The fact that the cell is not a well-mixed solution tells us, as is well-known, that it must not be in thermodynamic equilibrium. Prigogine's work [48] is deeply significant because it showed that ordered structures form in open systems under conditions of disequilibrium – maintained as such by a constant energy flow. Thus, although the phenomena he studied (e.g. the toroidal vortices of Rayleigh- Benard convection) are much simpler than what happens inside a cell, his insights give us a relatively concrete example of what a 'dynamical structure' might look like. The dynamic stability of cellular processes then constitutes a generalisation of Prigogine's ordered structures. Therefore, treating cellular processes as automata, or discrete low-dimensional dynamical systems, appears to be the most appropriate level of abstraction and entry point to understand biological construction of order in a way that is relatively easy to transfer to computer science.

3.3 Structure and Function in Biology and Computer Science

To make progress in this direction, we take as a starting hypothesis that the dynamically stable operation of the cell is critically dependent on two additional forms of structure that are more abstract than physical structure and that can be formalised mathematically as follows (see Figure 4):

- Time-independent algebraic structure of the automata modelling the cellular pathways. Algebraic structure gives rise to what we are calling static symmetries.
- Time-dependent Lie group structure of the dynamical systems modelling the same cellular pathways. This form of structure is formalised through a mix- ture of algebra and geometry and gives rise to what we are calling dynamic symmetries.

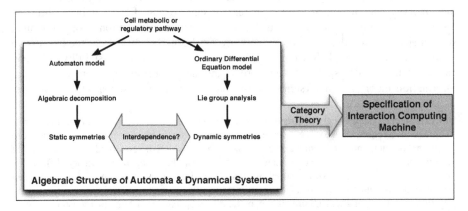

Fig. 4. Mathematical analysis workow to uncover biological symmetries

The relevance of the relationship between structure and function to all types of engineering and applied thinking motivates us to investigate how these two kinds of mathematical structure are related. The benefit of such a relationship would be the ability to specify desired behavioural properties and derive the corresponding structural properties.

In its simplest form, a finite-state automaton is a nite set of states acted upon by a semigroup of transformations. Until the 1960s the general consensus was that semigroups were too unstructured for anything useful to be done with them. This perception was changed by one of the landmark theorems in this field, the Krohn-Rhodes prime decomposition theorem for finite semigroups [38], which proved the existence of a much greater amount of structure in semigroups. The relevance of semigroups to automata has then made this mathematical theory of increasing interest to computer science over the past 40 years. Furthermore, the non-linear character of automata ([36]) suggests that they are the right instrument to model the enormously intricate feedback loops of discrete cellular processes. This observation is greatly strengthened by the current research of the Biocomputation Laboratory at the University of Hertfordshire, UK ([47, 26, 25, 24, 27]), in which several examples of cell regulatory and metabolic pathways are shown to be formalisable as finite-state automata. The application of Krohn- Rhodes decomposition to the corresponding semigroups then reveals the presence of a rich algebraic structure in the form of permutation groups and non-invertible components (flip-flops) at different levels of their hierarchical decomposition.

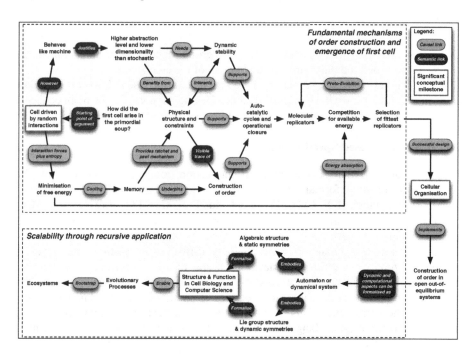

Fig. 5. Causal-semantic workflow summarising a part of the research rationale

The algebraic structure of automata does not account for their time-dependent or dynamic behaviour. Therefore, a significant challenge we face is how to make sense of the often non-integrable dynamical behaviour of non-linear systems. Systems biology, in fact, relies heavily on the numerical solution of the ordinary differential equations (ODEs) derived from the chemical rate equations modelling the cellular pathways, simply because no analytical solutions exist. However, as we mentioned above it is well-known that in many cases systems of coupled non-linear ODEs embody so-called global symmetries obtainable through Lie groups analysis [49]. Although global symmetries are quite constraining and are correspondingly difficult to find, this is not necessarily a drawback since biological systems exhibit ordered behaviour only within certain ranges of their parameters (e.g. temperature). In other words, Lie groups can help us solve mathematical models that are clearly very idealised approximations to how real systems work. However, the important point is that they do capture and formalise the concept of order in dynamical behaviour, which we have loosely called 'dynamic stability' above. It is not unreasonable to claim, therefore, that the symmetries corresponding to 'local' or parameter-limited ordered biological behaviour could be found through an extension of Lie's theory to less rigidly defined mathematical structures such as groupoids, as well as to discrete dynamical systems due to their closer relevance to automata:

> There are plenty of objects which exhibit what we clearly recognize as symmetry, but which admit few or no nontrivial automorphisms. It turns out that the symmetry, and hence much of the structure, of such objects can be characterized algebraically, if we use groupoids and not just groups. ([64]; quoted in [31])

Groupoids are like groups except that the group operation (usually functional composition) is defined only for some and not all of the elements.

Figure 5 gives an overall summary of the rationale of the research workflow and of some of the concepts we have discussed so far. Having summarised the main concepts of the mathematical theory, we now start building a bridge towards computer science.

3.4 Behaviour-Based Specification

It appears obvious that several parts of interaction computing systems could be described by existing formal specification frameworks or formal system, such as VDM [3], Z notation [60], CCS [45], π-calculus [46], CSP [33], LOTOS [4], ACP_τ [2], etc. While there are languages which are very similar to our approach, and Aspect-Oriented Programming (AOP) is certainly one of them, the reason for developing a new language is fundamentally different. Interaction computing is highly different from existing systems in terms of its concurrency, its interdependability, its realisation of functionality, its non-deterministic and probabilistic computation, and its modularity. Modifications of some specification languages may support all these properties. This has been shown in the past, for example, for Z. Step after the step the original language was extended with new features, such as non-determinism or the full support of temporal logic. This valuable engineering process extends a language such that it fits a certain need. However, this requires that the actual problem the language describes is similar.

Our problem is interaction computing and instead of trying to describe interaction computing using an existing language, adapting it to our needs, we take the opposite approach and start with analysing the problem first, i.e. its dynamical and structural properties. In the course of our research we will learn about this structure and identify basic functional components inspired by biology. This will also determine the primitives of the language. On top of that, our language will be based on behaviour the system to be described should exhibit. Here, the internal structure of the components realising this behaviour is not essential. They are hidden from the specification as they are far too complex. This is in strong contrast with existing formal specification methods which try to describe the actual functionality but not the behaviour. Here we define functionality as the actual functions which have to be executed to implement a certain behaviour.

Thus, the functionality of an interaction machine describes in detail the internal states and transitions the machine has to go through in order to achieve its desired behaviour, i.e. the specification would follow a white box characteristic approach. In contrast, the behaviour describes the observable or expected effects of a black box. Thus, behaviour strongly abstracts from the internal structure and gives a wider flexibility to its implementation. This takes the established high-level programming and specification languages one step further. While they already abstract from the hardware level and use higher-order programming language constructs, the biologically-inspired interaction computing specification language even abstracts from functionality and lifts programming and specification to the behavioural level. In our work we are studying how the two concepts of machine structure and its behaviour are strongly linked in categorical terms [40]. In particular, we show how a category of behaviour is directly linked to a category of machines realising this behaviour.

Additionally, to be able to transform an existing specification into an executable form, the specification language requires some operational semantics which allows us to translate a behaviour specification into interaction machines and their execution steps. Similar to functional or logical specification languages, the realisation of such an approach in an executable instance includes several implicit steps wihch are not explicitly stated in a machine specification. In interaction computing this process is even more complex because even simple operations are realised by multiple interactions between multiple machines. Adapting the operational semantics of an existing language becomes infeasible. Thus, we follow the general design process which tries to develop a language which actually fits best our needs.

Finally, we do not refuse the use of existing formal systems. In fact, our work already uses mechanisms [1] which allow us to transform one logic into a comparable one, to recognise the well-established correspondence between coalgebras and temporal logics (see also BIONETS deliverable D2.2.4 [22]), or which compare their internal structures. If we find that our systems possess properties which are describable by existing formal systems, we will opt for them, of course.

Thus, this work aims to develop the basis of an 'environment specification' language, which can be seen as a higher-abstraction software engineering specifi-cation language addressing both the structure and content of bio-inspired digital systems. Figure 6 shows at a high level how category theory can enable a mapping from algebraic and coalgebraic structures to algebraic and coalgebraic logic, as an initial step in this direction. This work is in progress ([20, 22, 21, 19, 59, 18, 40]) and elaborates concepts which map algebraic structure corresponding to automata into categories of behaviour.

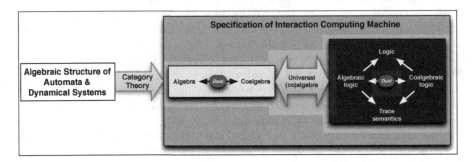

Fig. 6. Mapping of algebraic structures to logic structures through category theory

3.5   Organisation in Biology and Computer Science

The reliance on category theory is further motivated by Rosen [51, 52] who, following Rashevsky's ideas [50], first applied category theory to cell biology to develop a theory of "relational biology" as an alternative to the reductionist analytical methods still prevalent to this day. His main result was to prove that the cell metabolism repair function performed by the DNA is invertible into a DNA repair function performed by the cell metabolism. Hence the cell is 'self-sufficient' in terms of information, it contains all the information it needs to repair all of its parts. Of course we already knew that the cell is able to repair its DNA, but for our purposes it is very good to know that the same mathematical theory that can map automata to logic and dynamical systems is also able to capture important properties of the cell. Rosen's result has more recently been interpreted ([14]) as the mathematical analogue of Maturana and Varela's "operational closure" (or organisational closure) within the theory of autopoiesis [44]. In spite of the fact that Rosen's subsequent generalisation of this proof into a much more ambitious 'theory of Life' [54] has recently been criticised and has been the subject of a lively debate ([11, 13, 12, 41, 65]), Rosen should be credited with a simple but insightful observation:

> ... systems of the utmost structural diversity, with scarcely a molecule in common, are nevertheless recognizable as cells. This indicates that the essential features of cellular organization can be manifested by a profusion of systems of quite different structure. [53]

In other words, all cells, regardless of their structure, share a similar organisation. However, depending on their function, cells can have very different structure. This suggests that **Structure, Function,** and **Organisation** are equally fundamental concepts in biology.

In computer science, on the other hand, things are a bit different. In analogue computer systems the computation to be performed (Function) was strictly dependent on the electronic components utilised and their wiring (Structure). Digital computers, by contrast, were developed as "general-purpose machines" through extensive use of abstraction/layering. In contrast to biology and analogue computers, there is very little interdependence between Structure and Function in digital computers – by design! However, Organisation does map well from biology to computer science, where it is called Architecture. An interesting example of the applicability of these concepts is provided by the "conscientious software" of Gabriel and Goldman [30], who identify software that performs some useful external function as "allopoietic", in symbiotic coexistence with software that keeps the system alive as "autopoietic". A related concept that is similar to operational closure and that is a current focus of our research is to wire different allopoietic components together in order to form an autopoietic whole. A more in-depth discussion of autopoiesis-inspired computing can be found in another paper being presented at this conference [9].

The complexity of the problem and of the theory that is emerging is making it difficult to keep the various analogies, metaphors, and models straight, partly because the concepts apply at very different levels of abstraction. Table 1 provides a possible mapping between how these three fundamental concepts apply in biology, mathematics and computer science.

Table 1. Examples of how the fundamental properties of biology might map to other domains

	Biology	Mathematics	Computer Science
Structure	Shape of nerve cell	Group structure of cellular pathways	sequential/ parallel/ concurrent
Function	Nerve signal conduction	Metabolic pathway	Algorithm Behaviour
Organisation	Operational closure	Group closure property	Autopoietic architecture

3.6 Gene Expression Computing, or Interaction Computing

In reference to Figure 5, proto-evolutionary mechanisms in the primordial soup bootstrapped resilient organisational forms such as hypercycles [28] and auto-catalytic cycles [37] from random physical interactions. After the membrane emerged as a structure that could delimit an 'inside' from an 'outside', these so-called molecular replicators eventually led to the emergence of the cell with its autopoietic properties (organisationally closed, recursively self-generating). As we argued above, cellular pathways today are still driven by the same interaction and entropic physical processes. Thus, if we wish to emulate, in software, principles from biology that can rightfully claim 'fundamental' status, in its most general form context-sensitivity must work both ways, which argues for a reciprocal and pervasive interaction model.

Our work is inspired by the observation that the computation performed by a biological ecosystem can be conceptualised as a theoretical limit characterised by the number of peers in a distributed P2P architecture approaching infinity, with the amount of traditional computation performed by each approaching zero. This analogy can also be extended to the 'computation' performed by the cell's cytoplasm. More precisely, the computation performed by biological systems always involves at least two entities, each of which is performing a different, and often independent, algorithm which can only be advanced to its next state by the interaction itself. This is the kernel of the concept of interaction computing or gene expression computing. We wish to explore the implications of such a 'vanishing CPU' scenario because by providing a mathematical foundation to building nested and recursively interacting structures we believe that it underpins a model of emergent computation that will lead to new insights in biology and computer science, in equal measure.

This explains why we are trying to develop an emergent model of computation by mapping the regulatory and metabolic biochemical pathways of the cell to interacting automata. Such a model of computation will both require and enable a shift from a reliance on human design as the only source of order in software towards a greater reliance on information and structures built into the environment. In fact, the complexity of the cell's interior suggests that in the cell 'interaction' can acquire significantly greater semantics than, for example, perfect collisions between point particles in an ideal gas. We then notice that the cell is itself surrounded by other cells with which it communicates, and all are embedded in a complex mixture of tissues and fluids that form organs. Organs, in turn, cooperate in the functioning of individuals, which interact to form biological ecosystems. Thus, interactions happen at all scales within the nested and recursively organised hierarchical structure of all biological systems.

3.7 Computational Medium and RESTful Architecture

Interaction signals in biological systems are mediated in physical space by the solid, liquid or gaseous media that fill it (with the exception of light, which does not need a medium). Software systems, by contrast, do not interact over continuous metric spaces, they interact over topological spaces, or networks. By 'network' we do not mean simply the IP layer or below, we mean the term in the most general possible sense, applicable as a medium of low-abstraction signals, of application layer protocols, or of semantic and knowledge networks. In order to provide a roadmap of applicability to instantiate the theoretical and mathematical results of the project into the software and web environments of the future we need to understand how distributed and networked systems can support the interaction or gene expression computing models and their recursive application.

Our starting point for the development of a run-time framework that is general enough to support the mathematical results and that is relevant to today's web computing environments is a RESTful architecture for the definition of a message-passing interaction model for distributed environments. REST (Representational State Transfer [29]) in general, and the REST over HTTP architecture of the web specifically, constitutes a language in which interaction can be considered a primitive element. The REST architectural style has been conceived to reflect the architecture of the web. Since the architecture of the web is constrained at the lowest levels to enable extensibility at higher levels, higher-order capabilities such as support for complex interactions that require transactional guarantees (e.g. in long-running service applications) and querying languages can be constructed on top of it.

4 Conclusion

The aim of this paper was to provide a broad research framework through which the rationale of more focussed research activities could be understood [62, 35, 23, 9]. Much work remains to be done. However we hope that the framework we have presented here will appear plausible enough to attract more computer scientists, mathematicians, and cell biologists in the development of a common and unified theory of bio-computing for autopoietic digital ecosystems.

Acknowledgements

The authors wish to thank Dr Sotiris Moschoyiannis of the University of Surrey for clarifying how the RESTful perspective could be connected to the concept of interaction computing presented here. The support for this work by the OPAALS (FP6-034824) and the BIONETS (FP6-027748) EU projects is gratefully acknowledged.

References

1. Andréka, H., Neméti, I., Sain, I.: Universal Algebraic Logic, 1st edn. Studies in Universal Logic. Springer, Heidelberg (to appear)
2. Bergstra, J.A., Klop, J.W.: ACPτ: a universal axiom system for process specification, pp. 447–463 (1989)
3. Bjorner, D., Jones, C.B. (eds.): The Vienna Development Method: The Meta-Language. LNCS, vol. 61. Springer, Heidelberg (1978)
4. Bolognesi, T., Brinksma, E.: Introduction to the ISO specification language LOTOS. Comput. Netw. ISDN Syst. 14(1), 25–59 (1987)
5. Briscoe, G.: D6.1-Entropy-Based Complexity Measure for the Evolution-Based Self-Organisation of Agent Populations. DBE Project (2004),
   http://files.opaals.org/DBE/deliverables
6. Briscoe, G., De Wilde, P.: D6.2-Self-Organisation of Evolving Service Populations. DBE Project (2005), http://files.opaals.org/DBE/deliverables
7. Briscoe, G., De Wilde, P.: D6.4-Intelligence, learning and neural networks in distributed agent systems. DBE Project (2005),
   http://files.opaals.org/DBE/deliverables
8. Briscoe, G., De Wilde, P.: D6.5-The effect of distributed intelligence in evolutionary dynamics. DBE Project (2006),
   http://files.opaals.org/DBE/deliverables
9. Briscoe, G., Dini, P.: Towards Autopoietic Computing. In: Proceedings of the 3rd OPAALS International Conference, Aracaju, Sergipe, Brazil, March 22-23 (2010)
10. Buckingham, E.: The principle of similitude. Nature 96, 396–397 (1915)
11. Chu, D., Ho, W.K.: A Category Theoretical Argument Against the Possibility of Artificial Life: Robert Rosens Central Proof Revisited. Artificial Life 12, 117–134 (2006)
12. Chu, D., Ho, W.K.: Computational Realizations of Living Systems. Artificial Life 13, 369–381 (2007)
13. Chu, D., Ho, W.K.: The Localization Hypothesis and Machines. Artificial Life 13, 299–302 (2007)

14. Cornish-Bowden, A., Cardenas, M.L.: Self-organization at the origin of life. Journal of Theoretical Biology 252, 411–418 (2008)

15. Dini, P.: D18.4-Report on self-organisation from a dynamical systems and computer science viewpoint. DBE Project (2007), http://files.opaals.org/DBE

16. Dini, P., Berdou, E.: D18.1-Report on DBE-Specific Use Cases. DBE Project (2004), http://files.opaals.org/DBE

17. Dini, P., Briscoe, G., Munro, A.J., Lain, S.: D1.1: Towards a Biological and Mathematical Framework for Interaction Computing. OPAALS Deliverable, European Commission (2008), http://files.opaals.org/OPAALS/Year_2_Deliverables/WP01/

18. Dini, P., Horvath, G., Schreckling, D., Pfeffer, H.: D2.2.9: Mathematical Framework for Interaction Computing with Applications to Security and Service Choreography. BIONETS Deliverable, European Commission (2009), http://www.bionets.eu

19. Dini, P., Schreckling, D.: More Notes on Abstract Algebra and Logic: Towards their Application to Cell Biology and Security. In: 1st OPAALS Workshop, Rome, November 26-27 (2007)

20. Dini, P., Schreckling, D.: On Abstract Algebra and Logic: Towards their Application to Cell Biology and Security. In: Altman, E., Dini, P., Miorandi, D., Schreckling, D. (eds.) D2.1.1 Paradigms and Foundations of BIONETS research (2007)

21. Dini, P., Schreckling, D.: Notes on Abstract Algebra and Logic: Towards their Application to Cell Biology and Security. In: 2nd International Conference on Digital Ecosystems and Technologies, IEEE-DEST 2008, February 26-29 (2008)

22. Dini, P., Schreckling, D., Yamamoto, L.: D2.2.4: Evolution and Gene Expression in BIONETS: A Mathematical and Experimental Framework. BIONETS Deliverable, European Commission (2008), http://www.bionets.eu

23. Egri-Nagy, A., Dini, P., Nehaniv, C.L., Schilstra, M.J.: Transformation Semigroups as Constructive Dynamical Spaces. In: Proceedings of the 3rd OPAALS International Conference, Aracaju, Sergipe, Brazil, March 22-23 (2010)

24. Egri-Nagy, A., Nehaniv, C.L.: Algebraic Properties of Automata Associated to Petri Nets and Applications to Computation in Biological Systems. BioSystems 94(1-2), 135–144 (2008)

25. Egri-Nagy, A., Nehaniv, C.L., Rhodes, J.L., Schilstra, M.J.: Automatic Analysis of Computation in Biochemical Reactions. BioSystems 94(1-2), 126–134 (2008)

26. Egri-Nagy, A., Nehaniv, C.L.: Hierarchical coordinate systems for understanding complexity and its evolution with applications to genetic regulatory networks. Artificial Life 14(3), 299–312 (2008) (Special Issue on the Evolution of Complexity)

27. Egri-Nagy, A., Nehaniv, C.L.: SgpDec - software package for hierarchical coordinatization of groups and semigroups, implemented in the GAP computer algebra system (2008), http://sgpdec.sf.net

28. Eigen, M., Schuster, P.: The Hypercycle. Naturwissenschaften 65(1) (1978)

29. Fielding, R.: Architectural Styles and the Design of Network-based Software Architectures. UC Irvine PhD Dissertation (2000), http://www.ics.uci.edu/fielding/pubs/dissertation/top.htm

30. Gabriel, R.P., Goldman, R.: Conscientious software. In: OOPSLA'06, Portland, Oregon, October 22-26 (2006)

31. Golubitsky, M., Stewart, I.: Nonlinear Dynamics of Networks: The Groupoid Formalism. Bulletin of the American Mathematical Society 43, 305–364 (2006)

32. Heistracher, T., Kurz, T., Marcon, G., Masuch, C.: D9.1-Report on Fitness Landscape. DBE Project (2005), http://files.opaals.org/DBE/deliverables

33. Hoare, C.A.R.: Communicating sequential processes. ACM Commun. 21(8), 666–677 (1978)
34. Horvath, G.: Functions and Polynomials over Finite Groups from the Computational Perspective. The University of Hertfordshire, PhD Dissertation (2008)
35. Horvath, G., Dini, P.: Lie Group Analysis of p53-mdm3 Pathway. In: Proceedings of the 3rd OPAALS International Conference, Aracaju, Sergipe, Brazil, March 22-23 (2010)
36. Kalman, R.E., Falb, P.L., Arbib, M.A.: Topics in Mathematical System Theory. McGraw-Hill, New York (1969)
37. Kauffman, S.: The Origins of Order: Self-Organisation and Selection in Evolution. Oxford University Press, Oxford (1993)
38. Krohn, K., Rhodes, J.: Algebraic Theory of Machines. I. Prime Decomposition Theorem for Finite Semigroups and Machines. Transactions of the American Mathematical Society 116, 450–464 (1965)
39. Kurz, T., Marcon, G., Okada, H., Heistracher, T., Passani, A.: D9.2-Report on Evolutionary and Distributed Fitness Environment. DBE Project (2006), http://files.opaals.org/DBE/deliverables
40. Lahti, J., Huusko, J., Miorandi, D., Bassbouss, L., Pfeffer, H., Dini, P., Horvath, G., Elaluf-Calderwood, S., Schreckling, D., Yamamoto, L.: D3.2.7: Autonomic Services within the BIONETS SerWorks Architecture. BIONETS Deliverable, European Commission (2009), http://www.bionets.eu
41. Louie, A.H.: A Living System Must Have Noncomputable Models. Artificial Life 13, 293–297 (2007)
42. Maeda, S.: The similarity method for difference equations. IMA Journal of Applied Mathematics 38, 129–134 (1987)
43. Manrubia, S.C., Mikhailov, A.S., Zanette, D.H.: Emergence of Dynamical Order. World Scientific, Singapore (2004)
44. Maturana, H., Varela, F.: Autopoiesis and Cognition, the Realization of the Living. D. Reidel Publishing Company, Boston (1980)
45. Milner, R.: A Calculus of Communication Systems. LNCS, vol. 92. Springer, Heidelberg (1980)
46. Milner, R., Parrow, J., Walker, D.: A Calculus of Mobile Processes, I. Inf. Comput. 100(1), 1–40 (1992)
47. Nehaniv, C.L., Rhodes, J.L.: The Evolution and Understanding of Hierarchical Complexity in Biology from an Algebraic Perspective. Artificial Life 6, 45–67 (2000)
48. Nicolis, G., Prigogine, I.: Self-Organization in Nonequilibrium Systems. Wiley, New York (1977)
49. Olver, P.: Applications of Lie Groups to Differential Equations. Springer, Heidelberg (1986)
50. Rashevsky, N.: Mathematical Biophysics and Physico-Mathematical Foundations of Biology, vol. II. Dover, New York (1960)
51. Rosen, R.: A Relational Theory of Biological Systems. Bulletin of Mathematical Biophysics 20, 245–260 (1958)
52. Rosen, R.: The Representation of Biological Systems from the Standpoint of the Theory of Categories. Bulletin of Mathematical Biophysics 20, 317–341 (1958)
53. Rosen, R.: Some relational cell models: The metabolism-repair systems. In: Rosen, R. (ed.) Foundations of Mathematical Biology. Cellular Systems, vol. II, Academic Press, London (1972)
54. Rosen, R.: Life Itself. Columbia University Press, New York (1991)

55. Rowe, J.E., Mitavskiy, B.: D8.1 - report on evolution of high-level software components (April 2005)
56. Sanchez, R., Grau, R., Morgado, E.: A novel Lie algebra of the genetic code over the Galois field of four DNA bases. Mathematical Biosciences 202, 156–174 (2006)
57. Sanchez, R., Morgado, E., Grau, R.: The genetic code boolean lattice. Communications in Mathematical and Computational Chemistry 52, 29–46 (2004)
58. Sanchez, R., Morgado, E., Grau, R.: Gene algebra from a genetic code algebraic structure. Journal of Mathematical Biology 51, 431–457 (2005)
59. Schreckling, D., Dini, P.: Distributed Online Evolution: An Algebraic Problem? In: IEEE 10th Congress on Evolutionary Computation, Trondheim, Norway, May 18-21 (2009)
60. Spivey, J.M.: The Z notation:a reference manual. Prentice-Hall, Inc., Upper Saddle River (1989)
61. Stewart, I.: Galois Theory, 2nd edn. Chapman and Hall, London (1989)
62. Van Leeuwen, I., Munro, A.J., Sanders, I., Staples, O., Lain, S.: Numerical and Experimental Analysis of the p53-mdm2 Regulatory Pathway. In: Proceedings of the 3rd OPAALS International Conference, Aracaju, Sergipe, Brazil, March 22-23 (2010)
63. Voss, R.F., Clarke, J.: 1/f noise in music: Music from 1/f noise. Journal of the Acoustical Society of America 63(1) (1978)
64. Weinstein, A.: Groupoids: unifying internal and external symmetry. Notices of the American Mathematical Society 43, 744–752 (1996)
65. Wolkenhauer, O.: Interpreting Rosen. Artificial Life 13, 291–292 (2007)
66. Wu, C.W.: Synchronization in Coupled Chaotic Circuits and Systems. World Scientific, Singapore (2002)

Transformation Semigroups as Constructive Dynamical Spaces

Attila Egri-Nagy[1], Paolo Dini[2], Chrystopher L. Nehaniv[1], and Maria J. Schilstra[1]

[1] Royal Society Wolfson BioComputation Research Lab
Centre for Computer Science and Informatics Research
University of Hertfordshire
Hatfield, Hertfordshire, United Kingdom
{a.egri-nagy,c.l.nehaniv,m.j.1.schilstra}@herts.ac.uk
[2] Department of Media and Communications
London School of Economics and Political Science
London, United Kingdom
p.dini@lse.ac.uk

Abstract. The informal notion of constructive dynamical space, inspired by biochemical systems, gives the perspective from which a transformation semigroup can be considered as a programming language. This perspective complements a longer-term mathematical investigation into different understandings of the nature of computation that we see as fundamentally important for the realization of a formal framework for interaction computing based on algebraic concepts and inspired by cell metabolism. The interaction computing perspective generalizes further the individual transformation semigroup or automaton as a constructive dynamical space driven by programming language constructs, to a constructive dynamical 'meta-space' of interacting sequential machines that can be combined to realize various types of interaction structures. This view is motivated by the desire to map the self-organizing abilities of biological systems to abstract computational systems by importing the algebraic properties of cellular processes into computer science formalisms. After explaining how semigroups can be seen as constructive dynamical spaces we show how John Rhodes's formalism can be used to define an Interaction Machine and provide a conceptual discussion of its possible architecture based on Rhodes's analysis of cell metabolism. We close the paper with preliminary results from the holonomy decomposition of the semigroups associated with two automata derived from the same p53-mdm2 regulatory pathway being investigated in other papers at this same conference, at two different levels of discretization.

1 Introduction

This expository paper has several goals and consists of three main parts. The first part concentrates on computer science and aims to show that transformation semigroups can provide a theoretical background for programming languages. The second part introduces the concept of interaction computing and discusses a possible architecture

F.A. Basile Colugnati et al. (Eds.): OPAALS 2010, LNICST 67, pp. 245–265, 2010.
© Institute for Computer Sciences, Social Informatics and Telecommunications Engineering 2010

for the Interaction Machine based on examples from cell biology. The third part streamlines the theory further and applies some of the algebraic results to the analysis of the p53-mdm2 regulatory pathway [24], as part of our on-going effort to understand and formalize the computation performed by the cell.

This paper is part of a research framework that is documented in the following four companion papers at this same conference:

- A Research Framework for Interaction Computing [7]
- Numerical and Experimental Analysis of the p53-mdm2 Regulatory Pathway [24]
- Lie Group Analysis of a p53-mdm2 ODE Model [16]
- Transformation Semigroups as Constructive Dynamical Spaces (this paper)
- Towards Autopoietic Computing [4]

1.1 The Programming Language Perspective

In the first part of the paper (Section 2) we would like to argue that:

1. Finite state automata, and thus transformation semigroups, have much wider applicability than it is thought traditionally.
2. A transformation semigroup is analogous to a programming language; therefore, whenever a piece of software can model some phenomenon, so does a semigroup.
3. There are different ways to achieve parallel computation, and they all fit naturally into the algebraic framework.
4. The presence of symmetry groups in a computational structure indicates a special kind of reversibility, which is not to be mistaken for the general idea of reversibility.

A *constructive dynamical space*[1] determines a set of possible processes (computations) and provides basic building blocks for these possible dynamics, equipped with ways of putting the pieces together. With these tools one can explore the space of possibilities; or, going in the other direction, given one particular dynamical system it is also possible to identify its components and their network of relations. A prime example is a programming language (complete with its runtime system): we build algorithmic (thus dynamical) structures using the language primitives in order to model or realize the dynamics that we are interested in. In this paper our main purpose is to show that finite state automata and transformation semigroups are other examples of constructive dynamical spaces, although they are not usually considered as such. The possible gain is that we can bring the algebraic results, mainly the hierarchical decomposition theories, into domains of applications outside mathematics. For example, currently programming languages do not include tools for automatic decomposition and reconstruction of the problem domain.

The term is used in artificial life research for artificial chemistries (e.g. [2]), and this does agree with our definition.

1.2 Interaction Computing

The second part of the paper (Section 3) extends the perspective on an individual automaton to two or more interacting automata, and explores the implications of such

---

[1] The term is used in artificial life research for artificial chemistries (e.g. [2]), and this does agree with our definition.

a generalization. Figure 1 shows the broader theoretical context within which this paper is situated. The study of semigroups as programming languages fits within the activity labelled as "Algebraic properties of automata structure" in the figure. The concept of interaction computing can be seen as an area of application that aims to combine *interacting* constructive dynamical spaces to create an environment analogous to the cell's cytoplasm. In particular, Section 3 considers the following points:

1. Interaction computing builds on ideas that have been around since Turing's 1936 paper [23] and that require a shift in how we think about computing.
2. The Interaction Machine can be given a formal foundation and a high-level architecture based on Rhodes's work [22].
3. The groups found in the hierarchical decomposition of the semigroups associated with the automata derived from cellular pathways suggest a form of parallel computation that relies on cyclic phenomena and on interdependent algorithms (i.e. symbiotic cohesion, and generally opposite to loose coupling).

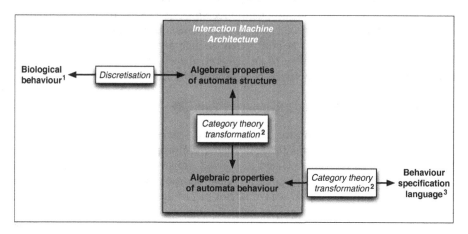

Fig. 1. Theoretical areas of relevance to interaction computing whose exploration is discussed in this and other papers: 1 = [24]; 2 = [8, 6, 21, 7]; 3 = future projects/papers; no superscript = this paper.

The term 'interaction computing' captures the essence of a particular form of biologically-inspired computing based on metabolic rather than evolutionary processes. Specifically, the concept of interaction computing is based on the observation that the computation performed by biological systems always involves at least two entities, each of which is performing a different, and often independent, algorithm which can only be advanced to its next state by interaction itself. The aim of the interaction computing approach is to reproduce in software the self-organizing properties of cellular processes.

For this concept to make sense in a biological context one needs to choose an appropriate level of abstraction. In particular, our perspective views the stochastic nature of cell biochemistry mainly as a mechanism of dimensional reduction that does not necessarily need to be emulated in any detail. For example, a gene expresses hundreds of mRNA molecules which, in turn, engage hundreds of ribosomes for no other reason than to maximize the probability that a particular, single genetic instruction will be carried out as a single step in a metabolic process, such as the synthesis of a particular enzyme. As a consequence of this dimensional reduction (hundreds to 1), a higher level of abstraction than that at which stochastic molecular processes operate is justified in the modelling approach – in particular, a formalization that retains, and builds on, the discrete properties of cell biology.

However, even the resulting lower-dimensional system can't plausibly be imagined to perform the complexity of a cell's functions driven simply by a uniform distribution of interaction probability between its (now fewer) components. Additional structure and constraints must be at play, for example as provided by molecular selectivity. Whereas the evolutionary processes that have led to molecular selectivity and the underlying physical processes that 'fold and hold' the relevant proteins and molecules together are fascinating phenomena that can be recognized as the ultimate and the material *causes* of order construction in biology, respectively, this does not entail that it is necessary to reproduce these mechanisms to arrive at self-organizing formal systems. We think it is sufficient to recognize the *effects* of these phenomena as embodying the essence of the cell's discrete behaviour as a kind of computation, whose ordered properties can be formalized through algebra. This is the motivation for our work in the development of an algebraic theory of interaction computing. A formal foundation for this theory has been provided by Rhodes's work [22].

1.3   Analysis of the p53-mdm2 Regulatory Pathway

The above ideas for how constructive dynamical spaces and interaction computing could be realized to improve the flexibility and self-* properties of software have only begun to be addressed. As explained in [7] and shown in Figure 1, using category theory we need to develop a mapping to relate the specification of behaviour to the algebraic structure(s) needed to realize it. But before this mapping can be developed we first need to understand in greater depth the algebraic structures that underpin the observed self-organizing properties of biological behaviour. Therefore, the third part of this paper (Section 4) looks in greater depth at the algebraic structure of the semigroup associated with the automata derived from a particular system, the p53-mdm2 regulatory pathway [24], for two levels of resolution in its discretization.

2   Theoretical Framework

2.1   Finite Automata as Computational Spaces

A finite state automaton and its algebraic counterpart, a transformation semigroup, is a computational object, i.e. something that computes. This is not at all surprising, especially from the extreme computationalist's viewpoint ("everything, that changes, computes") and from a classical source in cybernetics [1], but the statement still needs further explanation: *In what sense does an abstract algebraic structure entail computation?*

Formal definition. A finite automaton can be defined as $A = (A, Q, \lambda)$ where A is the set of *input symbols*, the *input alphabet*; Q is the set of *states*, the *state space* ; and λ is the *state transition function* $\lambda : Q \times A \rightarrow Q$. Since we are talking about a finite state automaton, all the objects involved are finite.

Static versus dynamical view. Traditionally, finite automata appear in computability and formal language theories [14]. For problems in those fields we need to extend the definition by giving initial and accepting states. Moreover, an output alphabet can be given together with an output function mapping a state and an input symbol pair to an output symbol. But here our main interest is not using the automaton for a particular purpose, e.g. recognizing certain languages, but to study the automaton itself as a *discrete dynamical system*. However, by looking at the minimalistic definition of finite automata, it is obvious that an automaton is not *dynamical*, i.e. nothing moves in it. Its dynamics, rather, is *potential*, the automaton creates room for possible 'movements'. Applying input symbols to states may move them to other states according to λ, producing a trajectory, but there is no predescribed dynamics of the automaton. The automaton itself rather is merely the setting for any one of an entire space of possible behaviours; therefore, we need to supply the machine with instructions, i.e. we have to provide a starting state and a sequence of input symbols in order to observe change, computation. We can combine the atomic transitions by concatenating input symbols that are interpreted by λ sequentially. This way the series of operations denoted by a sequence of symbols becomes an algorithm, and λ gives meaning to the symbols, thus λ acts as an interpreter. A computational problem can be formalized generally as getting to state r from state q, or producing output r from input q (the states q and r can be arbitrarily complex and r may not be known when starting the algorithm). An algorithmic solution for this problem is

$$q \cdot a_1 \cdots a_n = r \tag{1}$$

where $a_1,...,a_n$ is the sequence of steps of the algorithm.[2] At this point we encounter some possible terminological confusion. So far the input of the automaton was the input symbol, but in the above example we used the state as the input. Looking at λ, the 'machinery' of the automaton reveals that indeed it has two kinds of inputs as a state transition function: the state and the input symbol. This is very important since this view departs from the traditional interpretation. So we can talk about states as

---

[2] To exploit automata computational capacity further, one might also generalize (1) so that in addition to a sequence of atomic operations, $a_1 ...a_n$, one allows a sequence of atomic operations and variables that each evaluate to some basic operation based an additional component of input (i.e. other inputs than state q). That is, one has

$$q \cdot t_1 \cdots t_n = r \tag{2}$$

where each t_i is either an input symbol a $\in$ A, or a variable v, ranging over A, whose value is determined by some 'environmental' input. The generalization allows one to make use of so-called *functional completeness* properties of simple non-abelian groups that allow them to compute any finitary function (similar to the two-element Boolean algebra). See [18, 15]. For simplicity, we do not pursue this promising variant further in this particular paper.

inputs, like data fed into the algorithm. This looks quite natural if we think in terms of a universal computer: we have to provide both the algorithm and the input data for the algorithm to work on.

This concludes the introduction of the metaphor

$$\text{finite state automaton} \equiv \text{programming language.}$$

Before continuing, let's summarize: Q, the set of states of the automaton A, is the problem domain in which we would like to solve problems. These problems include going from a start state to a target, a 'desired' state, or a solution. We can also consider states as bits of data and do calculations on them, so the 'number crunching' aspect of computation also appears here. A, the set of input symbols, contains the language primitives, i.e. the basic commands that can be combined into longer sequences to form programs. The transition function λ is then the machinery: the compiler, interpreter, runtime system with libraries, processor, etc. In short,

states Q	problem domain
input symbols A	basic commands
input strings A^+	algorithms
$w : Q \to Q, \ w \in A^+$	computational function
transition function λ	runtime system

We need to differentiate between the automaton and the mathematical function (machine) that the automaton implements (realizes), since the same computational function (same state mapping) can be expressed by different algorithms (sequences). In other words, following Rhodes's terminology, the machine is the behaviour while the automaton or circuit is what implements it.

2.2 Problems, Critique

Problem: "Finite state automata are too constrained compared to general-purpose programming languages". At first sight an automaton as a programming language seems to be quite constrained. We show that the only constraint is finiteness, and this poses no practical problems. Usually programming languages are all considered to be Turing-complete, thus being in a completely different computational class than the one formed by finite automata. This is only true if we assume unbounded resources for the machine on which the programming language is implemented. But if we consider a physically existing (hence finite) computer, then we have a finite state automaton. Its state set may be enormous but still finite. For example, one state of a computer with 1 gigabyte memory can be described by 8796093022208 bits, therefore it has $2^{8796093022208}$ distinct states, so the number of possible state transformations on this computer is

$$\left(2^{8796093022208}\right)^{2^{8796093022208}},$$

which is quite a big number. Of course this large collection of possible computational functions contains the workings of all possible operating systems, with all possible applications as substructures. The situation is reminiscent of the Library of Babel [3].

One of the great capabilities of the human mind is that it can look at things from different perspectives, ignoring the view in which they normally appear. This is really needed here as we usually meet finite automata only of moderate size and usually in textbooks (probably the largest ones in natural language processing), but still not in the magnitude mentioned above. Thus, *finiteness is the only constraint*, and that also applies to programming languages implemented on computers.

"Finite state automata-based programming is a specialized technique mainly for implementing lexical and syntactic analysers". Our approach is not about one particular technique that can be applied in a programming situation (like using state transition tables in an algorithm), but a more general conceptual framework to enable us to see the applications of automata in more general settings, in dynamical systems.

2.3 A More Comprehensive View: Transformation Semigroups

A^+ forms a semigroup under the operation of concatenation, and its elements can be interpreted, using λ recursively, as different mappings of the state set to itself. The finite set of these mappings is called the semigroup $S(A)$ corresponding to the automaton A and these mappings can be combined by function composition (i.e. following one by another). In (1), a computation was just one run of the algorithm on a particular input x (in (1) this input is the state q). The same algorithm can be applied to other inputs (states) as well. Therefore, each $a \in A$ is an (atomic) algorithm which yields a (single) mapping $Q \rightarrow Q$; the latter is a (generating) element of the corresponding transformation semigroup $S(A)$. So semigroup elements realize algorithms defined for all possible inputs; therefore, a semigroup is a "constructive space of total algorithms", which is basically a programming language.

Built-in primitives $\equiv$ generator sets. Programming languages are equipped with different sets of built-in basic tools; thus the same algorithm requires longer code in a low-level language than in a high-level one. Analogously, a transformation semigroup with a smaller set of generators will have longer words to express the same semigroup element. Different semigroups can have common elements, just like different programming languages can share the same tools and techniques. The extreme case is the automaton where we have a symbol for each semigroup element, corresponding to a language in which there is a language primitive for each expressible algorithm.

Self-modifying algorithms. This powerful idea is incorporated in some languages (e.g. LISP), but it is not a mainstream property. In the algebraic settings it is natural as semigroup elements can be multiplied by each other, thus algorithms can act on each other (algorithms as inputs and outputs of other algorithms), thereby modifying themselves. This is just a consequence of the generalization of Cayley's theorem to semigroups.

2.4 Going Parallel: The Role of Permutation Groups

Once we start talking about transformation semigroups, the special class of permutation groups naturally comes into the picture. As a consequence, we need to give an account of symmetries in a computational context.[3]

Reversibility. A permutation of the state set (or a subset of it) as a totally defined algorithm has the peculiar property that, when applied to the whole set in parallel, it gives back the whole set: no two states/inputs are collapsed into the same output. This is like maintaining the set of possible future states, not losing any possibility. This prompts us to believe that algebraic groups of algorithms coincide with the notion of reversibility. This is certainly true if we consider the parallel movement of all states, but it becomes subtle if we study individual inputs: permutations are not the only way to achieve reversibility. The following examples will show the different kinds of reversibility.

Example: Reversibility by resets. Let's consider state 1 of the state set $\{1, 2\}$ and two transformations $t_1 = \left(\begin{smallmatrix}1 & 2\\ 1 & 1\end{smallmatrix}\right)$, $t_2 = \left(\begin{smallmatrix}1 & 2\\ 2 & 2\end{smallmatrix}\right)$. t_2 takes 1 to 2, and we can reverse this movement since we can go back to 1 by t_1. t_1, t_2 are called constant maps, or resets.

Example: Cyclic reversibility. Similarly, by applying the permutation $(1, 2)$ (or $p = \left(\begin{smallmatrix}1 & 2\\ 2 & 1\end{smallmatrix}\right)$) we can go from 1 to 2: $1 \cdot p = 2$, but we can get back to 1 by applying p again: $1 \cdot pp = 1$. So we achieve reversibility by 'going forward', by a kind of "cyclic reversibility". This can be achieved only by permutations.

Cycles. Another interpretation of the symmetries that algebraic groups symptomize is the presence of periodic cycles of reactions in the metabolic systems from which the automata are derived, the invariant feature being the existence of a stable cycle that is preserved by the elements of the group. This too is compatible with the parallel interpretation of the algorithm arising from a group. In fact, conceptually, a group element can be conceived to act on all the states at once, and this can be implemented as a set of parallel automata (as many as there are states), all transitioning simultaneously and in parallel. Such a model is entirely consistent with periodic pathways such as the Krebs cycle, whose equilibrium operation is characterized by all the reactions taking place in parallel, at the same average rate. In such a case a group element can be considered as realizing a parallel algorithm on several inputs. This can be viewed on different levels:

1. acting on the whole Q
2. acting on the subsets of Q
3. acting on multisets of elements of Q

More generally, there are other natural possibilities too (for any positive integer n): n-tuples of elements of Q, n-tuples of subsets of Q, or n-tuples multisets of elements of

---

[3] This is because of two facts: (1) a symmetry of an object is a transformation that leaves some aspect of that object invariant; and, (2) the set of invertible transforma- tions that leave some aspect of a mathematical object invariant form a group with the operation of functional composition.

Q, not to mention coproduct structures, etc. As transformations are the computational functions realized by totally defined algorithms, the definitions of these actions can be derived in a natural way from the action of transitions symbols of the automaton.

2.5 The Gain: Algebraic Results

After we presented the analogy, finally we can show the benefits of thinking about the automata in a different way.

Hierarchical coordinatization. When dealing with huge automata we need some tools for organizing their structure. Just as, when analysing software, we find modules and subroutines that are combined in a hierarchical way, so too we can decompose semigroups with hierarchical coordinates [20].

The disappearance of memory–processor duality. The algebraic theory of machines gives us a level of abstraction at which we can consider both processing and memory units having the same nature, namely they are semigroups. For storing single bits of information the semigroup of the flip-flop automaton can be used and for calculations the permutation groups can be used. In the hierarchical construction of automata this allows us to dynamically change the ratio of computational power and storing capacity on demand, as in real computational problems either the power of the processor or the insufficient amount of memory is the bottleneck. For exploiting this idea in practical computing we would need some physical implementation, e.g. reconfigurable hardware that is capable of dynamically allocating automata structures.

3 The Interaction Machine Concept

The above discussion provides a theoretical backdrop against which we can now propose some novel ideas about computing, in particular interaction computing. The concept of interaction computing was originally inspired by a physics perspective on biological systems ([5, 7] and references therein); in other words, in looking for general principles that appear to govern self-organizing behaviour in cell biology, the role of interactions appeared to be such a fundamental feature that it seemed indispensable to replicate it in computational systems in order to develop an 'architecture of self-organization' in software along analogous principles.

The addition of an algebraic automata theory perspective to the above stream of work has opened the possibility to develop a formalism that can express the behaviour of biological systems seen as dynamical systems in a manner that is consistent with the mathematical foundations of computer science. We now discuss a few examples that are helping us understand how the algebraic properties of automata discussed so far can be interpreted in the context of 'biological computation', and how this mapping between algebra and biology is helping us imagine the architectural requirements of the 'Interaction Machine'.

3.1 Sequential Machines and Their Realizations

In this section we now consider not only computations inside individual transformation semigroups (as in Section 2), but also (1) their behaviours (abstracting away states) as formalized by *sequential machines*, which is closely related to software specification, and (2) a higher-level constructive dynamical 'meta- space', one in which sequential machines are basic entities that may interact and combine in various ways to create new kinds of computational structures; in particular, they extend to *interacting machines*, whose deployment together can create complex interaction structures.

Many years after proving the prime decomposition theorem for semigroups and machines [19], John Rhodes published a book that he had started working on in the 1960s and that has come to be known as the 'Wild Book' [22]. In this book he provides a very clear discussion of an alternative to Turing machines, which we believe to be a very promising starting point for a model of interaction computing.

As we know, an algorithm implementable with a Turing machine is equivalent to the evaluation of a mathematical function. As Wegner and co-workers argued in a series of papers over the last 20 years ([13] and references therein), the evaluation of a mathematical function can afford to take place by following an 'internal clock', i.e. the Turing machine is isolated from its environment while it evaluates the algorithm. Biological systems, on the other hand, are continually interrupted by external inputs and perturbations. As an example of this class of computations Golding and Wegner used 'driving home from work', which they described as a non-Turing-computable problem. Turing himself had foreseen this possibility in his original 1936 paper as the 'choice machine' [23], although he did not pursue it further.

Similarly, Rhodes starts from the familiar definition of a sequential machine. The sequential machine accepts an input at each discrete point in time and generates an output once all the inputs have been received:

> Let A be a non-empty set. Then $A^+ = \{(a_1, ..., a_n) : n \geq 1$ and $a_j \in A\}$. A *sequential machine* is by definition a function $f : A^+ \rightarrow B$, where A is the basic input set, B is the basic output set, and $f(a_1, ..., a_n) = b_n$ is the output at time n if a_j is the input at time j for $1 \leq j \leq n$.

This is clearly related to the formal definition of the finite automaton given at the beginning of Section 2, but there is a twist: Rhodes prefers to make a sharp distinction between a *machine*, which he equates to a *mathematical function*, and the *realization of that machine*, which he calls a *circuit* and that is essentially an *automaton*:

Mathematical concept...	...and its realization
Machine or mathematical function	Circuit or automaton
Automata behaviour	Automata structure

Due to the need to maintain the development of a theory of interaction computing on firm mathematical grounds, we follow his approach. More specifically, the separation between a machine and its realization matches well the distinction between the description (formalization) of *behaviour* and the automaton *structure* necessary to achieve it. It is essential for us to maintain this distinction in light of a parallel thread of research [7, 6] which applies categorical morphisms to automata behaviour in order to derive a specification language. To understand better what we might be aiming to specify, let's develop the idea further.

We are going to use a generalization of the sequential machine, also by Rhodes, which produces an output for each input it receives and not just in correspondence of the most recent input. We are going to call this generalization an interact**ing** machine:

> Let $f : A^+ \to B$ be a sequential machine. Then an *interacting machine* f^+ :
> $A^+ \to B^+$ is defined by $f^+(a_1, ..., a_n) = (f(a_1), f(a_1, a_2), ..., f(a_1, ..., a_n))$.

Thus, an Interact**ion** Machine (IM) can be built by joining two or more interact**ing** machines. Such an IM will still accept inputs from outside itself ("the environment") and will produce outputs for the environment. The realization of either machine is achieved through a finite-state automaton that Rhodes calls a "circuit", C , but that in the literature is more commonly called a Mealy automaton:

> C = (A, B, Q, λ, δ) is an *automaton* with basic input A, basic output B, states Q, next-state function λ, and output function δ iff A and B are finite non-*empty sets, Q is a non-empty set*, $\lambda : Q \times A \to Q$, and $\delta : Q \times A \to B$.

Having established then that the problem of computation is posed in two parts, a mathematical function and its realization, we continue to rely on Rhodes to define a few more concepts related to the latter, in order to develop a relatively concrete working terminology. The next concept we need is the realization of an algorithm, as follows. Let $C = (A, B, Q, \lambda, \delta)$ be an automaton. Let $q \in Q$. Then $C_q : A^+ \to B$ is the *state trajectory associated with state q* and it is defined inductively by

$$C_q(a_1) = \delta(q, a_1)$$
$$C_q(a_1, ..., a_n) = C_{\lambda(q, a_1)}(a_2, ..., a_n), \text{ for } n \geq 2.$$

We say that C *realizes* the machine $f : A^+ \to B$ *iff* $\exists q \in Q : C_q = f$. By a simple extension of the above definitions it is fairly easy to see that the output of an algorithm can be associated with a sequential machine when the output corresponds to the result after the last input, whereas it is associated with an interacting machine when there are as many outputs as there are inputs:

$$C_q(a_1, ..., a_k) = b_k, \qquad \text{for } k = 1, ..., n$$
$$C^+{}_q(a_1, ..., a_n) = (b_1, ..., b_n).$$

Rhodes then introduces more formalism to define precisely a particular automaton that realizes a function f as C (f), and goes on to prove that C (f) is the unique minimal automaton that realizes f. He goes on to say

> The reason why Turing machine programs to realize a computable f are not unique and the circuit which realizes the (sequential) machine f (namely C (f)) is unique is not hard to fathom. In the sequential machine model we are given much more information. It is 'on-line' computing; we are told what is to happen at each unit of time. The Turing machine program is 'off-line' computing; it just has to get the correct answer – there is no time restraint, no space restraint, etc. ([22]: 58)

Rather than constructing dynamical behaviour through a sequential algorithm expressed in a programming language, which can be realized by a single automaton or Turing machine, we are talking about constructing dynamical behaviour through the interaction of two or more *finite*-state automata. This is because, if the possibility that Q be infinite is left open as the above definition does, "then the output function could be a badly non-computable function, with all the interesting things taking place in the output map δ, and we are back to recursive function theory" ([22]: 59). Therefore, we note the interesting conclusion that, for a tractable approach, Q must be finite and that the realization of an Interaction Machine must be made up of interacting finite-state automata. Thus, from the mathematical or behavioural perspective, we will build the Interaction Machine by using sequential (or interacting) machines as basic units and by combining them in various ways.

The mathematical and computer science challenge, therefore, is to develop a formalism that is able to capture the non-linear character of the dynamics of an arbitrary number of coupled metabolic systems, so that when this mathematical structure is mapped to our yet-to-be-developed specification language we will be able to specify software *environments* capable of supporting self-organizing behaviour through interaction computing, driven by external stimuli from users or other applications.

In the next subsection we switch the perspective from synthetic, i.e. *constructing* automata and models of computation, to analytical, i.e. *analysing* cellular pathways and interpreting the role and function of the algebraic structures they harbour.

3.2 Permutation Groups from the Biological Point of View

The presence of permutation groups in the hierarchical decomposition of the semigroups associated with automata derived from cellular pathways presents a difficult puzzle. This is because a permutation group can be interpreted in different ways: the fact that groups are defined up to isomorphism means that different 'implementations' could be derived from the same abstract group. In this subsection we explore some of the possibilities. Ultimately, if we are not able to resolve this puzzle theoretically, *ex ante*, it will be resolved *ex post* by building different kinds of interaction machines, based on different interpretations of these groups, and by seeing which kind behaves in the most convincingly 'biological' way.

Based on the analysis of the Krebs metabolic cycle and the p53-mdm2 regulatory pathway, it appears that the permutation groups may be associated with cyclic phenomena in biochemical processes. The challenge is to relate such cyclic or oscillatory behaviour to the computational properties of its mathematical discretization.

It is helpful, first, to clarify the connection between cycles and oscillations by resorting to an idealized system in the form of the simple harmonic oscillator of elementary physics. As shown in Figure 2, for a simple harmonic oscillator periodic cycles (in some parameter space, which could include also Euclidean space) are mathematically indistinguishable from oscillations (in time, at a fixed point in space). It is possible, therefore, that the same kind of algebraic structure, i.e. non-trivial permutation groups, could result from different kinds of cyclic biochemical phenomena. For example, the discrete algebraic analysis discussed in the next section indicates that biochemical processes such as the Krebs cycle, the concentration levels of whose metabolites can be assumed to remain constant over time, have similar 'algebraic signature' (i.e. permutation groups) to processes such as the p53-mdm2 pathway [24, 16], in which the concentrations of the compounds can oscillate (for some parameter values) as a function of time.[4]

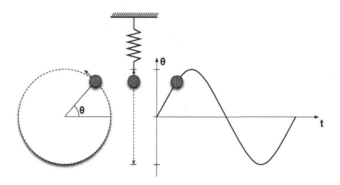

Fig. 2. Periodic behaviour of the simple harmonic oscillator

The reason may be found in the fact that the models that give rise to these signatures describe what will happen (or what could happen, if there is a choice) to individual instances of classes of molecules or molecular complexes, but do not distinguish between instances of the same class. In simulations of the synchronized Krebs cycle and the p53-mdm2 pathway under conditions that promote sustained oscillations, indistinguishable instances of particular classes, such as citrate and active p53, appear and disappear periodically.

What could be the computational significance of such periodic structures? We see at least two possibilities.

Cyclic interpretation. If a element g of a permutation group G is interpreted as a function $g : X \rightarrow X$ from a subset X of the state set Q to itself, then if g is non-trivial, then some element $x \in X$ is actually moved by g, so when the system reaches state $x \in X$ and is acted upon by g it will transition to state $y \in X$, i.e. $xg = y$ (where the group action is taken as multiplication on the right). Now, if g is applied again, this time to y, another state $z \in X$ will be reached. This process could be repeated such that

$$x \rightarrow xg \rightarrow xg^2 \rightarrow xg^3 \rightarrow \cdots \rightarrow xg^{(n-1)} \rightarrow x, \qquad (3)$$

---

[4] We hasten to add that the permutation groups found in these two systems are *not* of the same kind. Therefore, the two systems are different in some important way. However, our first goal is to find a plausible explanation for the presence of permutation groups in the first place, and it is only in this specific sense that we mean that these two systems have the same algebraic signature.

where the arrow "→" means "multiply on the right by g", for some $n > 1$, where n divides the order of g.[5] This is periodic behaviour, or a cycle of the system's states, that the system can potentially exhibit.

Parallel interpretation. Although we are still far from pinpointing the characteristics of 'cellular computation', as discussed in [6] and consistently with the discussion in Section 2 there is one aspect that is hard to overlook: parallelism. It is obvious that biochemical events take place in parallel in solution. We also know that a group element acting on a set can be conceived of as permuting *all* the elements of that set, at once. In the language of automata, this corresponds to the parallel transition of *every* state of the automaton to its next state as determined by the transition function, in parallel. This is not a single machine, clearly. We are talking about an ensemble of machines that are computing in parallel, each of which is at a different location along the *same* algorithm.

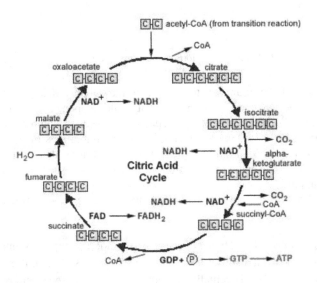

Fig. 3. Schematic of the Krebs or citric acid cycle [17]

A specific example could help elucidate the above point. Take, for instance, the Krebs or citric acid cycle, shown in Figure 3. It is clear that all the reactions around the cycle can be happening simultaneously and in parallel. But if we assume that the average concentrations of the metabolites in each stage around the cycle are constant then, strictly speaking, this system is always in the same state. Thus, in this case the discretization of this system into a finite-state automaton must be based on the history of an individual molecule as it travels around the circle. Such a molecule will indeed experience different 'states' as it is transformed at each stage. The hierarchical

---

[5] For example, $g = (123)(45)$ has order 6, and g acting repeatedly on the state 4 is $4 \to 4g = 5 \to 5g = 4g^2 = 4$, i.e. $n = 2$ in this case and 2 divides 6.

decomposition of the semigroup associated with such an automaton contains non-trivial permutation groups, e.g. C_2 and C_3 [9, 22]. This could indicate the parallel transition of 2 and 3 (abstracted or 'macro-') states, respectively under the same metabolic program (sequence of atomic transition operations). Once the concept of parallel transitions has been accepted as a possibility the generalization of the same mechanism to the simultaneous transition of all the states and macro-states permuted by a given subgroup, as we are proposing here, is equally possible and worth investigating further.

3.3   A Conceptual Architecture for the IM

Building on the above discussion, let us now simplify the system such that it is closed and let us neglect the irreversible (semigroup) components for the moment. Rather than performing an isolated algorithm to evaluate a mathematical function, as a Turing machine does, we are talking about an 'open' or 'permeable' algorithm, which can be coupled to other algorithms at each of its steps. A non-trivial question concerns the form such a coupling could or should take.

As we saw in Section 2 an algorithm is realized by one or more generating semigroup elements concatenated in a sequence or string of symbols. Because each semigroup element is a mapping $Q \rightarrow Q$, knowledge of the algorithm and of the state space implies knowledge of the state transition function λ. Therefore, an algorithm is a part of the realization of a sequential machine or mathematical function in the sense of Rhodes. It may therefore be more useful and appropriate to talk about the coupling between automata rather than between algorithms.

Automata can be coupled in different ways, for example by overlapping on one or more states. Such a case seems important for modelling symbiosis but presents some difficulties, since a common state for two automata would seem to imply that the two automata are actually associated with the same physical system and simply drawing on different sets of its states, with one in common. Alternatively, the output alphabet of an automaton could have some symbols in common with the input alphabet of another automaton, and vice versa. In this way, each automaton reaching certain states would trigger inputs to the other, causing it to transition. The same arrangement could be generalized to 3 or more automata coupled in different ways. Thus, state transitions taking place in one automaton could trickle through the system, causing other automata to advance their own algorithms.

What remains to be seen is whether such an architecture of computation will yield more powerful qualities (in the sense of non-functional requirements), e.g. greater efficiency, where efficiency could be defined as

$$\frac{\text{number of functional requirements satisfied}}{\text{number of state transitions required}} \tag{4}$$

This is in fact what the map of metabolic pathways looks like: a complex network of automata intersecting or communicating at several steps, which therefore means that the chemical reactions taking place at each of these intersection points are 'serving' multiple algorithms at the same time. This is one of the points of inspiration of the concepts of interaction computing, symbiotic computing, and multifunctionality [7].

From an evolutionary point of view, the emergence of an architecture that seems to violate just about all the accepted principles of software engineering (modularization, separation of concerns, loose coupling, encapsulation, well-defined interfaces, etc) might be explainable simply in terms of efficiency. In the presence of finite energy resources and multiple survival requirements, the organisms and the systems that survived were those that could optimize the fulfilment of such requirements for the most economical use of materials and energy; hence the overloading, the multifunctionality, and so forth. This in fact is not an unfamiliar concept in software engineering: where resources are constrained, for example in specialized drivers for real-time systems, it is common to code directly in assembly language and to develop code that is incredibly intricate and practically impossible for anyone but the engineer who wrote it to understand. Evolution has achieved the same effect with biological systems, but to an immensely greater extent.

Whether or not the above concepts will withstand closer scrutiny as we continue to develop the theory, it is interesting to show a visualization of the interaction machine according to Rhodes, in Figure 4 ([22]: 198). Rhodes does not use this term, but he does talk about the DNA and the cytoplasmic processes as interacting automata. His description of the cell as a system of interacting automata comes after the formal development of the sequential machine and its realization (that we presented above) all the way to the Krohn-Rhodes prime decomposition theorem ([22]: 62), and after a lengthy analysis and discussion of the Krebs cycle. Therefore, we expect Rhodes's work to continue to provide valuable inputs to our emerging theory of interaction computing.

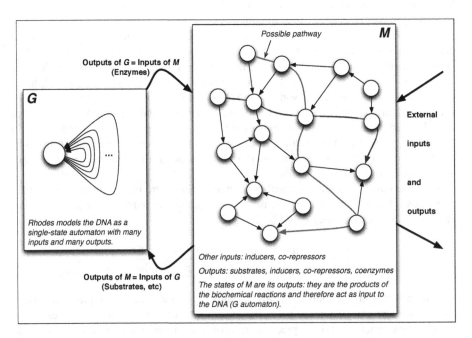

Fig. 4. Conceptual architecture of the biological realization of the Interaction Machine, based on Rhodes ([22]: 198)

4 Some Results from the Algebraic Analysis of the p53-mdm2 Regulatory Pathway

By using the now available SgpDec computer algebra software package [12] we can start applying these algebraic methods to real-world biological problems.

4.1 Petri Net of p53-mdm2 System

Here we briefly study the algebraic structure of the p53-mdm2 system. The p53 protein is linked to many other proteins and processes in the cell, but its coupling to the mdm2 protein appears to be particularly important for understanding cancer. Depending on the concentration level of p53, the cell can (in order of increasing concentration): (1) operate normally; (2) stop all functions to allow DNA repair to take place; (3) induce reproductive senescence (disable cellular reproduction); and (4) induce apoptosis instantly (cell 'suicide'). Therefore, p53 is a very powerful and potentially very dangerous chemical that humans (and most other animals) carry around in each cell, whose control must be tuned very finely indeed. Roughly 50% of all cancers result from the malfunction of the p53- mdm2 regulatory pathway in damaged cells that should have killed themselves.

P53 levels are controlled by a fast feedback mechanism in the form of the mdm2 protein. P53 is synthesized all the time, at a fairly fast rate; but the presence of p53 induces the synthesis of mdm2, which binds to p53 and causes it to be disintegrated. When the DNA is damaged (for instance by radiation in radiotherapy) the cell responds by binding an ATP molecule to each p53, bringing it to a higher energy level that prevents its destruction and causes its concentration to rise. Thus in the simplest possible model there are in all 4 biochemical species: p53 (P), mdm2 (M), p53-mdm2 (C), and p53* (R), whose concentrations are modelled by 4 coupled and non-linear ordinary differential equations (ODEs) [24]. Whereas [16] analyzes the algebraic structure of this pathway from the point of view of the Lie group analysis of the ODEs, in this paper we look at the algebraic structure of the same pathway from the point of view of the discrete finite-state automata that can be derived from it.

4.2 Algebraic Decomposition

It is now a common practice to model biological networks as Petri nets. Petri nets can easily be transformed into finite automata, though this operation should be carried out with some care (for a detailed explanation see [11]). Depending on the capacity allowed in the places of the Petri net, the automaton's state set can be different in size, thus we can get different resolutions during discretization.

When we distinguish only between the presence and absence (in sufficient concentration) of the molecules, the derived automaton has only 16 states (see Fig. 5).[6] This may be considered as a very rough approximation of the original process, but still group components do appear in this simple model: S_3, the symmetric group on 3 points is among the components. The existence of these group components is clearly connected to oscillatory behaviours, however the exact nature of this relation still needs to be investigated.

---

[6] 4 places each taking on 2 possible values (0 or 1) makes $2^4 = 16$ states.

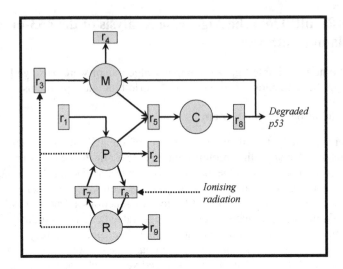

Fig. 5. Petri net for the p53-mdm2 regulatory pathway [10]. P = p53, M = mdm2, C = p53-mdm2, R = p53*.

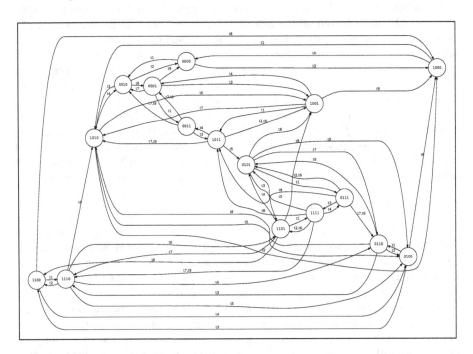

Fig. 6. Automaton derived from 2-level Petri net of the p53 system (16 states). The labels on the nodes encode the possible configurations for M, C, P and R (in this order). 0 denotes the absence, 1 the presence of the given molecule. For instance, 0101 means that C and R are present.

If we in addition distinguish two levels of significant concentration, i.e. absence (or very low, sub-threshold concentration), low, and high levels concentrations, then the corresponding automaton has 81 states ($3^4 = 81$) and the algebraic decomposition reveals that it contains S_5 among its components. The appearance of S_5 is particularly interesting as it contains A_5, the alternating group on 5 points, which is the smallest SNAG (simple non-abelian group). The SNAGs are considered to be related to error-correction [22] and functional completeness [18, 15, 7]. The latter property of SNAGs makes them a natural candidate for realizing an analogue of 'universal computation' within the finite realm. The presence of a SNAG in the decompostion of the p53-mdm2 path- way suggests that the cell could potentially be performing arbitrarily complex finitary computation using this pathway. Moreover, how the activity of the rest of the cell interacts with the p53-mdm2 pathway to achieve such fine regulatory control could depend on other components (interacting machines) in the interaction architecture of the cell harbouring similar group structures.

5 Conclusions

In this paper we have tried to paint a broad-brush picture of the possibilities for the roles of algebraic structures in enabling open-ended constructive computational systems, based on recent work and on our current results. The most immediate example is to regard a transformation semigroup as a constructive dynamical space in which computation can be carried out, analogous to programming languages on present-day computers. Although practical implementations of automata in real computers cannot help being finite, the size of the state space of a modern computer is well beyond our human ability to count without losing all sense of scale (many times the number of molecules in the known universe) and so 'approximates' the infinite size of a Turing machine reasonably well for practical purposes.

The work of Rhodes suggests, however, a much stronger claim, that sequential and interacting machines inspired by biological systems must be finite in a much more limiting sense, and that algebraic structure is the key to finitary computation and to understanding how computation can proceed via interaction. This leads us to a meta-space of constructive dynamical systems in which the interacting parts that synergetically conspire to achieve complex computation are individual machines or transformation semigroups themselves. We are combining this insight with the results of the algebraic analysis of the semigroups derived from cellular pathways and with more synthetic tentative arguments about the architecture of an Interaction Machine that could replicate biological computation. We have not yet reached actionable conclusions, but we feel we are uncovering interesting mathematical and computational properties of cellular pathways and are finding intriguing correspondences between biochemical systems, dynamical systems, algebraic systems, and computational systems.

Acknowledgments

Partial support for this work by the OPAALS (FP6-034824) and the BIONETS (FP6-027748) EU projects is gratefully acknowledged.

References

1. Ashby, W.R.: An Introduction to Cybernetics. Chapman & Hall Ltd., London (1956), http://pespmci.vub.ac.be/books/IntroCyb.pdf
2. Banzhaf, W.: Artificial chemistries towards constructive dynamical systems. Solid State Phenomena, 97–98, 43–50 (2004)
3. Borges, J.L.: La biblioteca de Babel (The Library of Babel). El Jardin de senderos que se bifurcan. Editorial Sur. (1941)
4. Briscoe, G., Dini, P.: Towards Autopoietic Computing. In: Proceedings of the 3rd OPAALS International Conference, Aracaju, Sergipe, Brazil, March 22-23 (2010)
5. Dini, P., Briscoe, G., Van Leeuwen, I., Munro, A.J., Lain, S.: D1.3: Biological Design Patterns of Autopoietic Behaviour in Digital Ecosystems. OPAALS Deliverable, European Commission (2009), http://files.opaals.org/OPAALS/Year_3_Deliverables/WP01/
6. Dini, P., Horvath, G., Schreckling, D., Pfeffer, H.: D2.2.9: Mathematical Framework for Interaction Computing with Applications to Security and Service Choreography. BIONETS Deliverable, European Commission (2009), http://www.bionets.eu
7. Dini, P., Schreckling, D.: A Research Framework for Interaction Computing. In: Proceedings of the 3rd OPAALS International Conference, Aracaju, Sergipe, Brazil, March 22-23 (2010)
8. Dini, P., Schreckling, D., Yamamoto, L.: D2.2.4: Evolution and Gene Expression in BIONETS: A Mathematical and Experimental Framework. BIONETS Deliverable, European Commission (2008), http://www.bionets.eu
9. Egri-Nagy, A., Nehaniv, C.L., Rhodes, J.L., Schilstra, M.J.: Automatic Analysis of Computation in BioChemical Reactions. BioSystems 94(1-2), 126–134 (2008)
10. Egri-Nagy, A., Nehaniv, C.L., Schilstra, M.J.: Symmetry groups in biological networks. In: Information Processing in Cells and Tissues, IPCAT'09 Conference, April 5-9 (2009) (Journal preprint)
11. Egri-Nagy, A., Nehaniv, C.L.: Algebraic properties of automata associated to petri nets and applications to computation in biological systems. BioSystems 94(1-2), 135–144 (2008)
12. Egri-Nagy, A., Nehaniv, C.L.: SgpDec - software package for hierarchical coordinatization of groups and semigroups, implemented in the GAP computer algebra system (2008), http://sgpdec.sf.net
13. Golding, D., Wegner, P.: The Church-Turing thesis: Breaking the myth. In: Computability in Europe (CiE) conference series (2005)
14. Hopcroft, J.E., Motwani, R., Ullman, J.D.: Introduction to Automata Theory, Languages, and Computation, 2nd edn. Addison-Wesley Longman Publishing Co., Inc., Amsterdam (2001)
15. Horvath, G.: Functions and Polynomials over Finite Groups from the Computational Perspective. The University of Hertfordshire, PhD Dissertation (2008)
16. Horvath, G., Dini, P.: Lie Group Analysis of p53-mdm3 Pathway. In: Proceedings of the 3rd OPAALS International Conference, Aracaju, Sergipe, Brazil, March 22-23 (2010)
17. ICT4US, http://ict4us.com/mnemonics/en_krebs.htm
18. Krohn, K., Maurer, W.D., Rhodes, J.: Realizing complex boolean functions with simple groups. Information and Control 9(2), 190–195 (1966)
19. Krohn, K., Rhodes, J.: Algebraic theory of machines. I. Prime decomposition theorem for finite semigroups and machines. Transactions of the American Mathematical Society 116, 450–464 (1965)

20. Krohn, K., Rhodes, J.L., Tilson, B.R.: The prime decomposition theorem of the algebraic theory of machines. In: Arbib, M.A. (ed.) Algebraic Theory of Machines, Languages, and Semigroups, ch. 5, pp. 81–125. Academic Press, London (1968)
21. Lahti, J., Huusko, J., Miorandi, D., Bassbouss, L., Pfeffer, H., Dini, P., Horvath, G., Elaluf-Calderwood, S., Schreckling, D., Yamamoto, L.: D3.2.7: Autonomic Services within the BIONETS SerWorks Architecture. BIONETS Deliverable, European Commission (2009), http://www.bionets.eu
22. Rhodes, J.L.: Applications of Automata Theory and Algebra via the Mathematical Theory of Complexity to Biology, Physics, Psychology, Philosophy, and Games. World Scientific Press, Singapore (2009); foreword by Hirsch, M.W. edited by Nehaniv, C.L. (Original version: University of California at Berkeley, Mathematics Library, 1971)
23. Turing, A.: On Computable Numbers, with an Application to the Entschei-dungsproblem. Proceedings of the London Mathematical Society 42(2), 230–265 (1936); a correction. ibid 43, 544–546 (1937)
24. Van Leeuwen, I., Munro, A.J., Sanders, I., Staples, O., Lain, S.: Numerical and Experimental Analysis of the p53-mdm2 Regulatory Pathway. In: Proceedings of the 3rd OPAALS International Conference, Aracaju, Sergipe, Brazil, March 22-23 (2010)

Numerical and Experimental Analysis of the p53-mdm2 Regulatory Pathway

Ingeborg M.M. van Leeuwen[1,2], Ian Sanders[1], Oliver Staples[1], Sonia Lain[1,2], and Alastair J. Munro[1]

[1] Surgery and Oncology, Ninewells Hospital, University of Dundee, DD1 9SY Dundee, UK
ingeborg@maths.dundee.ac.uk
[2] Microbiology, Tumor and Cell Biology, Karolinskat Institute, SE-17177 Stockholm, Sweden

Abstract. The p53 tumour suppressor plays key regulatory roles in various fundamental biological processes, including development, ageing and cell differentiation. It is therefore known as "the guardian of the genome" and is currently the most extensively studied protein worldwide. Besides members of the biomedical community, who view p53 as a promising target for novel anti-cancer therapies, the complex network of protein interactions modulating p53's activity has captivated the attention of theoreticians and modellers due to the possible occurrence of oscillations in protein levels in response to stress. This paper presents new insights into the behaviour of the p53 network, which we acquired by combining mathematical and experimental techniques. Notably, our data raises the question of whether the discrete p53 pulses in single cells, observed using fluorescent labelling, could in fact be an artefact. Furthermore, we propose a new model for the p53 pathway that is amenable to analysis by computational methods developed within the OPAALS project.

Keywords: Mathematical modelling, Systems Biology, Oscillations, Pulses, Cell Cycle, Cancer.

1 Introduction

Oscillating systems in living organisms are mathematically tractable and, through analysis and understanding of their behaviour, might provide useful insights into the design of computer systems. It therefore seemed appropriate for us, as biologists within the OPAALS project, to look critically at oscillating systems in biology to see whether or not they might be used to inform the computer science aspects of the project. This paper is part of a research framework that is documented in the following four companion papers at this same conference:

- A Research Framework for Interaction Computing [15]
- Lie Group Analysis of a p53–mdm2 ODE Model [24]
- Transformation Semigroups as Constructive Dynamical Spaces [17]
- Towards Autopoietic Computing [10]

F.A. Basile Colugnati et al. (Eds.): OPAALS 2010, LNICST 67, pp. 266–284, 2010.
© Institute for Computer Sciences, Social Informatics and Telecommunications Engineering 2010

2   Oscillations in Biology

Since autopoiesis is one of OPAALS's key areas of interest, and since the archetypical autopoietic entity is the cell, we have looked systematically at sub-cellular systems that oscillate. An oscillating system can be defined as one whose attributes rise and fall in a regular fashion over a sustained period. Such systems have interesting implications for computer science: each oscillation could be regarded as the tick of a digital clock, a pulse that might be counted.

Probably the most obvious role for self-sustaining cellular oscillators in biology is as timekeepers, controlling events such as circadian rhythms [45] and embryonic development [7]. However cellular oscillators are not limited to time-keeping. Experimental as well as theoretical work has demonstrated that cellular oscillators might be capable of performing a wide variety of important functions [20]. Among them are making decisions concerning the fate of a cell [2]; controlling of calcium-dependent signalling pathways; facilitating cellular responses to changes in environment; and regulating cellular energy production [8].

2.1   The p53/mdm2 Network as an Oscillating (?) System

We have chosen to investigate the behaviour of the p53/mdm2 system. Mainly because it is a critical element in determining the fate of a cell [29] – with obvious analogies to the success (or failure) of an enterprise – and because it is one with which we are somewhat familiar. We also believed, on the basis of previously published work [4, 19, 28, 27], that oscillatory behaviour was characteristic of the system and that, by further exploring these oscillations, we could usefully contribute to the mathematical and computer science aspects of the OPAALS project. P53 is a protein that is activated in response to stress or damage to a cell and has variable effects on cellular behaviour: cells may rest and repair the damage; they may die via programmed cell death (apoptosis); they may continue to behave as normal but, later, prove incapable of division (premature replicative senescence). P53 is negatively regulated by mdm2 (murine double minute 2) – mdm2 is an enzyme that targets p53 for degradation and elimination via a process termed ubiquitination. By convention, the human homologue of mdm2 is termed hdm2.

As a first step we sought to justify some of the assumptions underlying the published work on sustained oscillations in the p53/mdm2 system. The single – cell experiments [28], upon which an elaborate mathematical superstructure has been built [1, 5, 12, 13, 19, 27, 32, 36, 38], used fluorescent labelling to monitor simultaneous changes in the levels of p53 and mdm2 following irradiation. A key assumption in any such experimental system is that the labelling does not have any effect upon the function of the protein. Our own experimental work shows that, somewhat unexpectedly, the labelling of mdm2 renders it functionally inactive. This raises the possibility that the oscillations observed in the single–cell fluorescence experiments might be an artefact caused by the labelling procedure.

2.2   Experimental Data

The Western blots shown in Figure 1 illustrate this point. A Western blot is a gel-based method for assessing the amounts of specific proteins present in an experimental system. The darker the band, the more protein is present. The gels in the top half of the figure show the levels of unlabelled hdm2 and p53 in cells transfected with different amounts of expression plasmids for wild-type p53 and hdm2. Normal hdm2 down-regulates p53 and, as expected, we observe lower p53 levels in the presence of hdm2 (right–hand blue box) than in the absence of hdm2 (left–hand blue box) . When the experiment is repeated with hdm2 labelled with yellow fluorescent protein (hdm2-YFP), however, the outcome is different. In the lower pair of gels in Figure 1 there is, if anything, more p53 expression within the blots enclosed by the right–hand box. This suggests strongly that labelling of hdm2 might impair its ability to down-regulate p53 expression.

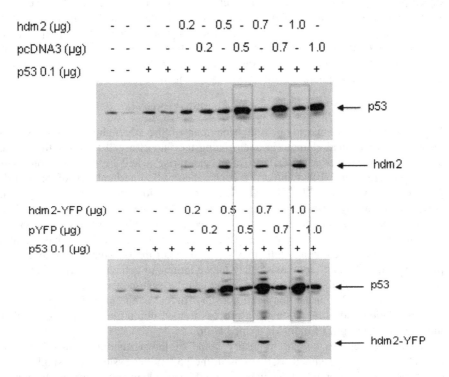

Fig. 1. MCF-7 cells were transiently transfected with hdm2 or hdm2-YFP (pU293), and as controls pcDNA3 or pYFP were also transfected in a gradient. In addition p53 wild-type (0.1µg) plasmid was transfected as indicated. In the lower gel there is abundant p53 (right–hand blue box) despite presence of hdm2. If the labelled hdm2 were functional we would expect to see inhibition of p53 but, if anything, p53 is induced.

It is possible that the sustained oscillations that have been observed in single– cell experiments may be due to artefacts induced by the abnormalities in mdm2 function introduced by the labelling procedure. This leaves us somewhat sceptical concerning the ability of regular sustained oscillations in the p53/mdm2 system, within individual cells, to act as a digital counting mechanism for determining the fate of a cell.

Nevertheless we have good evidence from our own work (e.g. Figure 2) and that of others [30] that the p53/mdm2 system can oscillate in response to irradiation – but that these oscillations are damped. There are several possible explanations for such damped oscillations – the most parsimonious is that p53 and MDM2 operate in a simple feedback loop but with a discrete time delay [34]. Another explanation – less favoured by us for the reasons given – is that the damping at the aggregate level is caused by the presence of a mixed population of cells.

Having set out to use sustained oscillations in the p53/mdm2 system as the basis of our mathematical models we were forced into an unexpected conclusion: close scrutiny of some so-called "oscillating systems" shows that the biological observations upon which they are based may be flawed and, furthermore, the models used to demonstrate oscillations may only show oscillatory behaviour under certain restricted and somewhat artificial conditions. The literature at the interface between mathematics and biology is not as robust as a superficial reading would lead us to believe. Perhaps this is a reflection of a more general problem in interdisciplinary research: it is difficult to achieve a balance of expertise. Mathematical sophistication can be undermined by biological naïveté, and *vice versa*.

Our disappointment with our initial exploration of oscillating systems raises a fairly fundamental question: when we are using natural science to influence mathematics, algebra and computer science, are we seeking to inspire using a model that accurately reflects the natural world, a robust template; or are we simply generating possibilities, initiating chains of thought and encouraging creative approaches? If our aim is the latter then the extent to which our models fit the biological observations is less than critical – indeed could be entirely unimportant; if our aim is the former, then the goodness–of–fit is crucial. At this point it seems entirely reasonable to leave the question rhetorical, unanswered.

Despite our concerns about *sustained* oscillations in the p53/mdm2 system we are confident that the p53/mdm2 system exhibits *damped* oscillations in cell populations, as shown in Figure 2. This information has been used to inform the mathematical models that are discussed in the following section.

3   Mathematical Modelling of the p53 Network

During the cell cycle, key proteins undergo cyclic synthesis and degradation, the associated changes in their levels triggering progression through the cell–cycle phases. In recent years, similar oscillatory behaviour has been shown in several other regulatory networks. The NF-κB transcription factor, for instance, presents sustained nucleo–cytoplasmic fluctuations after stimulation with TNFα [3, 25, 35]. Similarly, the levels of Hes1 protein and mRNA oscillate in cell culture upon serum treatment [6, 23]. In the mouse segmentation clock, not only the levels of Hes1 but also of components of the Wnt and Notch pathways have been shown to oscillate *in vivo* [14, 49]. Given the biomedical importance of these pathways, the molecular basis and biological implications of their oscillations have attracted extensive attention from experimentalists and theoreticians alike. Although the signalling pathways above differ in their components and biological outcomes, the oscillations are believed to

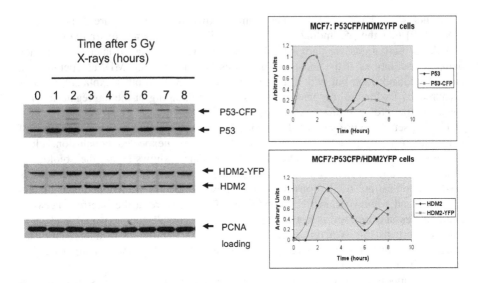

Fig. 2. Levels of p53 and mdm2 after irradiation of cells in suspension. There is a prominent peak for p53 at 2 hours post exposure followed by a second (lower) peak at 6 hours; for mdm2 the peaks are at 3 hours and 7 hours. Findings consistent with a delay between the rise in p53 and the induction of mdm2. Each observation is based on a pooled estimate from > 100, 000 cells; the error bars (s.e.m.) are very narrow and are obscured by the data points themselves.

share a common underlying mechanism: a negative feedback loop (NFL) combined with a transcriptional delay [34, 37, 44]. Notably, such a mechanism is present in the p53–mdm2 network (Figure 3), as p53 promotes the synthesis of its main negative regulator, mdm2 (Figure 4). The possibility that p53 levels oscillate in response to stress has therefore been explored both experimentally and theoretically.

The p53 pathway has been subject to extensive modelling efforts, the first model being published by Bar-Or and co-workers in 2000. The existing models can be classified into two categories based on their purpose. On the one hand, relatively simple models describing the role of p53 in cancer have been developed as part of multiscale models for the disease [11, 16, 31, 40]. Ribba et al. [40], for instance, used p53 as a switch adopting values 0 and 1 in the absence and presence of DNA damage, respectively, whereas in the cellular automaton model developed by Alarcon and co-workers [11] cells possess different automaton features depending on whether their p53 status is wild-type or mutant. On the other hand, various kinetic models have been built to investigate whether the system can give rise to damped oscillations in p53 and mdm2 in cell populations and un- damped pulses in individual cells [1, 5, 9, 13, 19, 28, 32, 33, 34, 30, 38, 39, 48, 51]. The vast majority assume that a cell consists of a single compartment in which the proteins of interest are abundant and evenly distributed; hence they undertake a continuum, deterministic approach, using either ordinary or delay differential equations (ODEs or DDEs) to describe the changes in protein levels. Alternative modelling approaches include stochastic, spatial and multiple compartment models. Proctor et al. [38], for instance, developed

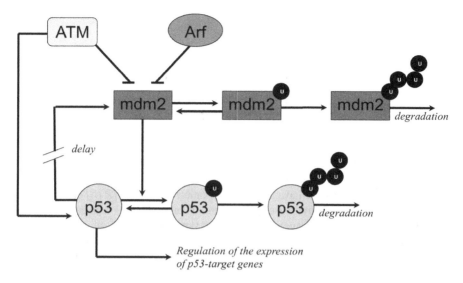

Fig. 3. Simple schematic of the p53 network. The p53 tumour suppressor – the guardian of the genome – acts as a transcription factor, regulating the expression of over a hundred target genes [18, 29, 47]. The mdm2 protein, a RING finger-dependent ubiquitin protein ligase, binds to p53 and targets it for proteasomal degradation [22]. Both p53 and mdm2 are highly regulated proteins [26].

stochastic models to account for the intercellular variation in the number of macromolecules. In contrast, Ciliberto *et al.* [13] formulated a deterministic, compartmental ODE model that distinguishes between nuclear and cytoplasmic mdm2. Finally, Gordon *et al.* [21] relaxed the intracellular homogeneity assumption further and used partial differential equations (PDEs) to describe intracellular protein density patterns.

To prevent the negative feedback loop linking p53 and mdm2 (Figure 4) from immediately inhibiting itself and, in particular, to enable oscillations to occur, a delay in the negative feedback loop is required [43]. Such a delay can be modelled in different ways. In some cases the delay has been included implicitly in mathematical models – Lev Bar-Or *et al.* [30], for instance, included an additional unknown component in the p53–mdm2 pathway, whereas Ciliberto *et al.* [13] combined positive and negative feedbacks. The alternative is to incorporate time delays explicitly. Srividya *et al.* [42] have shown that discrete delay terms can help reduce the number of variables and parameters required to describe a molecular system by replacing one or more intermediate reactions. Concerning spatial effects, [34] demonstrated that the waiting times associated with transcription and translation can be fused into a single time delay without altering the dynamical properties of the system. Hence, several models [5, 21, 34, 36] calculate the rate of mdm2 syntesis at time t as a function of the amount of p53 present in the system at time $t - \tau$.

Below we discuss in detail three existing theoretical approaches to the p53 pathway (i.e. [30, 34, 13]), while in the next section we present a new model. To facilitate

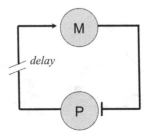

Fig. 4. Negative feedback loop with transcriptional delay linking p53 and mdm2

comparison of the models, we have unified the notation as follows: *[P (t)]* and *[M (t)]* are the total intracellular concentrations of p53 and mdm2 at time t; S (t) represents a transient stress stimulus (e.g. DNA damage); $s_\star$ are *de novo* synthesis rates; $k_\star$ are production rates (e.g. phosphorylation); $j_\star$ are reverse reactions (e.g. desphosphorylation); $d_\star$ are degradation rates; $\sigma_\star$ are transport rates; $K_\star$ are saturation coefficients and $c_\star$ are additional constants.

3.1 Model I: Lev Bar-Or et al. (2000)

The first model for the p53-mdm2 network, formulated by Lev Bar-Or *et al.* [30], describes the interaction between p53 and mdm2. The time–lag between p53 activation and p53-mediated induction of mdm2 synthesis is modelled by the presence of a hypothetical intermediary, X. The proposed kinetic equations are [30]:

$$
\mathrm{d}[P]/\mathrm{d}t = s_p - \left(d_p + \widehat{d}_{pm}(t)[M] \right)[P],
$$

$$
\mathrm{d}[M]/\mathrm{d}t = s_{m0} + \frac{s_{mx}[X]^n}{[X]^n + K_{mx}^n} - d_m[M], \tag{1}
$$

$$
\mathrm{d}[X]/\mathrm{d}t = \frac{s_s \mathcal{S}(t)[P]}{1 + c_x[M][P]} - d_x[X],
$$

with initial conditions [P (0)] = Po, [M (0)] = Mo, and [X (0)] = 0. The stress stimulus S is assumed to influence the behaviour of the system in two ways, namely by promoting p53's transcriptional activity and by downregulating mdm2-mediated degradation of p53. The former mechanism is modelled by making the synthesis of X an increasing function of S, while the latter is incorporated by making $\widehat{d}_{pm}$ a decreasing function of S. Numerical analyses and simulations revealed that, for certain parameter values, the system can show damped oscillations in which mdm2 and p53 levels peak out of phase.

3.2 Model II: Monk (2003)

In contrast to [30], the model proposed by Monk [34] not ony characterizes the dynamics of the mdm2 and p53 proteins, but also the changes in the level of mdm2 mRNA. Morever, it explicitly accounts for a transcriptional delay as follows:

$$\mathrm{d}[P]/\mathrm{dt} = s_p - \left(d_{p0} + \frac{d_{pm2}[M]^2}{[M]^2 + K_{pm}^2} \right)[P],$$

$$\mathrm{d}[R_M]/\mathrm{dt} = s_{rm0} + \frac{s_{rm1}[P(t-\tau)]^n}{[P(t-\tau)]^n + K_{rm}^n} - d_{rm}[R_M], \qquad (2)$$

$$\mathrm{d}[M]/\mathrm{dt} = s_{rt}[R_M] - d_m[M],$$

where both p53's transcriptional activity and mdm2's ubiquitin-ligase activity are assumed to be saturating functions. The system demonstrates oscillatory behaviour for certain parameter values, the period of the oscillations depending on the transcriptional delay and the protein and mRNA half-lives.

3.3 Model III: Ciliberto et al. (2005)

Compared with the approaches above, Ciliberto *et al.* [13] incorporated substantially more biological detail in their model, to arrive at a relatively more sophisticated description of the p53–mdm2 network that accounts for two subcellular compartments, namely the nucleus and the cytoplasm. As it is assumed that mdm2 has to be phosphorylated in order to enter the nucleus, the model includes three molecular forms of mdm2: nuclear mdm2, and both unphosphorylated and phosphorylated cytoplasmic mdm2. Moreover, ubiquitination is modelled as a multistep process (Figure 3), involving three molecular forms of p53 (i.e., non-ubiquitinated, mono-ubiquitinated and poly-ubiquitinated protein). The dynamics of the six molecular components is expressed by:

$$\mathrm{d}[P]/\mathrm{dt} = s_p - d_p[P] - d_{pu}[P_{UU}],$$

$$\mathrm{d}[P_U]/\mathrm{dt} = k_u([P] - [P_U] - [P_{UU}])[M_N] + j_u[P_{UU}]$$
$$\qquad\qquad\qquad - (j_u + d_{p1} + k_u[M_N])[P_U],$$

$$\mathrm{d}[P_{UU}]/\mathrm{dt} = k_u[M_N][P_U] - (j_u + d_{pu} + d_{p1})[P_{UU}],$$

$$\mathrm{d}[M_N]/\mathrm{dt} = (\sigma_{cn}[M_{PC}] - \sigma_{nc}[M_N])v - \left(d_{m0} + \frac{d_{m1}\mathcal{S}(t)}{\mathcal{S}(t) + K_m} \right)[M_N], \qquad (3)$$

$$\mathrm{d}[M_C]/\mathrm{dt} = s_{m0} + \frac{s_{m1}[P]^n}{[P]^n + K_{pm}^n} + k_{pc}[M_{PC}] - \left(d_{m0} + \frac{k_{pc}}{[P] + K_{pc}} \right)[M_C],$$

$$\mathrm{d}[M_{PC}]/\mathrm{dt} = \frac{k_{pc}[M_C]}{[P] + K_{pc}} + \sigma_{nc}[M_N] - (j_{pc} + \sigma_{cn} + d_{m0})[M_{PC}],$$

The state variables above represent the concentrations of total p53, P; monoubiquitinated p53, P_U; poly-ubiquitinated p53, P_{UU}; nuclear mdm2, M_N; unphosphorylated cytoplasmic mdm2, Mc ; and phosphorylated cytoplasmic mdm2, M_{PC}. The parameter v denotes the nuclear-cytoplasmic volume ratio. Unlike in the previous models, the value of the stress function S depends on the level of p53, as it is assumed that p53 plays a role in DNA repair:

$$\mathrm{d}\mathcal{S}/\mathrm{dt} = \widehat{k}_s(I(t)) - \frac{d_s\mathcal{S}(t)[P]}{\mathcal{S}(t) + K_s},$$

where $\widehat{k}_s$ is the DNA damage production rate as a function of time and dose of irradiation. Based on numerical analyses and simulations, the authors suggest that the discrete pulses observed by Lahav *et al.* [28] might be the result of a combination of positive and negative feedbacks. In the model, *the positive feedback originates from two opposing negative effects: nuclear mdm2 induces p53 degradation, while p53 inhibits nuclear entry of mdm2, by inhibiting phosphorylation of mdm2 in the cytoplasm. The negative feedback loop is the well-known fact that p53 induces the synthesis of mdm2* [13].

Eqns (3) also predict that the level of mdm2 can increase in the absence of mdm2 synthesis, which indicates a mass balance problem. This can be solved by rewriting the expressions for the changes in $[M_N]$ and $[M_{PC}]$:

$$d[M_N]/dt = v\sigma_{cn}[M_{PC}] - \sigma_{nc}[M_N] - \left(d_{m0} + \frac{d_{m1}\mathcal{S}(t)}{\mathcal{S}(t) + K_m}\right)[M_N],$$

$$d[M_{PC}]/dt = \frac{k_{pc}[M_C]}{[P] + K_{pc}} + \frac{\sigma_{nc}[M_N]}{v} - (j_{pc} + \sigma_{cn} + d_{m0})[M_{PC}].$$

(4)

An analogous modification has been introduced in [1], which presents a four ODE model inspired by [13]. The new system of ODEs based on (4) does not oscillate for the parameter values used in [13].

4 Formulation of the New Model

To show that there is yet another possible biological explanation for the occurrence of oscillations in p53 levels in response to stress, we will now introduce a new model for the p53 pathway, which we will refer to as Model IV. The structure of our simple network is shown in Figure 6. It describes the interactions between four molecular components: P_I, the p53 tumour suppressor, M, p53's main negative regulator, mdm2, C, the p53–mdm2 complex and, P_A, an *active* form of p53 that is resistant against mdm2-mediated degradation. The model accounts for the following phenomena: (1) basal p53 synthesis; (2) basal (i.e. mdm2-independent) p53 degradation; (3) mdm2 synthesis; (4) basal mdm2 degradation; (5) p53–mdm2 complex assembly; (6) p53–mdm2 complex dissociation; (7) mdm2-mediated p53 ubiquitination and subsequent elimination; (8) stress-induced p53 activation; (9) p53 inactivation; and (10) basal degradation of active p53. According to the reaction scheme shown in Figure 6, the changes in the concentrations of the four molecular components are given by:

$$d[P_I]/dt = r_1 - r_2 - r_5 + r_6 - r_8 + r_9,$$
$$d[M]/dt = r_3 - r_4 - r_5 + r_6 + r_7,$$
$$d[C]/dt = r_5 - r_6 - r_7,$$
$$d[P_A]/dt = r_8 - r_9 - r_{10}.$$

(5)

where r_i (t), for $i = 1, \ldots, 10$, is the rate of reaction i at time t and $[X]$ denotes the concentration of molecular component X, for $X = P_I$, P_A M, and C.

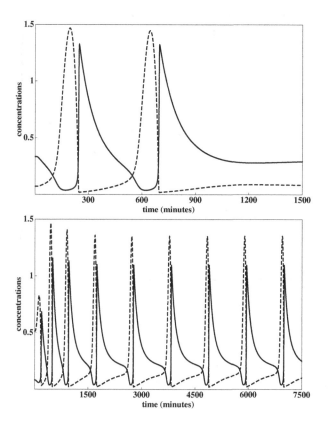

Fig. 5. Simulations with the original model by Ciliberto *et al.* [13] (Eqns (3)) showing two pulses *(left)* and sustained oscillations *(right)* in the levels of total p53, *[P]* (dashed lines), and nuclear mdm2, *[M_N]* (solid lines). For the parameter values, see the original paper [13]. In the left and right panels, the value of the cytoplasm:nucleus ratio, v, is 15 and 11, respectively.

4.1 Simplifying Assumptions

For simplicity, we assume that the basal p53 synthesis rate, r_1, remains constant in time and that basal degradation rates, r_2, r_4 and r_{10}, are proportional to the corresponding substrate concentrations. The mdm2 protein, a RING finger-dependent ubiquitin protein ligase, is known to bind to p53 and target it for proteasomal degradation [22]. In the model, the binding of mdm2 to p53 is assumed to be reversible with assembly and dissociation rates r_5 and r_6, respectively. Furthermore, experimental evidence has shown that mdm2's downregulation of p53 is inhibited under DNA damage [41]. We have incorporated this observation into the model by assuming that ionising radiation induces the phosphorylation of p53, which prevents mdm2 binding. That is, r_8 is an increasing function of the level of radiation exposure. Finally, reaction 3 represents the negative feedback loop in which p53 transactivates expression of the *MDM2* gene [50]. As we expect this expression rate to reach a maximum value when there is negligible delay between the binding of two successive

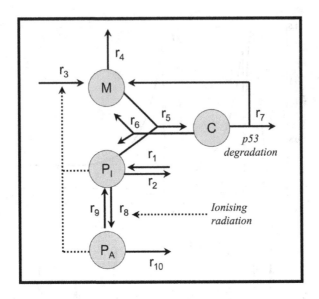

Fig. 6. Schematic of the new p53-mdm2 interaction model. M = mdm2 protein; P_I = *inactive* p53 protein, C = mdm2-p53 complex; P_A = *active* p53 protein.

p53 molecules to the *MDM2* promoter region, we assume that r_3 is a saturating function of the total level of p53. Given the assumptions above, the reaction rates can be calculated as follows:

$$r_1(t) = s_p,$$
$$r_2(t) = d_p[P_I(t)],$$
$$r_3(t) = s_{m0} + \frac{s_{m1}[P_I(t)] + s_{m2}[P_A(t)]}{[P_I(t)] + [P_A(t)] + K_m},$$
$$r_4(t) = d_m[M(t)],$$
$$r_5(t) = k_c[P_I(t)][M(t)],$$
$$r_6(t) = j_c[C(t)],$$
$$r_7(t) = k_u[C(t)],$$
$$r_8(t) = k_a S(t)[P_I(t)],$$
$$r_9(t) = j_a[P_A(t)],$$
$$r_{10}(t) = d_p[P_A(t)].$$

Substitution of the expressions above into Eqns (5) yields:

$$d[P_I]/dt = s_p + j_a[P_A] - (d_p + k_a S(t))[P_I] - k_c[P_I][M] + j_c[C],$$
$$d[M]/dt = s_{m0} + \frac{s_{m1}[P_I] + s_{m2}[P_A]}{[P_I] + [P_A] + K_m} + k_u[C] + j_c[C] - (d_m + k_c[P_I])[M],$$
$$d[C]/dt = k_c[P_I][M] - (j_c + k_u)[C],$$
$$d[P_A]/dt = k_a S(t)[P_I] - (j_a + d_p)[P_A].$$

4.2 Model IV: Dimensionless Equations

The state variables in Figure 6 can be scaled as

$$P_i = k_u[P_I]/s_p,$$
$$M = k_u[M]/s_p,$$
$$C = k_u[C]/s_p,$$
$$P_a = k_u[P_A]/s_p \text{ and}$$
$$\tau = k_u t.$$

In terms of these new variables, the model equations reduce to:

$$\mathrm{d}P_i/\mathrm{d}\tau = 1 + \beta_a P_a - (\beta_p + \alpha_a \mathcal{S}(\tau))P_i \\ - \alpha_c P_i M + \beta_c C, \tag{6}$$

$$\mathrm{d}M/\mathrm{d}\tau = \alpha_{m0} + \frac{\alpha_{m1} P_i + \alpha_{m2} P_a}{P_i + P_a + \kappa_m} + C \\ - (\beta_m + \alpha_c P_i)M + \beta_c C, \tag{7}$$

$$\mathrm{d}C/\mathrm{d}\tau = \alpha_c P_i M - (1 + \beta_c)C, \tag{8}$$

$$\mathrm{d}P_a/\mathrm{d}\tau = \alpha_a \mathcal{S}(\tau)P_i - (\beta_a + \beta_p)P_a. \tag{9}$$

where the dimensionless parameters are defined as:

$$\alpha_a = k_a/k_u,$$
$$\beta_a = j_a/k_u,$$
$$\alpha_c = k_c s_p/k_u^2,$$
$$\beta_c = j_c/k_u,$$
$$\alpha_{m0} = s_{m0}/s_p,$$
$$\alpha_{m1} = s_{m1}/s_p,$$
$$\alpha_{m2} = s_{m2}/s_p,$$
$$\beta_m = d_m/k_u,$$
$$\beta_p = d_p/k_u \text{ and}$$
$$\kappa_m = k_u K_m/s_p.$$

In the simulations depicted in Figure 7, two modes of stress have been considered, namely a discrete pulse insult at time zero and long-term exposure to a constant stressful stimulus. The former is expressed as:

$$\mathcal{S}(t) = \mathrm{e}^{-c_s t} \qquad \text{and} \qquad \mathcal{S}(\tau) = \mathrm{e}^{-\gamma \tau}, \tag{10}$$

in dimensional and dimensionless form, respectively (the dimensionless stress coefficient is $\gamma = c_s/k_u$). In contrast, the latter is modelled as:

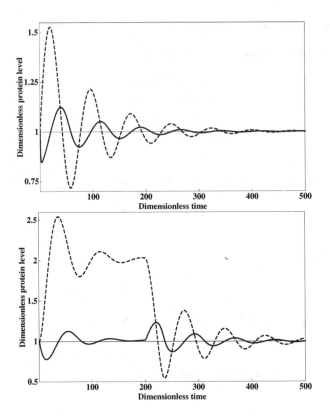

Fig. 7. Simulations with the dimensionless system (6–9) showing damped oscillations in the levels of total p53, P_i+P_a+C (dashed lines), and total mdm2, $M+C$ (solid lines). The values have been normalised with respect to the corresponding concentrations in the absence of stress (i.e., $P_{i0}+C_0$ and M_0+C_0, respectively). The parameter values used in the simulations are provided in Table 1. One dimensionless time unit corresponds to 2.5 minutes. *Left panel*: response to a single pulse insult at time $\tau = 0$ (Eqn (10) with $\gamma = 2.5$). *Right panel*: response to a constant, long-term insult (Eqn (11) with $\gamma = 2.5$, $\tau_i = 200$ and $S_0 = 0.05$).

$$\begin{aligned} \mathcal{S}(\tau) &= \mathcal{S}_0 && \text{for} && 0 \leq \tau \leq \tau_i \\ &= \mathcal{S}_0 e^{-\gamma(\tau-\tau_i)} && \text{for} && \tau > \tau_i. \end{aligned} \tag{11}$$

Notably, the response of the system to the two kinds of insult is very different.Exposure to a single, pulse insult at time zero results in damped oscillations in the levels of p53 and mdm2 around their steady-state values in the absence of stress. In contrast, exposure to a long-term signal causes the system to move to a new steady-state, fluctuating transiently. This new steady-state has a higher p53 level, which depends on the strength of the signal. When the stimulus ends, the system returns to its original steady-state, displaying a second round of damped oscillations.

Table 1. Parameter values used in the model simulations shown in Figure 7

$[P_I(0)] = 9.42$ nM	$k_a = 20$ min^{-1}
$[P_A(0)] = 0$ nM	$k_c = 4$ min^{-1} nM^{-1}
$[C(0)] = 3.49$ nM	$k_u = 0.4$ min^{-1}
$[M(0)] = 0.037$ nM	$j_a = 0.2$ min^{-1}
$s_{m0} = 2 \times 10^{-3}$ nM/min	$j_c = 2 \times 10^{-3}$ min^{-1}
$s_{m1} = 0.15$ nM/min	$s_p = 1.4$ nM/min
$s_{m2} = 0.2$ nM/min	$d_p = 2 \times 10^{-4}$ min^{-1}
$d_m = 0.4$ min^{-1}	
$K_m = 100$ nM	
$P_{i0} = 2.69$	$\alpha_a = 50$
$P_{a0} = 0$	$\alpha_c = 35$
$C_0 = 1.0$	
$M_0 = 0.01$	
$\alpha_{m0} = 0.00143$	$\beta_a = 0.5$
$\alpha_{m1} = 0.107$	$\beta_c = 5 \times 10^{-3}$
$\alpha_{m2} = 0.143$	
$\beta_m = 1$	$\beta_p = 5 \times 10^{-4}$
$\kappa_m = 28.57$	

5 Discussion

In the above section we have visited four alternative mathematical descriptions of the p53 network. Their specific features are highlighted in Table 2. These models, which represent only a small sample of the models available in the literature, were chosen to illustrate the use of differential equations in systems biology and, in particular, to show how very different mechanisms can succeed in explaining the same data. The modelling efforts addressed here were motivated mainly by two experimental studies. First, Lev Bar-Or et al. [30] exposed mouse fibroblasts NIH 3T3 cells expressing wild-type p53 and mdm2 to 5 Gy of irradiation and then measured the protein levels at several time points after exposure. Their Western Blots showed two peaks in the level of p53, each followed by a peak in mdm2 approximately one hour later. Second, Lahav and co-workers [28] created a cell line expressing p53 and mdm2 tagged with fluorescent proteins to study the dynamics of these molecules in single cells. In response to γ-radiation, they observed digital pulses in both p53 and mdm2 levels (for a critical discussion on this approach, see Section 2.2). Unifying the two experimental observations, Ma et al. [32] suggested that *the damped oscillations previously observed in cell populations can be explained as the aggregate behaviour of single cells.* Comparing the four models above, models I [30] and II [34] are the most similar, as they both produce an oscillatory behaviour based on the mdm2–p53 NFL (Figure 4) alone and both account for an intermediary component linking of the level of p53 to the rate of mdm2 synthesis. The main difference is that model II also includes an explicit transcriptional delay, which implies a shift of the mathematical approach from ODEs to DDEs. According to model III [13], however, the mechanism underlying the oscillations is more complex and involves both negative and positive

Table 2. Summary of the mathematical models described in this chapter

	FEATURES	MECHANISM(S) FOR OS-CILLATIONS
MODEL I Lev Bar-Or *et al* [30]	• 3 ODEs + 1 expression for the stress function • Includes an unknown intermediary	• Implicit delay, in the form of an unknown intermediary, in the mdm2–p52 negative feedback loop
MODEL II Monk [34]	• 2 ODEs + 1 DDE • Includes mdm2 mRNA dynamics	• Implicit time delay, in the form of a known intermediary (i.e, mdm2 mRNA), in the mdm2-p53 negative feedback loop • Explicit time delay for gene transcription
MODEL III Ciliberto *et al* [13]	• 6 ODEs + 1 ODE for the stress function • Distinguishes between nuclear and cytoplasmic mdm2 • Accounts for 3 molecular forms of p53 and 2 forms of mdm2 • DNA damage enhances mdm2 degradation • p53 promotes DNA repair	• Combination of positive and negative feed back loops between mdm2 and p53 • Implicit time delay (mdm2 has to be phosphorylated and then shuttled into the nucleus before it can degrade p53)
MODEL IV New model	• 4 ODEs + 1 expression for the stress function • Characterises of the dynamics of the p53-mdm2 complex • Accounts for a mdm2-resistant form of p53	• mdm2-p53 negative feedback loop • p53 binding protects mdm2 from proteosomal degradation

feedbacks. In response to ionising radiation, DNA damage *increases abruptly as does [. . .] the rate of [. . .] degradation of [M$_N$]. As [M$_N$] decreases, [P] increases, which causes an initial drop in [M$_{PC}$] and a steady increase in [M$_C$]. When a sufficient amount of mdm2 accumulates in the cytoplasm, it initiates a change of regime: phosphorylated mdm2 enters the nucleus, causing increased degradation of p53, which relieves the inhibition of mdm2 phosphorylation in the cytoplasm, allowing more mdm2 to enter the nucleus. The positive feedback loop causes the abrupt drop in [P] and rise in [M$_N$]. The drop in [P] cuts off the synthesis of [M$_C$], and consequently [M$_C$] and [M$_N$] drop. The system is back to the original state, [P] starts to accumulate again due to the low level of [M$_N$] and a new oscillation starts [13].*

While model III incorporates substantially more biologically-relevant information than models I and II, its main weaknesses are the strong dependence of the behaviour of the system on the relative size of the nucleus (see the simulations in Figure 3) and the mass balance problems discussed above (see corrected Eqns (4)).

Model III predicts a decrease in nuclear mdm2 in response to stress (Figure 5), and a subsequent reduction in mdm2-mediated p53 degradation. This is also the case in the context of model IV (Figure 7). The reason behind the drop in mdm2 is different, though. According to model III, the degradation rate of nuclear mdm2 is an increasing function of the level of DNA damage. In contrast, under model IV, p53-binding protects mdm2 from proteasomal degradation and, therefore, any decrease in [P$_I$] translates naturally in an increased mdm2 elimination rate. Hence, when radiation promotes the transformation of P$_I$ into P$_A$, mdm2 has less chance to bind to P$_I$ and is thus at a higher risk of being rapidly degraded. This model prediction highlights the importance of accounting for the dynamics of the mdm2–p53 complex, as suggested by Proctor *et al.* [38] *since the regulation of p53 is dependent on its interaction with mdm2, we would expect that the oscillatory behaviour of the system would be strongly affected by the binding affinity of mdm2 to p53. Therefore any mechanistic model of the sytem should include the mdm2–p53 complex.* The reduction in mdm2 levels in response to stress in models III and IV plays a role equivalent to the transcriptional delay in models I and II: it enables the levels of mdm2 and p53 to peak out of phase, thereby allowing oscillations to occur.

The purpose of the experimental and numerical work is to gain better insight into the dynamics of the chosen cellular regulatory pathway and of the ODE models that represent it most faithfully. The next step is to derive the corresponding automaton and extract the algebraic structure of its semigroup (in OPAALS), whilst simultaneously performing the Lie group analysis of the ODE system (BIONETS). Once these more powerful analytical tools are in place, we can re-examine the models discussed in this chapter, in order to correlate algebraic structure to observed or numerically calculated behaviour.

Acknowledgements

The partial support for this work by the OPAALS (FP6-034824) EU project is gratefully acknowledged. The experimental work summarised in Figures 1 and 2 was supported by other sources (CRUK, The Ninewells Cancer Campaign, a bequest from the estate of Mrs D.B. Miller). Individuals contributing to the data presented include Elizabeth Sinclair and Johanna Campbell.

References

1. Abou-Jaude, W., Ouattara, D.A., Kaufman, M.: From structure to dynamics: Frequency tuning in the p53-mdm2 network. i. Logical approach. J. Theor. Biol. 258, 561–577 (2009)
2. Agrawal, S., Archer, C., Schaffer, D.V.: Computational models of the notch network elucidate mechanisms of context-dependent signaling, PLoS. Comput. Biol. 5, e1000390 (2009)
3. Ankers, J.M., Spiller, D.G., White, M.R.H., Harper, C.V.: Spatio-temporal protein dynamics in single living cells. Curr. Opinion Biotechnol. 19, 375–380 (2009)
4. Batchelor, E., Loewer, A., Lahav, G.: The ups and downs of p53: understanding protein dynamics in single cells. Nat. Rev. Cancer 9, 371–377 (2009)
5. Batchelor, E., Mock, C.S., Bhan, I., Loewer, A., Lahav, G.: Recurrent initiation: A mechanism for triggering p53 pulses in response to DNA damage. Mol. Cell. 30, 277–289 (2008)

6. Bernard, S., Cajavec, B., Pujo-Menjouet, L., Mackey, M.C., Herzel, H.: Modelling transcriptional feedback loops: the role of Gro/Tle1 in Hes1 oscillations. Phil. Trans. R. Soc. A 364, 1155–1170 (2006)

7. Bessho, Y., Kageyama, R.: Oscillations, clocks and segmentation. Curr. Opinion Gen. Dev. 13, 379–384 (2003)

8. Bier, M., Teusink, B., Kholodenko, B.N., Westerhoff, H.: Control analysis of gly-colytic oscillations. Biophys. Chem. 62, 15–24 (1996)

9. Bottani, S., Grammaticos, B.: Analysis of a minimal model for p53 oscillations. J. Theor. Biol. 249, 235–245 (2007)

10. Briscoe, G., Dini, P.: Towards Autopoietic Computing. In: Proceedings of the 3rd OPAALS International Conference, Aracaju, Sergipe, Brazil, March 22-23 (2010)

11. Byrne, H.M., van Leeuwen, I.M.M., Owen, M.R., Alarcon, T.A., Maini, P.K.: Multiscale modelling of solid tumour growth. In: Bellomo, N., Chaplain, M.A.J., de Angelis, E. (eds.) Selected Topics on Cancer Modelling: Genesis, Evolution, Immune Competition and Therapy. Modelling and Simulation in Science, Engineering and Technology, Birkhauser, Boston (2008)

12. Chickarmane, V., Nadim, A., Ray, A., Sauro, H.: A p53 oscillator model of dna break repair control (2006), http://arxiv.org/abs/q-bio.mn/0510002

13. Ciliberto, A., Novak, B., Tyson, J.J.: Steady states and oscillations in the p53/mdm2 network. Cell Cycle 4, 488–493 (2005)

14. Dequeant, M.L., Glynn, E., Gaudenz, K., Wahl, M., Chen, J., Mushegian, A., Pourquie, O.: A complex oscillating network of signaling genes underlies the mouse segmentation clock. Science 314, 1595–1598 (2006)

15. Dini, P., Schreckling, D.: A Research Framework for Interaction Computing. In: Proceedings of the 3rd OPAALS International Conference, Aracaju, Sergipe, Brazil, March 22-23 (2010)

16. Dionysiou, D.D., Stamatakos, G.S.: Applying a 4D multiscale in vivo tumour growth model to the exploration of radiotherapy scheduling: the effects of weekend treatment gaps and p53 gene status on the response of fast growing solid tumors. Cancer Informatics 2, 113–121 (2006)

17. Egri-Nagy, A., Dini, P., Nehaniv, C.L., Schilstra, M.J.: Transformation Semigroups as Constructive Dynamical Spaces. In: Proceedings of the 3rd OPAALS International Conference, Aracaju, Sergipe, Brazil, March 22-23 (2010)

18. Farmer, G., Bargonetti, J., Zhu, H., Friedman, P., Prywes, R., Prives, C.: Wild-type p53 activates transcription in vivo. Nature 358, 83–84 (1992)

19. Geva-Zatorsky, N., Rosenfeld, N., Itzkovitz, S., Milo, R., Sigal, A., Dekel, E., Yarnitzky, T., Liron, Y., Polak, P., Lahav, G., Alon, U.: Oscillations and variability in the p53 system. Mol. Systems Biol. 2 (2006)

20. Goldbeter, A.: Computational approaches to cellular rhythms. Nature 420, 238–245 (2002)

21. Gordon, K.E., Van Leeuwen, I.M.M., Lain, S., Chaplain, M.A.J.: Spatio-temporal modelling of the p53-mdm2 oscillatory system. Math. Model. Nat. Phenom. 4, 97–116 (2009)

22. Haupt, Y., Maya, R., Kazaz, A., Oren, M.: Mdm2 promotes the rapid degradation of p53. Nature 387, 296–299 (1997)

23. Hirata, H., Yoshiura, S., Ohtsuka, T., Bessho, Y., Harada, T., Yoshikawa, K., Kageyama, R.: Oscillatory expression of the Bhlh factor Hes1 regulated by a negative feedback loop. Science 298, 840–843 (2002)

24. Horvath, G., Dini, P.: Lie Group Analysis of p53-mdm3 Pathway. In: Proceedings of the 3rd OPAALS International Conference, Aracaju, Sergipe, Brazil, March 22-23 (2010)

25. Krishna, S., Jensen, M.H., Sneppen, K.: Minimal model of spiky oscillations in NF-KB. Proc. Natl. Acad. Sci. USA 103, 10840–10845 (2006)
26. Kruse, J.P., Gu, W.: Modes of p53 regulation. Cell 137, 609–622 (2009)
27. Lahav, G.: Oscillations by the p53-mdm2 feedback loop. Adv. Exp. Med. Biol. 641, 28–38 (2008)
28. Lahav, G., Rosenfield, N., Sigal, A., Geva-Zatorsky, N., Levine, A.J., Elowitz, M.B., Alon, U.: Dynamics of the p53-mdm2 feedback loop in individual cells. Nat. Gen. 36, 147–150 (2004)
29. Lane, D.P.: p53, guardian of the genome. Nature 358, 15–16 (1992)
30. Lev Bar-Or, R., Maya, R., Segel, L.A., Alon, U., Levine, A.J., Oren, M.: Generation of oscillations by the p53-mdm2 feedback loop: a theoretical and experimental study. Proc. Natl. Acad. Sci. USA 97, 11250–11255 (2000)
31. Levine, H.A., Smiley, M.W., Tucker, A.L., Nilsen-Hamilton, M.: A mathematical model for the onset of avascular tumor growth in response to the loss of p53 function. Cancer Informatics 2, 163–188 (2006)
32. Ma, L., Wagner, J., Rice, J.J., Hu, W., Levine, A.J., Stolovitzky, G.A.: A plausible model for the digital response of p53 to dna damage. Proc. Natl. Acad. Sci. USA 102, 14266–14271 (2005)
33. Mihalas, G.I., Simon, Z., Balea, G., Popa, E.: Possible oscillatory behaviour in p53-mdm2 interaction computer simulation. J. Biol. Syst. 8, 21–29 (2000)
34. Monk, N.A.M.: Oscillatory expression of hes1, p53, and NF-KB driven by transcriptional time delays. Curr. Biol. 13, 1409–1413 (2003)
35. Nelson, D.E., Ihekwaba, A.E., Elliott, M., Johnson, J.R., Gibney, C.A., Foreman, B.E., Nelson, G., See, V., Horton, C.A., Spiler, D.G., Edwards, S.W., McDowell, H.P., Unitt, J.F., Sullivan, E., Grimley, R., Benson, N., Broomhead, D., Kell, D.B., White, M.R.: Oscillations in NF-KB signaling control de dynamics of gene expression. Science 306, 704–708 (2004)
36. Ogunnaike, B.A.: Elucidating the digital control mechanism for dna damage repair with the p53-mdm2 system: single cell data analysis and ensemble modelling. J. R. Soc. Interface 3, 175–184 (2006)
37. Pigolotti, S., Krishna, S., Jensen, M.H.: Oscillation patterns in negative feedback loops. Proc. Natl. Acad. Sci. USA 104, 6533–6537 (2007)
38. Proctor, C.J., Gray, D.A.: Explaining oscillations and variability in the p53-mdm2 system. BMC Syst. Biol. 2, 75 (2008)
39. Puszynski, K., Hat, B., Lipniacki, T.: Oscillations and bistability in the stochastic model of p53 regulation. J. Theor. Biol. 254, 452–465 (2008)
40. Ribba, B., Colin, T., Schnell, S.: A multiscale mathematical model of cancer, and its use in analyzing irradiation therapies. Theor. Biol. Med. Model. 3, 7 (2006)
41. Shieh, S.Y., Ikeda, M., Taya, Y., Prives, C.: DNA damage-induced phosphorylation of p53 alleviates inhibition by mdm2. Cell 91, 325–334 (1997)
42. Srividya, J., Gopinathan, M.S., Schnells, S.: The effects of time delays in a phosphorylation-dephosphorylation pathway. Biophys. Chem. 125, 286–297 (2007)
43. Tiana, G., Jensen, M.H., Sneppen, K.: Time delay as a key to apoptosis induction in the p53 network. Eur. Phys. J. B 29, 135–140 (2002)
44. Tiana, G., Krishna, S., Pigolotti, S., Jensen, M.H., Sneppen, K.: Oscillations and temporal signalling in cells. Phys. Biol. 4, R1–R17 (2007)
45. Tigges, M., Marquez-Lago, T.T., Stelling, J., Fussenegger, M.: A tunable synthetic mammalian oscillator. Nature 457, 309–312 (2009)

46. Vogelstein, B., Lane, D.P., Levine, A.J.: Surfing the p53 network. Nature 408, 307–310 (2007)
47. Wagner, J., Ma, L., Rice, J.J., Hu, W., Levine, A.J., Stolovitzky, G.A.: P53-mdm2 loop controlled by a balance of its feedback strength and effective dampening using atm and delayed feedback. IEE Proc. Syst. Biol. 152, 109–118 (2005)
48. Wawra, C., Kuhl, M., Kestler, H.A.: Extended analyses of the wntβ-cateni pathway: robustness and oscillatory behaviour. FEBS Lett. 581, 4043–4048 (2007)
49. Zauberman, A., Flusberg, D., Haupt, Y., Barak, Y., Oren, M.: A functional p53-response intronic promoter is contained within the human mdm2 gene. Nucleic Acids Res. 23, 2584–2592 (1995)
50. Zhang, T., Brazhnik, P., Tyson, J.J.: Exploring mechanisms of DNA damage response: p53 pulses and their possible relevance to apoptosis. Cell Cycle 6, 85–94 (2007)

Lie Group Analysis of a p53-mdm2 ODE Model

Gábor Horváth and Paolo Dini

Department of Media and Communications
London School of Economics and Political Science
London, United Kingdom
{g.horvath,p.dini}@lse.ac.uk

Abstract. This paper presents a symmetry analysis based on Lie groups of a system of ordinary differential equations (ODEs) modelling the p53-mdm2 regulatory pathway. This pathway is being investigated across several research groups as a biological system from which to extract dynamical and algebraic characteristics relevant to the emerging concept of Interaction Computing. After providing a conceptual motivation for the approach and some biological background for the choice of pathway, the paper gives an intuitive introduction to the method of Lie groups for a non-mathematical audience. This is followed by a general statement of the problem of finding the symmetries of a general system of four 1st-order ODEs, and then by the analysis of one such system modelling the p53-mdm2 pathway. The system chosen does not appear to harbour any symmetries, and therefore the effectiveness of the Lie group method cannot be demonstrated on this particular example. The symmetry analysis, however, helped reduce the system to a single Riccati equation for a specific choice of parameters, whose oscillatory behaviour appears to be relevant to the bio-computing perspective being discussed in a companion paper.

Keywords: Lie group, p53, symmetry, oscillation.

1 Introduction

Over the past few years, and across the DBE, OPAALS and BIONETS projects, we have mobilised a growing number of researchers and institutions in order to deepen and extend our study of the formal structure of cellular processes as the principal source of inspiration in our approach to biologically-inspired computing [5]. The rationale for this approach derives from the observation that the cell performs extremely complex functions by executing an intricate tangle of intersecting and interdependent metabolic and regulatory biochemical pathways, each of which can be considered analogous to an algorithm. Such behaviour is not planned and emerges spontaneously from the more elementary physical and chemical interactions at lower scales of description. What makes it possible is an immensely complex causal chain that links a stable periodic table of the elements to biological system behaviour through a stable set of physical laws. Thus, physical laws have bootstrapped an ecosystem that interacts with the

F.A. Basile Colugnati et al. (Eds.): OPAALS 2010, LNICST 67, pp. 285–304, 2010.
© Institute for Computer Sciences, Social Informatics and Telecommunications Engineering 2010

biological systems living in it in such a way as to support their existence and reproduction.

Whereas the global and pervasive order construction mechanism that has enabled the bootstrapping of the ecosystem to happen is ultimately explainable through the memory-dependent dynamics of Darwinian evolution acting across many generations, a process called phylogeny, the same evolutionary mechanism has 'discovered' ways in which biological systems and their sub-systems can interact, internally and with their environments, in order to construct order during the life of the individual. These interactions are able to construct recursively nested complex structures at multiple scales, a process called ontogeny or morphogenesis, and then keep the metabolism of the adult organism running throughout its lifetime. Rather than attempting to reproduce or emulate this overwhelmingly complex and physical order construction process, our approach has been to focus on the *output* of this process, in the form, for example, of a stable metabolic or regulatory pathway. The key concept is stability: in spite of the unpredictability of the biochemical inputs to the cell, appropriate cellular functions will be performed reliably, as long as the cell or the individual is healthy.

As discussed more extensively in [5], the concept of regularity (over space, time, or different scales of description) is best formalised through algebraic symmetries. Thus, our research programme has been to analyse the hierarchical algebraic structure of the mathematical models derived from metabolic pathways, to interpret what their dynamic and computational function might be, and to then apply these algebraic structures as constraints on automata in order to obtain a model of computation that we are calling Interaction Computing. The evidence so far suggests that a stable pathway is not stable of its own accord, but is kept within the analogue of a stable 'potential well' by the pathways it is biochemically coupled to, which therefore act as constraints. As a consequence, it appears that in order to achieve the Interaction Computing vision (so far developed only conceptually) we will have to understand how multiple threads, that are performing different algorithms, need to be coupled so that they can aid or constrain each other, as the case may be.

In [5] we refer to this concept as the kernel of Symbiotic Computing, a model of computing that is meant to apply to the interaction of higher-level software constructs, such as whole services. The predominant mathematical model used to analyse the interdependence of biochemical pathways is a set of coupled, and generally non-linear, ordinary differential equations (ODEs) derived from the chemical reaction equations. The set of dependent variables in such a set of ODEs is made up of the concentrations of compounds participating in the chemical reactions. Starting from these same chemical reaction equations, the system dynamics can be discretised as a Petri net, from which a finite-state automaton can be derived. The justification for this discretisation process is discussed in [5]. Our work has then focussed on looking for algebraic structures in both mathematical models of the same pathway: Lie group dynamical symmetries in the system of ODEs, discussed in this paper, and discrete computational symmetries in the automata [6].

This paper is part of a research framework that is documented in the following four companion papers at this same conference:

- A Research Framework for Interaction Computing [5]

  - Numerical and Experimental Analysis of the p53-mdm2 Regulatory Pathway [10]
  - Lie Group Analysis of a p53-mdm2 ODE Model (this paper)
  - Transformation Semigroups as Constructive Dynamical Spaces [6]
  - Towards Autopoietic Computing [1].

2   Oscillations in Biochemical Systems

We have focussed on the p53-mdm2 regulatory pathway because, in addition to its central role in the regulation of the cell cycle, it seemed that the oscillatory behaviour its constituents exhibit under particular conditions was too important a signature of the dynamics of the pathway to be overlooked, and felt that it could lead to useful insights for biologically-inspired algorithms and architectures. In the work discussed in [6] we are taking a close look at how oscillations can be modelled from a discrete mathematics perspective. It appears that cyclic phenomena in biochemical processes give rise to permutation groups in the hierarchical decomposition of the semigroups associated with the automata derived from such pathways. An example of a cyclic biochemical phenomenon is the Krebs or citric acid cycle, shown in Figure 1.

It is helpful to clarify the connection between cycles and oscillations by resorting to an idealised system in the form of the simple harmonic oscillator of

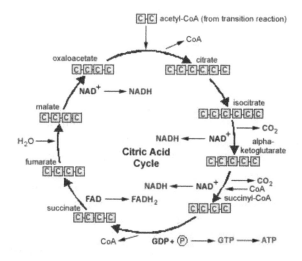

Fig. 1. Schematic of the Krebs or citric acid cycle [8]

elementary physics. As shown in Figure 2, for a simple harmonic oscillator periodic cycles (in some parameter space, which could include also Euclidean space) are mathematically indistinguishable from oscillations (in time, at a fixed point in space). For a biochemical system, the discrete algebraic analysis discussed in [6] indicates that biochemical processes such as the Krebs cycle, the concentration levels of whose metabolites can be assumed to remain constant over time, have the same 'algebraic signature' (i.e. permutation groups) as processes such as the p53-mdm2 pathway [10], in which the concentrations of the compounds oscillate as a function of time. The reason may be found in the fact that the models that give rise to these signatures describe what will happen (or what could happen, if there is a choice) to individual instances of classes of molecules or molecular complexes, but do not distinguish between instances of the same class. In simulations of the synchronized Krebs cycle and the p53-mdm2 pathway under conditions that promote sustained oscillations, indistinguishable instances of particular classes, such as citrate and active p53, appear and disappear periodically. Thus, the preliminary analytical results are confirming our initial hunch and the choice of problem.

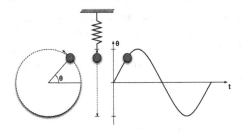

Fig. 2. Periodic behaviour of the simple harmonic oscillator

3   Background for p53-mdm2 System

A good discussion of the p53-mdm2 system is provided in [10,3], so here we summarise the highlights. Although the p53 protein participates in many pathways in the cell, we focussed on a small subset as a starting point. The p53 protein is linked to many other proteins and processes in the cell, but its coupling to the mdm2 protein appears to be particularly important for understanding cancer. Depending on the concentration level of p53, the cell can (in order of increasing concentration): (1) operate normally; (2) stop all functions to allow DNA repair to take place; (3) induce replicative senescence (disable cellular replication); and (4) induce apoptosis instantly (cell "suicide"). Therefore, p53 is a very powerful and potentially very dangerous chemical that humans (and most other animals) carry around in each cell, whose control must be tuned very finely indeed. Roughly 50% of all cancers result from the malfunction of the p53-mdm2 regulatory pathway in damaged cells that should have killed themselves.

P53 levels are controlled by a fast feedback mechanism in the form of the mdm2 protein. P53 is synthesised all the time, at a fairly fast rate; but the presence of p53 induces the synthesis of mdm2, which binds to p53 and causes it to be disintegrated. When the DNA is damaged (for instance by radiation in radiotherapy) the cell responds by binding an ATP molecule to each p53, bringing it to a higher energy level that prevents its destruction and causes its concentration to rise. Thus there are in all 4 biochemical species: p53, mdm2, p53-mdm2, and p53*, whose concentrations are modelled by 4 coupled and non-linear ordinary differential equations (ODEs).

As discussed in [10], the p53-mdm2 regulatory system is characterised by oscillatory and non-oscillatory regimes. A simplification of the p53-mdm2 system we have been working with assumes that in the absence of DNA damage with zero p53* initially the response of the system to non-equilibrium starting values of p53 and mdm2 is to create a peak of p53-mdm2 until enough p53 is destroyed and its level is brought back to equilibrium, without oscillations. On the other hand, if p53* is present because of DNA damage, then the system responds with a damped oscillation in its p53 level, until the damage is fixed. There are two ways in which Lie groups can help us with this problem: (1) by helping us solve the system of equations, as reported here in e.g. deriving a single first-order Riccati equation from the original system of 4 ODEs; or (2) providing additional information on the symmetry structure of the problem. This second aspect has not so far yielded directly usable results for this problem.

Figure 3 shows a subset of the chemical reactions associated with this pathway. Arrow heads indicate that a particular compound stimulates the production of the compound it points to, whereas a flat segment at the end of a line indicates inhibition. The small black circles with a 'u' inside indicate 'ubiquitination', which means labelling a particular compound for degradation. This system has been modelled by neglecting ATM and Arf for the moment.

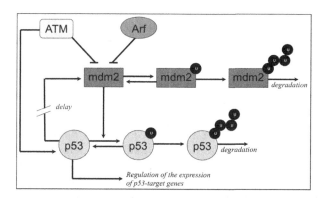

Fig. 3. Simplified interaction network for the p53-mdm2 regulatory pathway [3]

In Figure 4 the different forms of the p53 and mdm2 proteins are identified as follows: P_I = p53, M = mdm2, C = p53-mdm2 compound, P_A = phosphorylated p53, also denoted by p53*. From this type of diagram a Petri net can be easily derived, and the interdependencies between the compounds can be checked intuitively. The dotted arrows indicate promotion of gene expression.

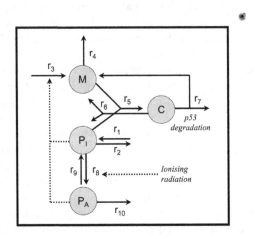

Fig. 4. Schematic of the simplified p53-mdm2 regulatory pathway [3]

The detailed description of the p53 model we use here can be found in [10]. We will use the dimensionless model described in that paper. Each of the four variables is a function of dimensionless time t. The equations are the following (subscript t indicates total derivative with respect to dimensionless time):

$$Pi_t = 1 + \beta_a Pa - (\beta_p + s(t)) Pi$$
$$- \alpha_c PiM + \beta_c C, \tag{1}$$

$$M_t = \alpha_{m0} + \frac{\alpha'_{m1} Pi + \alpha'_{m2} Pa}{Pi + Pa + \kappa_m} + (1 + \beta_c) C$$
$$- (\beta_m + \alpha_c Pi) M, \tag{2}$$

$$C_t = \alpha_c PiM - (1 + \beta_c) C, \tag{3}$$

$$Pa_t = s(t) Pi - (\beta_a + \beta_p) Pa, \tag{4}$$

with $Pi(0) = Pi_0$, $M(0) = M_0$, $C(0) = C_0$, $Pa(0) = Pa_0$. Here, all greek letter parameters are constants, and Pa_0 is normally zero since its presence results from radiation damage. The radiation damage is modelled by the function $s(t)$. If there is no radiation, then $s(t) = 0$. If radiation is kept on a constant level, then $s(t)$ is constant. If the stimulation is a discrete pulse, then it can be modelled as $s(t) = \alpha_a e^{-\gamma t}$, with some parameters α_a and γ.

In the next section we give a conceptual introduction to why Lie groups are useful in solving differential equations. This will hopefully make the subsequent section, which is quite mathematical, more accessible to a broader audience.

4   Overview of Lie's Method

Symmetries are transformations that move a solution of the system of differential equations into another solution of the system. Understanding symmetries of a system of differential equations can be a milestone in their investigation. They can be useful for several reasons: first and most importantly if a 'trivial solution' is known (e.g. the constant function is a solution of many physical systems) then new, possibly less trivial solutions of the system can be generated. A typical example is the heat equation, where determining the symmetry group and knowing that the constant function is a solution enables us to recover its highly nontrivial fundamental solution. This example is discussed in full detail in [4], so in this section we only recapitulate the main concepts behind the method and then move to the analysis of the p53-mdm2 pathway in the next section. Second, understanding symmetries of a system of differential equations might enable us to create new methods for solving the particular system. A typical example for this application is the well-known method of solving an ODE with integrating factor. Another application of this type is explained in [2].

The fundamental idea of the method is to find a way to treat the differential equation as an algebraic equation in order to understand the underlying structure of the system better [9,7]. For this, it is helpful to first understand the basic notions of differential equations. Every differential equation consists of two different types of variables: *independent variables* and *dependent variables*. As the names suggest, the independent variables do not depend on any other variables, while the dependent variables are functions of the independent variables. One tries to find these functions when seeking a solution for a system of differential equations. What makes systems of differential equations different to systems of algebraic equations is that various differentials or derivatives of the dependent variables occur in the system. The first idea is to treat these differentials as 'new' dependent variables, which will turn the system into an algebraic system. These new dependent variables are called *prolongated variables*. Now, using the prolongated variables, the system can be treated as an algebraic system of equations, which defines a manifold in a higher-dimensional space. Every solution curve of the original system appears on the surface of this manifold.

Manifolds are topological spaces such that every point has a neighbourhood isomorphic to $\mathbb{R}^n$ for some n. In other words a manifold is a set of points which locally resemble $\mathbb{R}^n$. It can also be described as a continuously parameterisable space. Knowing these diffeomorphisms ("differentiable morphisms", also called coordinate charts) one can generalise the different notions (e.g. smooth function, vector field, tangent space, etc.) from $\mathbb{R}^n$ to manifolds and use them to perform calculations. We do not want to go deeply into the general theory of manifolds, but we want the reader to appreciate that this part of differential geometry has been heavily investigated and therefore is quite well understood. From our point of view this means that considering the manifold constructed from the system of differential equations gives us powerful tools to investigate the structural properties of the system.

As discussed in [2], if a first-order ODE is given, then there is only one independent variable (e.g. x or, perhaps more pertinent to biology and automata, time t), only one dependent variable (the function we are looking for, e.g. $u = u(x)$) and only one prolongated variable, which is the derivative of the dependent variable (e.g. $u^{(1)} = \frac{\partial u}{\partial x}$). Now, the variables $x, u, u^{(1)}$ span a three-dimensional space, which is called the *jet space*. The differential equation defines a two-dimensional surface in this three-dimensional space, and every solution curve of the original equation is a curve on this surface. Figure 5, reproduced from [2], shows an example for the differential equation

$$\frac{dy}{dx} = e^{-x}y^2 + y + e^x \tag{5}$$

Symmetries are transformations that move a solution (green curve on Figure 5) into another solution (blue curve on Figure 5). Lie's key observation was that these transformations are not discrete as in many situations (e.g. as discussed in [6]), but rather continuous, i.e. the symmetries continuously move a solution into another solution. These (the symmetries) are shown as black curves lying in the surface defining the ODE in Figure 5. Thus, for every such symmetry there exists a corresponding vector field (the *infinitesimal generator* of the symmetry), which at every point tells us the 'speed' vector of how the point is moved continuously by the symmetry. The speed at every point is shown as a red arrow tangent to the black curves in the same figure. Since solutions lie in the manifold determined by the abovementioned algebraic system, every such infinitesimal generator has to be tangent to the manifold. This is a (differential) condition, which enables us to compute the symmetries. In general the higher is the order of a system, the more conditions have to be satisfied by an infinitesimal generator, and thus the easier it is to find the complete symmetry group (in many situations knowing only a

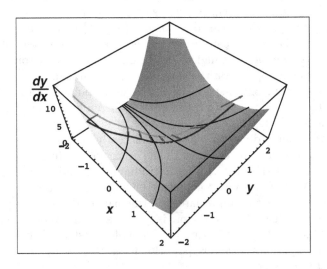

Fig. 5. Visualisation of solutions of a differential equation in the jet space

handful of symmetries can help enormously in solving a system of differential equations). After determining the infinitesimal generators, one can compute the symmetries with the so-called method of *exponentiation*, which is nothing more than the integral of a vector field to obtain a flow.

Thus, continuing the example of a first-order ODE, every symmetry moves a solution curve continuously into another solution curve. The speed of this movement is a vector field, which has to be tangent to the surface, since the solution curves lie on (in) the surface. The requirement to be tangent places a condition on the vector field. If a vector field satisfying this particular condition is found, the corresponding symmetry can be obtained by exponentiating the vector field.

Every symmetry corresponds to a vector field (the infinitesimal generator of the symmetry), such that if we move a solution curve along the flow corresponding to this vector field, then we again obtain solution curves. These symmetries form a (continuous) group called a *Lie group*. It is at first a bit difficult to see that Lie groups also have a manifold structure, and therefore that they have a tangent space at every point. However, looking again at Figure 5, we can see that the green solution curve is 'carried along' by the black curves, which act as a 'flow', into the blue solution curve. Thus, in this case the Lie group (technically this is a Lie subgroup, see below) is a transformation from the plane to the plane that can also be 'lifted' onto the same surface that defines the 'shape' of the differential equation. Moreover, for each point on the solution curve the Lie subgroup can be visualised as a 1-dimensional manifold, i.e. as one of the black curves.[1]

For the example shown in Figure 5, the Lie subgroup is [2]:

$$x'(x, y, \varepsilon) = x + \varepsilon$$
$$y'(x, y, \varepsilon) = ye^{\varepsilon}. \tag{6}$$

Here ε is the continuous parameter of the Lie subgroup (i.e. $\varepsilon \in \mathbb{R}$). Although the above transformation maps the whole xy-plane to itself, it is convenient for the sake of this discussion to identify (x, y) with the green curve and (x', y') as the blue curve. Hence, we can easily see that the red arrows correspond to a finite value of ε (and this is indeed how they were generated in this figure). As $\varepsilon \to 0$, we approach the identity element of the subgroup, the green solution will therefore map to itself, and each of the red arrows becomes infinitesimal and tangent to its black curve at its intersection with the green solution. The tangent space at the identity is called the corresponding *Lie algebra* of the Lie group. The elements of a Lie algebra are vector fields, i.e. every element is a collection of vectors.

In Figure 5 the red arrows form a vector field, which is an element of the Lie algebra. This vector field corresponds to a symmetry, which is an element of the Lie group. We can consider the 1-dimensional manifold generated by this symmetry as one of the Lie subgroups of this equation's Lie group. It can be

---

[1] More precisely, the Lie subgroup is composed of the whole family of black curves, which corresponds to a particular transformation or symmetry.

visualised by the 'flow' shown in Figure 5, with the red arrows being the speed of the flow at every point. Of course, there could be more than one symmetry in any given system, and that is in fact generally the case. Almost all information of the Lie group is contained in the Lie algebra. Via this correspondence, one can replace non-linear conditions describing a symmetry by relatively simple infinitesimal conditions describing the corresponding Lie algebra element. Formally, the Lie algebra is formed by the vector fields of the infinitesimal generators of the symmetries lying in the Lie group. Every symmetry uniquely determines its infinitesimal generator and, by the exponentiating method, for every vector field there exists a corresponding symmetry. Thus knowing the Lie algebra of the symmetries is equivalent to knowing the symmetries themselves.

Since by definition a symmetry is a transformation that leaves something invariant, what is it that these symmetries leave invariant? It turns out that an equivalent definition to what has already been given is to say that a transformation of the dependent and independent variables of a system, as defined above, is a symmetry if it leaves *the functional form* of the system invariant; in other words, if the system of (ordinary or partial) differential equations looks *identical* when expressed in the new variables as when it is expressed in the old variables. We might be happy to believe that at face value, but it turns out that we can again use Figure 5 to improve our understanding. The transformation given by Eq. (6) is such that, when substituted into the original ODE, Eq. (5), the resulting ODE in the new variables will look identical. Therefore, the graph of the new ODE will look identical to Figure 5. Therefore, the geometrical visualisation of what a symmetry preserves is *the shape of the surface corresponding to the system when plotted in the jet space.* Of course, for anything but a single, first-order ODE it is not possible to 'see' this actual 'surface', but hopefully Figure 5 is clear enough to provide an intuitive understanding of the more general case. So now we can probably also see why symmetries map solutions to solutions. Since they map the ODE surface to itself, any set of points such as the green curve will be mapped to another set of points on the same surface. But all the points on the surface satisfy the ODE, by construction. Hence solutions are necessarily mapped to solutions, and this statement generalises to higher-dimensional systems.

It is instructive to think briefly about the converse case, i.e. what happens when we transform the system with a map that is *not* a symmetry. Quite simply, what will happen is that the functional form of the ODE will look different to Eq. (5), and hence the original surface will be mapped to a *new and different* surface. Hence, not all of the red arrows will necessarily be tangent to the original surface as instead they are in Figure 5.

The invertible transformations of a mathematical object that leave some feature of its structure invariant always form a group. This is the case here as well, as *all* Lie symmetries together, and not only each single one as discussed two paragraphs above, can be shown to satisfy the axioms of *the same* group. Thus, as we have already stated, each set of symmetries written as Eq. (6) (i.e. $\exp(\varepsilon \mathbf{v})$ for all $\varepsilon \in \mathbb{R}$) is more properly called a Lie *subgroup* of the system.

5   Lie Symmetry Analysis of p53-mdm2 System

5.1   The General Symmetry Conditions

The general discussion in this subsection builds on the work reported in [4]. The methods described in this report can help understand the analysis of the system of differential equations modelling the p53 network described in [3]. As discussed above, the version of this model that we are currently analysing uses four variables, each depending on the time t. Therefore, we examine first a general system of four ODEs in four variables:

$$
\begin{aligned}
u_t &= f\left(t, u, v, y, z\right), \\
v_t &= g\left(t, u, v, y, z\right), \\
y_t &= p\left(t, u, v, y, z\right), \\
z_t &= q\left(t, u, v, y, z\right).
\end{aligned}
\tag{7}
$$

As in Section 3, the subscript t denotes derivation with respect to dimensionless time. In Section 4 we explained that a symmetry of a system of differential equations is a mapping which moves solution curves to solution curves. Every symmetry has an infinitesimal generator, which is a vector field $\mathbf{v}$ of the form

$$
\begin{aligned}
\mathbf{v} = {}&\tau\left(t, u, v, y, z\right)\partial_t + \phi\left(t, u, v, y, z\right)\partial_u \\
&+ \psi\left(t, u, v, y, z\right)\partial_v + \mu\left(t, u, v, y, z\right)\partial_y \\
&+ \nu\left(t, u, v, y, z\right)\partial_z.
\end{aligned}
\tag{8}
$$

Here, $\partial_t = \frac{\partial}{\partial t}$, $\partial_u = \frac{\partial}{\partial u}$, $\partial_v = \frac{\partial}{\partial v}$, $\partial_y = \frac{\partial}{\partial y}$, $\partial_z = \frac{\partial}{\partial z}$ denote the corresponding basis vectors, or equivalently the corresponding partial differential operators. If $\mathbf{v}$ is an infinitesimal generator of a symmetry, then the flow of the symmetry (i.e. the mapping $(t, u, v, y, z) \rightarrow \Psi\left(\varepsilon, t, u, v, y, z\right)$) can be computed by exponentiating:

$$
\exp\left(\varepsilon \mathbf{v}\right)\left(t, u, v, y, z\right) = \Psi\left(\varepsilon, t, u, v, y, z\right),
$$

where

$$
\Psi\left(0, t, u, v, y, z\right) = \left(t, u, v, y, z\right),
$$

$$
\frac{d}{d\varepsilon}\Psi\left(\varepsilon, t, u, v, y, z\right) = \mathbf{v}\,\big|_{\Psi(\varepsilon, t, u, v, y, z)}.
$$

Let $\mathbf{v}$ be an infinitesimal generator of a symmetry of (7) in the form of (8). Moreover, for an arbitrary function $\zeta\left(t, u, v, y, z\right)$ let us define the following operator:

$$
S\left(\zeta\right) = \zeta_t + f\zeta_u + g\zeta_v + p\zeta_y + q\zeta_z.
\tag{9}
$$

Then the symmetry conditions for the system (7) are the following:

$$
\begin{aligned}
- f_t \tau - f_u \phi - f_v \psi - f_y \mu - f_z \nu \\
- f S(\tau) + S(\phi) = 0,
\end{aligned}
\tag{10}
$$

$$
\begin{aligned}
- g_t \tau - g_u \phi - g_v \psi - g_y \mu - g_z \nu \\
- g S(\tau) + S(\psi) = 0,
\end{aligned}
\tag{11}
$$

$$
\begin{aligned}
- p_t \tau - p_u \phi - p_v \psi - p_y \mu - p_z \nu \\
- p S(\tau) + S(\mu) = 0,
\end{aligned}
\tag{12}
$$

$$
\begin{aligned}
- q_t \tau - q_u \phi - q_v \psi - q_y \mu - q_z \nu \\
- q S(\tau) + S(\nu) = 0.
\end{aligned}
\tag{13}
$$

Let us observe that for arbitrary $\tau(t, u, v, y, z)$ the following coefficients always satisfy the symmetry conditions (10–13):

$$
\begin{aligned}
\phi(t, u, v, y, z) &= f(t, u, v, y, z) \cdot \tau(t, u, v, y, z), \\
\psi(t, u, v, y, z) &= g(t, u, v, y, z) \cdot \tau(t, u, v, y, z), \\
\mu(t, u, v, y, z) &= p(t, u, v, y, z) \cdot \tau(t, u, v, y, z), \\
\nu(t, u, v, y, z) &= q(t, u, v, y, z) \cdot \tau(t, u, v, y, z).
\end{aligned}
$$

This is equivalent to the vector $\mathbf{v}_\tau$ always being an infinitesimal generator of a symmetry:

$$
\mathbf{v}_\tau = \tau \partial_t + f\tau \partial_u + g\tau \partial_v + p\tau \partial_y + q\tau \partial_z.
$$

In fact, taking a closer look at this infinitesimal generator we can observe that it corresponds to the symmetry where the flows are exactly the solution curves of the system (7). Thus $\mathbf{v}_\tau$ does not move a solution curve into another solution curve, but rather moves *along* the solution curves. (For a more detailed explanation, see Section 3.3.2 of [4].) This symmetry does not give us any 'new' or useful information on the system (7). Therefore let us call the infinitesimal generator $\mathbf{v}_\tau$ (for arbitrary $\tau(t, u, v, y, z)$) a *trivial* infinitesimal generator.

The infinitesimal generators form a Lie algebra: i.e. if $\mathbf{v}_1$ and $\mathbf{v}_2$ are infinitesimal generators of some symmetries, then $\mathbf{v}_1 + \mathbf{v}_2$ and $[\mathbf{v}_1, \mathbf{v}_2]$ are infinitesimal generators, as well. We define an equivalence relation on the Lie algebra of infinitesimal generators. We call two infinitesimal generators equivalent, if their difference is a trivial infinitesimal generator, i.e. $\mathbf{v}_1$ and $\mathbf{v}_2$ are equivalent (denoted by $\mathbf{v}_1 \sim \mathbf{v}_2$) if and only if $\mathbf{v}_1 - \mathbf{v}_2 = \mathbf{v}_\tau$ for some $\tau(t, u, v, y, z)$. It is easy to see that this relation is indeed an equivalence relation, which captures the nontrivial symmetries.

If $\mathbf{v} = \tau \partial_t + \phi \partial_u + \psi \partial_v + \mu \partial_y + \nu \partial_z$ is an infinitesimal generator, then $\mathbf{v} \sim \mathbf{v} - \mathbf{v}_\tau = (\phi - f\tau)\partial_u + (\psi - g\tau)\partial_v + (\mu - p\tau)\partial_y + (\nu - q\tau)\partial_z$. Thus every infinitesimal generator is equivalent to one with coefficient $\tau = 0$. Thus without loss of generality we can assume that $\tau = 0$. Then the symmetry conditions of the system (7) are

$$-f_u\phi - f_v\psi - f_y\mu - f_z\nu + S\left(\phi\right) = 0,$$
$$-g_u\phi - g_v\psi - g_y\mu - g_z\nu + S\left(\psi\right) = 0,$$
$$-p_u\phi - p_v\psi - p_y\mu - p_z\nu + S\left(\mu\right) = 0,$$
$$-q_u\phi - q_v\psi - q_y\mu - q_z\nu + S\left(\nu\right) = 0,$$

where S is defined by (9).

5.2   The General p53 Model

In Eqs. (1)-(4), $s\left(t\right)$ represents the radiation stimulus, which induces the creation of Pa-type p53 molecules from Pi-type p53 molecules. The function $s(t)$ can have three different interesting forms depending on three situations:

1. When there is no stimulus at all, i.e. $s(t) = 0$. This can be considered as initial situation, when $Pa = 0$, and Pa will stay 0, as $Pa_t = 0$. This extremely simplifies the system by basically eliminating the variable Pa and the function $s(t)$.
2. When there is no stimulus (i.e. $s(t) = 0$), but there is some initial Pa value.
3. When it is kept at a constant level, i.e. $s(t) = \alpha_a$ is constant.
4. When the stimulation is a discrete pulse 'insult' at time zero, which can be modelled as $s(t) = \alpha_a e^{-\gamma t}$, with some parameters α_a and γ.

There is another simplification that can be done, which corresponds to the saturation of the system, i.e. to the expression

$$\frac{\alpha'_{m1} Pi + \alpha'_{m2} Pa}{Pi + Pa + \kappa_m}. \tag{14}$$

When κ_m is much bigger than Pi and Pa, we simply neglect the latter two in the denominator and replace (14) by

$$\alpha_{m1} Pi + \alpha_{m2} Pa,$$

where $\alpha_{m1} = \alpha'_{m1}/\kappa_m$ and $\alpha_{m2} = \alpha'_{m2}/\kappa_m$. Thus we consider the following system:

$$Pi_t = 1 + \beta_a Pa - \left(\beta_p + s\left(t\right)\right) Pi$$
$$- \alpha_c PiM + \beta_c C, \tag{15}$$
$$M_t = \alpha_{m0} + \alpha_{m1} Pi + \alpha_{m2} Pa + \left(1 + \beta_c\right) C$$
$$- \left(\beta_m + \alpha_c Pi\right) M, \tag{16}$$
$$C_t = \alpha_c PiM - \left(1 + \beta_c\right) C, \tag{17}$$
$$Pa_t = s\left(t\right) Pi - \left(\beta_a + \beta_p\right) Pa. \tag{18}$$

5.3   No Stimulus

In this situation there is no radiation stimulus at all, i.e. $s(t) = 0$. Moreover, as it is an initial situation, $Pa = 0$, as well. Incorporating it to the equations (15–18) and renaming the variables to u, v and w yields

$$u_t = 1 - \beta_p u - \alpha_c uv + \beta_c w,$$
$$v_t = \alpha_{m0} + \alpha_{m1} u - \beta_m v - \alpha_c uv + (1 + \beta_c) w,$$
$$w_t = \alpha_c uv - (1 + \beta_c) w.$$

Now, $\mathbf{v} = \phi(t, u, v, w) \partial_u + \psi(t, u, v, w) \partial_u + \rho(t, u, v, w) \partial_w$ is a symmetry if the functions ϕ, ψ and ρ satisfy the symmetry conditions:

$$(\beta_p + \alpha_c v) \phi + \alpha_c u \psi - \beta_c \rho + S(\phi) = 0, \qquad (19)$$
$$(-\alpha_{m1} + \alpha_c v) \phi + (\beta_m + \alpha_c u) \psi$$
$$- (1 + \beta_c) \rho + S(\psi) = 0, \qquad (20)$$
$$-\alpha_c v \phi - \alpha_c u \psi + (1 + \beta_c) \rho + S(\rho) = 0, \qquad (21)$$

where the operator S is defined by

$$S(\zeta) = \zeta_t + (1 - \beta_p u - \alpha_c uv + \beta_c w) \zeta_u$$
$$+ (\alpha_{m0} + \alpha_{m1} u - \beta_m v - \alpha_c uv$$
$$+ (1 + \beta_c) w) \zeta_v$$
$$+ (\alpha_c uv - (1 + \beta_c) w) \zeta_w.$$

Adding (21) to equations (19) and (20) and using the additivity of S yields

$$\beta_p \phi + \rho + S(\phi + \rho) = 0,$$
$$-\alpha_{m1} \phi + \beta_m \psi + S(\psi + \rho) = 0.$$

Now, if the parameters are $\beta_p = 1 = \alpha_{m1} = \beta_m$, then these equations are further simplified to

$$\phi + \rho + S(\phi + \rho) = 0, \qquad (22)$$
$$-\phi + \psi + S(\psi + \rho) = 0. \qquad (23)$$

From (22) we immediately obtain that

$$\phi = -\rho + \xi_1,$$

where ξ_1 is a solution to $\xi_1 + S(\xi_1) = 0$. Substituting it into (23) we obtain that

$$\psi = -\rho + \xi_2,$$

where ξ_1 and ξ_2 satisfy

$$\xi_1 + S(\xi_1) = 0, \qquad (24)$$
$$-\xi_1 + \xi_2 + S(\xi_2) = 0. \qquad (25)$$

To find the general solutions to equations (24) and (25) can be really hard, but we were able to find solutions by chance:

$$\xi_1 = c_1 \cdot e^{-t},$$
$$\xi_2 = c_1 \cdot t \cdot e^{-t} + c_2 \cdot e^{-t}.$$

Inspired by the fact that we can find nice nontrivial examples for the functions $\xi_1 = \phi + \rho$ and $\xi_2 = \psi + \rho$, we introduce the following coordinate change:

$$y(t) = u(t) + w(t), \tag{26}$$
$$z(t) = v(t) + w(t). \tag{27}$$

With these new coordinates, $u = y - w$, $v = z - w$ and equations (19–21) become

$$y_t - w_t = 1 - y + w - \alpha_c (y - w) \cdot (z - w)$$
$$+ \beta_c w,$$
$$z_t - w_t = \alpha_{m0} + y - w - (z - w)$$
$$- \alpha_c (y - w) \cdot (z - w) + (1 + \beta_c) w,$$
$$w_t = \alpha_c (y - w) \cdot (z - w) - (1 + \beta_c) w,$$

which is equivalent to the system

$$y_t = 1 - y, \tag{28}$$
$$z_t = \alpha_{m0} + y - z, \tag{29}$$
$$w_t = \alpha_c w^2 - (1 + \beta_c + \alpha_c y + \alpha_c z) w + \alpha_c y z. \tag{30}$$

Here, the solution to (28) is almost the same function as ξ_1:

$$y(t) = 1 + c_1 \cdot e^{-t}.$$

Substituting this solution into (29) we obtain a solution for $z(t)$ similar to ξ_2:

$$z(t) = 1 + \alpha_{m0} + c_1 \cdot t \cdot e^{-t} + c_2 \cdot e^{-t}.$$

Thus the original system (19–21) reduces to the Riccati equation

$$
\boxed{
\begin{aligned}
w_t = {} & \alpha_c w^2 - (1 + \beta_c + \alpha_c(2 + \alpha_{m0} \\
& + (c_1 + c_2) \cdot e^{-t} + c_1 \cdot t \cdot e^{-t})) w \\
& + \alpha_c \cdot (1 + c_1 \cdot e^{-t}) \\
& \cdot (1 + \alpha_{m0} + c_1 \cdot t \cdot e^{-t} + c_2 \cdot e^{-t}).
\end{aligned}
}
\tag{31}
$$

Had we found a solution to equation (31), $u(t)$ and $v(t)$ could be expressed in the following way:

$$u(t) = 1 + c_1 \cdot e^{-t} - w(t),$$
$$v(t) = 1 + \alpha_{m0} + c_1 \cdot t \cdot e^{-t} + c_2 \cdot e^{-t} - w(t).$$

The constants c_1 and c_2 are determined by the initial conditions:

$$c_1 = u(0) + w(0) - 1,$$
$$c_2 = v(0) + w(0) - 1 - \alpha_{m0}.$$

Unfortunately equation (31) cannot be solved in general, unless at least one solution can be found. Nevertheless, it can be solved numerically. Figure 6 shows the solution curve for the following choice of parameters [3]:

$$u(0) = 3.5 \qquad v(0) = 0.008 \qquad w(0) = 0.993$$
$$\alpha_{m0} = 0.00005 \qquad \alpha_c = 35 \qquad \beta_c = 0.005.$$

One can read from the figure that w increases in time until it reaches a maximum, then it decreases, which is consistent with its decaying factor.

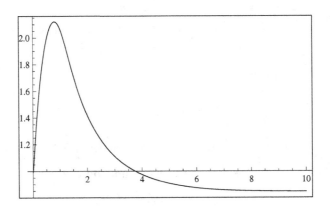

Fig. 6. Numerical solution to (31). The horizontal axis is the dimensionless time t, the vertical axis is w.

5.4   No Stimulus with Initial Pa Value

Let us assume now that there is no radiation, i.e. $s(t) = 0$, but unlike before there exists some active p53 initially, i.e. $Pa(0) = a_0 > 0$. Incorporating it into Equations (15–18) and renaming the variables to u, v, w and x yields

$$u_t = 1 - \beta_p u + \beta_a x - \alpha_c uv + \beta_c w,$$
$$v_t = \alpha_{m0} + \alpha_{m1} u + \alpha_{m2} x - \beta_m v - \alpha_c uv$$
$$+ (1 + \beta_c) w,$$
$$w_t = \alpha_c uv - (1 + \beta_c) w,$$
$$x_t = -(\beta_a + \beta_p) x.$$

Again, we can introduce the same coordinate change (26) and (27) inspired by the examination of the symmetry conditions of the system:

$$y(t) = u(t) + w(t),$$
$$z(t) = v(t) + w(t).$$

Using this coordinate change we obtain

$$x_t = -\left(\beta_a + \beta_p\right)x, \tag{32}$$

$$y_t = 1 + \beta_a x - \beta_p y + \left(\beta_p - 1\right)w, \tag{33}$$

$$z_t = \alpha_{m0} + \alpha_{m2}x + \alpha_{m1}y - \beta_m z$$
$$+ \left(\beta_m - \alpha_{m1}\right)w, \tag{34}$$

$$\boxed{w_t = \alpha_c w^2 - \left(1 + \beta_c + \alpha_c y + \alpha_c z\right)w + \alpha_c yz.} \tag{35}$$

Again, we observe that if $\beta_p = 1$ and $\alpha_{m1} = \beta_m$, then the equations (32–34) can be solved after each other. In particular the solution for (32) is

$$x(t) = c_1 \cdot e^{-(1+\beta_a)t},$$

for an arbitrary constant c_1. Substituting this result into (33) we obtain a solution:

$$y(t) = 1 - x(t) + c_2 \cdot e^{-t}$$
$$= 1 - c_1 \cdot e^{-(1+\beta_a)t} + c_2 \cdot e^{-t},$$

for an arbitrary constant c_2. Substituting this result into (34) we obtain a solution for $z(t)$. If $\beta_m \neq 1$ and $\beta_m \neq 1 + \beta_a$, then

$$z(t) = \frac{\alpha_{m0}}{\beta_m} + \frac{1}{(1-\beta_m)}$$
$$+ \frac{\alpha_{m2}\beta_m - \alpha_{m2} - \beta_a\beta_m}{(\beta_m - 1 - \beta_a)(\beta_m - 1)} \cdot x(t)$$
$$+ \frac{\beta_m}{\beta_m - 1} \cdot y(t) + c_3 \cdot e^{-\beta_m t}$$
$$= 1 + \frac{\alpha_{m0}}{\beta_m} + \frac{\beta_m - \alpha_{m2}}{1 + \beta_a - \beta_m} \cdot c_1 \cdot e^{-(1+\beta_a)t}$$
$$+ \frac{\beta_m}{\beta_m - 1} \cdot c_2 \cdot e^{-t} + c_3 \cdot e^{-\beta_m t},$$

for an arbitrary constant c_3. If $\beta_m = 1 + \beta_a$, then

$$z(t) = 1 + \frac{\alpha_{m0}}{1 + \beta_a}$$
$$+ (\alpha_{m2} - 1 - \beta_a) \cdot c_1 \cdot t \cdot e^{-(1+\beta_a)t}$$
$$+ \frac{1 + \beta_a}{\beta_a} \cdot c_2 \cdot e^{-t} + c_3 \cdot e^{-(1+\beta_a)t},$$

for an arbitrary constant c_3. If $\beta_m = 1$, then

$$z(t) = 1 + \alpha_{m0} + \frac{1 - \alpha_{m2}}{\beta_a} \cdot c_1 \cdot e^{-(1+\beta_a)t}$$
$$+ c_2 \cdot t \cdot e^{-t} + c_3 \cdot e^{-t},$$

for an arbitrary constant c_3. If $\beta_m = 1$ and $\beta_a = 0$, then

$$z(t) = 1 + \alpha_{m0} + (1 - \alpha_{m2}) \cdot c_1 \cdot e^{-2t}$$
$$+ c_2 \cdot t \cdot e^{-t} + c_3 \cdot e^{-t}.$$

for an arbitrary constant c_3.

Substituting these solutions into (35) the original system (19–21) reduces to a rather complicated Riccati equation. Had we found a solution to this equation, we could express $u(t)$ and $v(t)$ in the following way:

$$u(t) = y(t) - w(t),$$
$$v(t) = z(t) - w(t).$$

The constants c_1, c_2, c_3 are determined by the initial conditions $x(0)$, $u(0)$, $v(0)$, $w(0)$:

$$c_1 = x(0),$$
$$c_2 = x(0) + u(0) + w(0) - 1,$$

and if $\beta_m \neq 1$ and $\beta_m \neq 1 + \beta_a$, then

$$c_3 = \left(\frac{\beta_m}{\beta_m - 1} + \frac{\beta_m - \alpha_{m2}}{\beta_m - 1 - \beta_a} \right) \cdot x(0)$$
$$+ \frac{\beta_m}{1 - \beta_m} \cdot u(0) + v(0) + \frac{1}{1 - \beta_m} w(0)$$
$$+ \frac{\beta_m}{\beta_m - 1} + \frac{\beta_a + 1 - \beta_m}{\beta_m - 1 - \beta_a} \cdot \left(1 + \frac{\alpha_{m0}}{\beta_m} \right).$$

If $\beta_m = 1 + \beta_a$, then

$$c_3 = - \left(1 + \frac{1}{\beta_a} \right) \cdot x(0) - \left(1 + \frac{1}{\beta_a} \right) \cdot u(0)$$
$$+ v(0) - \frac{1}{\beta_a} \cdot w(0) + \frac{1}{\beta_a} - \frac{\alpha_{m0}}{1 + \beta_a}.$$

If $\beta_m = 1$, then

$$c_3 = \frac{\alpha_{m2} - 1}{\beta_a} \cdot x(0) + v(0) + w(0) - 1 - \alpha_{m0}.$$

If $\beta_m = 1$ and $\beta_a = 0$, then

$$c_3 = (\alpha_{m2} - 1) \cdot x(0) + v(0) + w(0) - 1 - \alpha_{m0}.$$

Figure 7 shows the solution curve for the following choice of parameters [3]:

$x(0) = 0$	$u(0) = 3.5$	$v(0) = 0.008$
$w(0) = 0.993$	$\alpha_{m0} = 0.00005$	$\alpha_c = 35$
$\beta_c = 0.005$	$\beta_m = 1$	$\beta_a = 0.5.$

It looks quite similar to the one shown in Figure 6, i.e. an overdamped system.

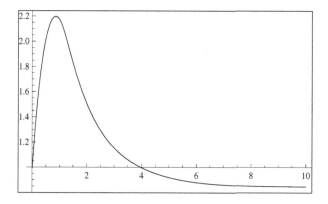

Fig. 7. Numerical solution to (35). The horizontal axis is the dimensionless time t, the vertical axis is w.

For other choices of parameters this system can indeed oscillate, as shown in [10]. In this paper our objective was to seek Lie symmetries and/or analytical solutions in order to gain insight into the analytical structure of the system, so we prioritised these activities over a study of the oscillations per se.

6 Conclusion

It is interesting that by making use of the transformation

$$w(t) = r'(t)/r(t)$$

the Riccati equations (31) and (35) can be transformed into linear ODEs in the new variable $r(t)$. However, because the coefficients of these ODEs are complicated functions of time, neither is integrable. Thus for this example we conclude that the Lie symmetry analysis has not been particularly useful, so far, beyond helping us see the possibility to reduce the original problem to a Riccati equation for a particular choice of parameters. The objective to gain insight into the analytical structure of the system has nonetheless been partially achieved, since the Riccati equation gives us a good starting point for further investigation.

Rather than an indication that there are no dynamical symmetries hiding in this regulatory pathway, this result is more likely a consequence of the simplicity of the p53-mdm2 ODE system chosen to represent it. More precisely, from the point of view of a mathematical model that is meant to capture the most important aspects of the physical phenomenon this model might be too simple. On the other hand, as a mathematical problem it is already rather complicated and practically impossible to solve analytically. So we need to look for a better model that is more expressive physically but simpler mathematically, which is not easy to do. More variables are in general used as part of this pathway, so that the next logical step is to enlarge the system to 5 or 6 equations. The present analysis has already helped us see quite a few features of the problem, and therefore provides a starting point upon which to build a more in-depth investigation.

Acknowledgements

The support for this work by the OPAALS (FP6-034824) and the BIONETS (FP6-027748) EU projects is gratefully acknowledged.

References

1. Briscoe, G., Dini, P.: Towards Autopoietic Computing. In: Proceedings of the 3rd OPAALS International Conference, Aracaju, Sergipe, Brazil, March 22-23 (2010)
2. Dini, P.: D18.4-Report on self-organisation from a dynamical systems and computer science viewpoint. DBE Project (2007), http://files.opaals.eu/DBE/
3. Dini, P., Briscoe, G., Van Leeuwen, I., Munro, A.J., Lain, S.: D1.3: Biological Design Patterns of Autopoietic Behaviour in Digital Ecosystems. OPAALS Deliverable, European Commission (2009),
   http://files.opaals.eu/OPAALS/Year_3_Deliverables/WP01/
4. Dini, P., Horváth, G., Schreckling, D., Pfeffer, H.: D2.2.9: Mathematical Framework for Interaction Computing with Applications to Security and Service Choreography. BIONETS Deliverable, European Commission (2009), http://www.bionets.eu
5. Dini, P., Schreckling, D.: A Research Framework for Interaction Computing. In: Proceedings of the 3rd OPAALS International Conference, Aracaju, Sergipe, Brazil, March 22-23 (2010)
6. Egri-Nagy, A., Dini, P., Nehaniv, C.L., Schilstra, M.J.: Transformation Semigroups as Constructive Dynamical Spaces. In: Proceedings of the 3rd OPAALS International Conference. Aracaju, Sergipe, Brazil, March 22-23 (2010)
7. Hydon, P.: Symmetry Methods for Differential Equations: A Beginner's guide. Cambridge University Press, Cambridge (2000)
8. ICT4US: http://ict4us.com/mnemonics/en_krebs.htm
9. Olver, P.: Applications of Lie Groups to Differential Equations. Springer, Heidelberg (1986)
10. Van Leeuwen, I., Munro, A.J., Sanders, I., Staples, O., Lain, S.: Numerical and Experimental Analysis of the p53-mdm2 Regulatory Pathway. In: Proceedings of the 3rd OPAALS International Conference, Aracaju, Sergipe, Brazil, March 22-23 (2010)

Author Index

Amritesh   44

Barretto, Saulo   100
Barros, Larissa   100
Botto, Francesco   76
Botvich, Dmitri   178
Bräuer, Marco   62
Briscoe, Gerard   199

Carrari R. Lopes, Lia   100, 109, 192
Chatterjee, Jayanta   44, 62
Colugnati, Fernando A. Basile   109, 118
Conboy, Kieran   20
Curran, Declan   31

Dini, Paolo   199, 224, 245, 285

Eder, Raimund   131
Egeraat, Chris van   31
Egri-Nagy, Attila   245

Heistracher, Thomas   131
Horváth, Gábor   285
Huhtamäki, Jukka   146

Kataishi, Rodrigo   1
Kiran, Yedugundla Venkata   76
Krause, Paul J.   213
Kurz, Thomas   131

Lain, Sonia   266
Lapteva, Oxana   62
Leeuwen, Ingeborg M.M. van   266

Lopes, Rodrigo Arthur de Souza Pereira   192
Lopes, Michelle   100

Malone, Paul   161, 178
McGibney, Jimmy   178
McLaughlin, Mark   161, 178
Miranda, Isabel   100
Morgan, Lorraine   20
Munro, Alastair J.   266
Mustaro, Pollyana Notargiacomo   192

Nehaniv, Chrystopher L.   245

Passani, Antonella   76
Piazzalunga, Renata   118

Rathbone, Neil   92
Razavi, Amir Reza   213
Rivera León, Lorena   1

Salonen, Jaakko   146
Sanders, Ian   266
Schilstra, Maria J.   245
Schreckling, Daniel   224
Serra, Fabio K.   213
Siqueira, Paulo R.C.   100, 213
Staples, Oliver   266
Steinicke, Ingmar   62

Zeller, Frauke   62